Mathematical Methods
in Robust Control of
Linear Stochastic Systems

MATHEMATICAL CONCEPTS AND METHODS IN SCIENCE AND ENGINEERING

Series Editor: Angelo Miele
Mechanical Engineering and Mathematical Sciences
Rice University

Mathematical Methods in Robust Control of Linear Stochastic Systems

Vasile Dragan
Institute of Mathematics of the Romanian Academy
Bucharest, Romania

Toader Morozan
Institute of Mathematics of the Romanian Academy
Bucharest, Romania

Adrian-Mihail Stoica
University Politehnica of Bucharest
Bucharest, Romania

 Springer

Vasile Dragan
Institute of Mathematics of
 the Romanian Academy
P.O. Box 1-764, Ro 70700
Bucharest, Romania
vdragan@rdslink.ro
Bucharest, Romania
amstoica@rdslink.ro

Toader Morozan
Institute of Mathematics of
 the Romanian Academy
P.O. Box 1-764, Ro 70700
Bucharest, Romania
toader.morozan@imar.ro

Adrian-Mihail Stoica
University Politehnica of Bucharest
Str. Polizu, No. 1, Ro-77206
Bucharest, Romania
amstoica@rdslink.ro

Mathematics Subject Classification (2000): 93EXX, 34F05

ISBN 978-1-4419-2143-7
e-ISBN 978-0-387-35924-3

Printed on acid-free paper.

Printed in the United States of America

9 8 7 6 5 4 3 2 1

springer.com

To our wives, Viorica, Elena and Dana
for their love, patience and support.

To our wives, Viorica, Elena and Dana
for their love, patience and support.

Contents

Preface

This monograph presents a thorough description of the mathematical theory of robust linear stochastic control systems. The interest in this topic is motivated by the variety of random phenomena arising in physical, engineering, biological, and social processes. The study of stochastic systems has a long history, but two distinct classes of such systems drew much attention in the control literature, namely stochastic systems subjected to white noise perturbations and systems with Markovian jumping. At the same time, the remarkable progress in recent decades in the control theory of deterministic dynamic systems strongly influenced the research effort in the stochastic area. Thus, the modern treatments of stochastic systems include optimal control, robust stabilization, and H^2- and H^∞-type results for both stochastic systems corrupted with white noise and systems with jump Markov perturbations.

In this context, there are two main objectives of the present book. The first one is to develop a mathematical theory of linear time-varying stochastic systems including both white noise and jump Markov perturbations. From the perspective of this generalized theory the stochastic systems subjected only to white noise perturbations or to jump Markov perturbations can be regarded as particular cases. The second objective is to develop analysis and design methods for advanced control problems of linear stochastic systems with white noise and Markovian jumping as linear-quadratic control, robust stabilization, and disturbance attenuation problems. Taking into account the major role played by the Riccati equations in these problems, the book presents this type of equation in a general framework. Particular attention is paid to the numerical aspects arising in the control problems of stochastic systems; new numerical algorithms to solve coupled matrix algebraic Riccati equations are also proposed and illustrated by numerical examples.

The book contains seven chapters. Chapter 1 includes some prerequisites concerning measure and probability theory that will be used in subsequent developments in the book. In the second part of this chapter, detailed proofs of some new results, such as the Itô-type formula in a general case covering the classes of stochastic systems with white noise perturbations and Markovian jumping, are given. The Itô-type formula plays a crucial role in the proofs of the main results of the book.

Chapter 2 is mainly devoted to the exponential stability of linear stochastic systems. It is proved that the exponential stability in the mean square of the considered class of stochastic systems is equivalent with the exponential stability of an appropriate class of deterministic systems over a finite-dimensional Hilbert space. Necessary and sufficient conditions for exponential stability for such deterministic systems are derived in terms of some Lyapunov-type equations. Then necessary and sufficient conditions in terms of Lyapunov functions for mean square exponential stability are obtained. These results represent a generalization of the known conditions concerning the exponential stability of stochastic systems subjected to white noise and Markovian jumping, respectively.

Some structural properties such as controllability, stabilizability, observability, and detectability of linear stochastic systems subjected to both white noise and jump Markov perturbations are considered in Chapter 3. These properties play a key role in the following chapters of the book.

In Chapter 4 differential and algebraic generalized Riccati-type equations arising in the control problems of stochastic systems are introduced. Our attention turns to the maximal, minimal, and stabilizing solutions of these equations for which necessary and sufficient existence conditions are derived. The final part of this chapter provides an iterative procedure for computing the maximal solution of such equations.

In the fifth chapter of the book, the linear-quadratic problem on the infinite horizon for stochastic systems with both white noise and jump Markov perturbations is considered. The problem refers to a general situation: The considered systems are subjected to both state and control multiplicative white noise and the optimization is performed under the class of nonanticipative stochastic controls. The optimal control is expressed in terms of the stabilizing solution of coupled generalized Riccati equations. As an application of the results deduced in this chapter, we consider the optimal tracking problem.

Chapter 6 contains corresponding versions of some known results from the deterministic case, such as the Bounded Real Lemma, the Small Gain Theorem, and the stability radius, for the considered class of stochastic systems. Such results have been obtained separately in the stochastic framework for systems subjected to white noise and Markov perturbations, respectively. In our book, these results appear as particular situations of a more general class of stochastic systems including both types of perturbations.

In Chapter 7 the γ-attenuation problem of stochastic systems with both white noise and Markovian jumping is considered. Necessary and sufficient conditions for the existence of a stabilizing γ-attenuating controller are obtained in terms of a system of coupled game-theoretic Riccati equations and inequalities. These results allow one to solve various robust stabilization problems of stochastic systems subjected to white noise and Markov perturbations, as illustrated by numerical examples.

The monograph is based entirely on original recent results of the authors; some of these results have been recently published in control journals and conferences proceedings. There are also some other results that appear for the first time in this book.

This book is not intended to be a textbook or a guide for control designers. We had in mind a rather larger audience, including theoretical and applied mathematicians and research engineers, as well as graduate students in all these fields, and, for some parts of the book, even undergraduate students. Since our intention was to provide a self-contained text, only the first chapter reviews known results and prerequisites used in the rest of the book.

The authors are indebted to Professors Gerhard Freiling and Isaac Yaesh for fruitful discussions on some of the numerical methods and applications presented in the book.

Finally, the authors wish to thank the Springer publishing staff and the reviewer for carefully checking the manuscript and for valuable suggestions.

October 2005

1

Preliminaries to Probability Theory and Stochastic Differential Equations

This first chapter collects for the readers' convenience some definitions and fundamental results concerning the measure theory and the theory of stochastic processes which are needed in the following developments of the book. Classical results concerning measure theory, integration, stochastic processes, and stochastic integrals are presented without proofs. Appropriate references are given; thus for the measure theory, we mention [27], [43], [55], [59], [95], [110]; for the probability theory we refer to [26], [55], [96], [104], [110] and for the theory of stochastic processes and stochastic differential equations we cite [5], [26], [55], [56], [69], [81], [97], [98]. However several results that can be found only in less accessible references are proved.

In Section 1.10 we prove a general version of the Itô-type formula which plays a key role in the developments of Chapters 3–5. The results concerning mean square exponential stability in Chapter 2 may be derived using an Itô-type formula which refers to stochastic processes that are solutions to a class of stochastic differential equations. This version of the Itô-type formula can be found in Theorem 39 of this chapter. Theorem 34, used in the proof of the Itô-type formula and also in Lemma 22 in Chapter 6 to estimate the stability radius, appears for the first time in this book.

1.1 Elements of measure theory

1.1.1 Measurable spaces

Definition 1. *A measurable space is a pair* $(\Omega, \mathcal{F})$, *where* Ω *is a set and* $\mathcal{F}$ *is a* σ-*algebra of subsets of* Ω; *that is,* $\mathcal{F}$ *is a family of subsets* $A \subset \Omega$ *with the properties*
   (i) $\Omega \in \mathcal{F}$;
   (ii) *if* $A \in \mathcal{F}$, *then* $\Omega - A \in \mathcal{F}$;
   (iii) *if* $A_n \in \mathcal{F}, n \geq 1$, *then* $\cup_{n=1}^{\infty} A_n \in \mathcal{F}$.

If $\mathcal{F}_1$ and $\mathcal{F}_2$ are two σ-algebras of subsets of Ω, by $\mathcal{F}_1 \vee \mathcal{F}_2$ we denote the smallest σ-algebra of subsets of Ω which contains the σ-algebras $\mathcal{F}_1$ and $\mathcal{F}_2$.

By $\mathcal{B}(\mathbf{R}^n)$ we denote the σ-algebra of Borel subsets of $\mathbf{R}^n$, that is, the smallest σ-algebra containing all open subsets of $\mathbf{R}^n$.

For a family C of subsets of Ω, $\sigma(C)$ will denote the smallest σ-algebra of subsets of Ω containing C; $\sigma(C)$ will be termed the σ-algebra generated by C.

If $(\Omega_1, \mathcal{G}_1)$ and $(\Omega_2, \mathcal{G}_2)$ are two measurable spaces, by $\mathcal{G}_1 \otimes \mathcal{G}_2$ we denote the smallest σ-algebra of subsets of $\Omega_1 \times \Omega_2$ which contains all sets $A \times B$, $A \in \mathcal{G}_1$, $B \in \mathcal{G}_2$.

Definition 2. *A collection C of subsets of Ω is called a π-system if*
   (i) $\phi \in C$, *and*
   (ii) *if $A, B \in C$, then $A \cap B \in C$.*

The next result proved in [118] is frequently used in probability theory.

Theorem 1. *If C is a π-system and $\mathcal{G}$ is the smallest family of subsets of Ω such that*
   (i) $C \subset \mathcal{G}$;
   (ii) *if $A \in \mathcal{G}$, then $\Omega - A \in \mathcal{G}$;*
   (iii) $A_n \in \mathcal{G}, n \geq 1$, *and* $A_i \cap A_j = \phi$ *for* $i \neq j$ *implies* $\cup_{n \geq 1}^{\infty} \in \mathcal{G}$, *then* $\sigma(C) = \mathcal{G}$.

Proof. Since $\sigma(C)$ verifies (i), (ii), and (iii) in the statement, it follows that $\mathcal{G} \subset \sigma(C)$.

To prove the opposite inclusion, we show first that $\mathcal{G}$ is a π-system.

Let $A \in \mathcal{G}$ and define $\mathcal{G}(A) = \{B; B \in \mathcal{G} \text{ and } A \cap B \in \mathcal{G}\}$.

Since $A - B = \Omega - [(A \cap B) \cup (\Omega - A)]$, it is easy to check that $\mathcal{G}(A)$ verifies the conditions (ii) and (iii), and if $A \in C$, then (i) is also satisfied. Hence for $A \in C$ we have $\mathcal{G}(A) = \mathcal{G}$; consequently, if $A \in C$ and $B \in \mathcal{G}$, then $A \cap B \in \mathcal{G}$. But this implies $\mathcal{G}(B) \supset C$ and therefore $\mathcal{G}(B) = \mathcal{G}$ for any $B \in \mathcal{G}$. Hence $\mathcal{G}$ is a π-system and now, since $\mathcal{G}$ verifies (ii) and (iii), it is easy to verify that $\mathcal{G}$ is a σ-algebra and the proof is complete. $\square$

1.1.2 Measures and measurable functions

Definition 3. (a) *Given a measurable space $(\Omega, \mathcal{F})$, a function $\mu : \mathcal{F} \rightarrow [0, \infty]$ is called a* measure *if:*
   (i) $\mu(\phi) = 0$;
   (ii) *if $A_n \in \mathcal{F}, n \geq 1$, and $A_i \cap A_j = \phi$ for $i \neq j$, then*

$$\mu(\cup_{n=1}^{\infty} A_n) = \sum_{n=1}^{\infty} \mu(A_n).$$

(b) *A triplet $(\Omega, \mathcal{F}, \mu)$ is said to be a* space with measure.

(c) *If $\mu(\Omega) = 1$, we say that μ is a* probability *on $\mathcal{F}$, and in this case the triplet $(\Omega, \mathcal{F}, \mu)$ is termed a* probability space.

A measure μ is said to be σ-finite *if there exists a sequence $A_n, n \geq 1, A_n \in \mathcal{F}$ with $A_i \cap A_j = \phi$ for $i \neq j$ and $\Omega = \cup_{n=1}^{\infty} A_n$ and $\mu(A_n) < \infty$ for every n.*

Definition 4. *Given a measurable space* $(\Omega, \mathcal{F})$, *a function* $f: \Omega \longmapsto \mathbf{R}$ *is said to be a* measurable function *if for every* $A \in \mathcal{B}(\mathbf{R})$ *we have* $f^{-1}(A) \in \mathcal{F}$, *where* $f^{-1}(A) = \{\omega \in \Omega; f(\omega) \in A\}$.

It is easy to prove that $f: \Omega \longmapsto \mathbf{R}$ is measurable if and only if $f^{-1}((-\infty, \alpha)) \in \mathcal{F}$ for every $\alpha \in \mathbf{R}$.

Remark 1. It is not difficult to verify that if $(\Omega_1, \mathcal{F}_1)$ and $(\Omega_2, \mathcal{F}_2)$ are two measurable spaces and if $f : \Omega_1 \times \Omega_2 \to \mathbf{R}$ is $\mathcal{F}_1 \otimes \mathcal{F}_2$ measurable, then for each $\omega_2 \in \Omega_2$ the function $\omega_1 \longmapsto f(\omega_1, \omega_2)$ is $\mathcal{F}_1$ measurable and for each $\omega_1 \in \Omega_1$ the function $\omega_2 \longmapsto f(\omega_1, \omega_2)$ is $\mathcal{F}_2$ measurable.

Definition 5. *A measurable function* $f : \Omega \longmapsto \mathbf{R}$ *is said to be a* simple measurable function *if it takes only a finite number of values.*

We shall write a.a. and a.e. for *almost all* and *almost everywhere*, respectively; $f = g$ a.e. means $\mu(f \neq g) = 0$.

Definition 6. *Let* $(\Omega, \mathcal{F}, \mu)$ *be a space with measure* $f_n : \Omega \to \mathbf{R}$, $n \geq 1$, *and* $f : \Omega \to \mathbf{R}$ *be measurable functions.*

(i) *We say that* f_n *converges to* f *for a.a.* $\omega \in \Omega$ *or equivalently* $\lim_{n \to \infty} f_n = f$ *a.e.* $(f_n \overset{a.e.}{\to} f)$ *if*

$$\mu \left\{\omega; \lim_{n \to \infty} f_n(\omega) \neq f(\omega)\right\} = 0.$$

(ii) *We say that the sequence* f_n *converges in measure to* f $(f_n \overset{\mu}{\to} f)$ *if for every* $\delta > 0$, *we have* $\lim_{n \to \infty} \mu\{\omega; |f_n(\omega) - f(\omega)| > \delta\} = 0$.

Theorem 2. *Assume that* $\lim_{n \to \infty} f_n = f$ *a.e. and that* $\mu(\Omega) < \infty$. *Then* $f_n \overset{\mu}{\to} f$. $\square$

Theorem 3. *(Riesz's theorem) If* $f_n \overset{\mu}{\to} f$, *then there exists a subsequence* f_{n_k} *of the sequence* f_n *such that* $\lim_{k \to \infty} f_{n_k} = f$ *a.e.* $\square$

Corollary 4. *Let* $(\Omega, \mathcal{F}, \mu)$ *be a space with measure such that* $\mu(\Omega) < \infty$. *Then the following assertions are equivalent:*

(i) $f_n \overset{\mu}{\to} f$;

(ii) *any subsequence of* f_n *contains a subsequence converging a.e. to* f. $\square$

As usual, in the measure theory two measurable functions f and g are identified if $f = g$ a.e. Moreover, if $f : \Omega \to \overline{\mathbf{R}} = [-\infty, \infty]$ is measurable, that is, $f^{-1}([-\infty, \alpha)) \in \mathcal{F}$ for every $\alpha \in \mathbf{R}$ and if $\mu(|f| = \infty) = 0$, then f will be identified with a function $g : \Omega \to \mathbf{R}$ defined as follows:

$$g(\omega) = \begin{cases} f(\omega) & \text{if } |f(\omega)| < \infty, \text{ and} \\ 0 & \text{if } |f(\omega)| = \infty. \end{cases}$$

Theorem 5. *If* $(\Omega_1, \mathcal{F}_1, \mu_1)$ *and* $(\Omega_2, \mathcal{F}_2, \mu_2)$ *are two spaces with* σ-*finite measures, then there exists a unique measure* $\mu : \mathcal{F}_1 \otimes \mathcal{F}_2 \to [0, \infty]$ *such that* $\mu(A \times B) = \mu_1(A)\mu_2(B)$ *for all* $A \in \mathcal{F}_1$ *and* $B \in \mathcal{F}_2$. *This measure* μ *will be denoted by* $\mu_1 \times \mu_2$. $\square$

1.1.3 Integration

Theorem 6. *Let $f \geq 0$ be a measurable function. Let us define*

$$f_n(\omega) = \sum_{i=1}^{2^n n + 1} \frac{i-1}{2^n} \chi_{A_{i,n}}(\omega),$$

where

$$A_{i,n} = \left\{ \omega; \frac{i-1}{2^n} \leq f(\omega) < \frac{i}{2^n} \right\}, i = 1, 2, \ldots, 2^n n,$$

$$A_{2^n n + 1,n} = \{\omega; f(\omega) \geq n\},$$

and $\chi_A(\omega)$ is the indicator function of the set A; that is, $\chi_A(\omega) = 1$ if $\omega \in A$ and $\chi_A(\omega) = 0$ if $\omega \in \Omega - A$. Then we have:

(i) $0 \leq f_n \leq f_{n+1}$ *and* $\lim_{n \to \infty} f_n(\omega) = f(\omega), \omega \in \Omega$;

(ii) $0 \leq a_n \leq a_{n+1}$, *where* $a_n = \sum_{i=1}^{2^n n + 1} \frac{i-1}{2^n} \mu(A_{i,n})$ *(with the convention that $0 \cdot \infty = 0$).* □

Definition 7. (i) *Let $f \geq 0$ be a measurable function on a space with measure $(\Omega, \mathcal{F}, \mu)$ and $f_n, a_n, n \geq 1$ be the sequences defined in the statement of Theorem 6. By definition $a_n = \int_\Omega f_n d\mu$ and $\int_\Omega f d\mu = \lim_{n \to \infty} a_n$.*

(ii) *A measurable function $f : \Omega \to \mathbf{R}$ is called an* integrable function *if $\int_\Omega |f| d\mu < \infty$, and in this case,*

$$\int_\Omega f d\mu = \int_\Omega f^+ d\mu - \int_\Omega f^- d\mu,$$

where

$$f^+ = \frac{1}{2}(|f| + f); \quad f^- = \frac{1}{2}(|f| - f).$$

(iii) *We say that the integral of a measurable function f exists if at least one of the integrals $\int_\Omega f^+ d\mu$ or $\int_\Omega f^- d\mu$ is finite; if $\int_\Omega f^+ d\mu = \infty$ and $\int_\Omega f^- d\mu < \infty$, then by definition, $\int_\Omega f d\mu = \infty$, and if $\int_\Omega f^+ d\mu < \infty$ and $\int_\Omega f^- d\mu = \infty$, by definition, $\int_\Omega f d\mu = -\infty$.*

Remark 2. It can be proved that the definition of the integral $\int_\Omega f d\mu$ in Definition 7(i) is not dependent upon the choice of the particular monotonic increasing sequence of simple measurable functions f_n converging to f. If f is a simple measurable function with values $c_1, c_2, \ldots, c_n$, then by definition

$$\int_\Omega f d\mu = \sum_{i=1}^{n} c_i \mu(\{\omega; f(\omega) = c_i\}).$$

It is known that

(i) $|\int_\Omega f d\mu| \leq \int_\Omega |f| d\mu$;

(ii) If $f = g$ a.e., then $\int_\Omega f d\mu = \int_\Omega g d\mu$; if $A \in \mathcal{F}$, by definition $\int_A f d\mu = \int_\Omega \chi_A f d\mu$.

By $L^p(\Omega)$, $p \geq 1$, we denote the space of all measurable functions $f : \Omega \to \mathbf{R}$ with $\int_\Omega |f|^p d\mu < \infty$.

Let us define

$$\|f\|_p = \left(\int_\Omega |f|^p d\mu \right)^{\frac{1}{p}} \text{ if } f \in L^p.$$

Regarding the integrable functions we recall the following useful results.

Theorem 7. *(Holder's inequality) If $f \in L^p(\Omega)$, $p > 1$, and $g \in L^q(\Omega)$ with $\frac{1}{p} + \frac{1}{q} = 1$, then $fg \in L^1(\Omega)$ and*

$$\|fg\|_1 \leq \|f\|_p \|g\|_q.$$ □

Taking, in the above theorem, $p = \frac{s}{r}$, $f = |h|^r$, $g = 1$, one obtains the following result.

Corollary 8. *If $\mu(\Omega) < \infty$ and $1 \leq r < s$, then $h \in L^s(\Omega)$ implies $h \in L^r(\Omega)$ and if $\mu(\Omega) = 1$, we have $\|h\|_r \leq \|h\|_s$.* □

Definition 8. *Let f_n, $f \in L^p$. We say that $f_n \to f$ in L^p or $f_n \xrightarrow{L^p} f$ if*

$$\lim_{n \to \infty} \int_\Omega |f_n - f|^p d\mu = 0.$$

Theorem 9. *If $f_n \xrightarrow{L^p} f$ then $f_n \xrightarrow{\mu} f$.* □

1.2 Convergence theorems for integrals

Let $(\Omega, \mathcal{F}, \mu)$ be a space with measure. The following results are well known in measure theory.

Theorem 10. *(Fatou's Lemma) Let $f_n \geq 0, n \geq 1$, be a sequence of measurable functions. Then*

$$\int_\Omega (\underline{\lim} f_n) d\mu \leq \underline{\lim} \int_\Omega f_n d\mu.$$ □

Theorem 11. *(Lebesgue's Theorem) Let f_n, f be measurable functions and $|f_n| \leq g$, $n \geq 1$, a.e. where g is an integrable function. If $\lim_{n\to\infty} f_n = f$ a.e., then $f_n \xrightarrow{L^1} f$, and therefore $\lim_{n\to\infty} \int_\Omega f_n d\mu = \int_\Omega f d\mu$.* □

Theorem 12. *Let f_n, f be measurable functions. If $|f_n| \leq g, n \geq 1$, for some integrable function g and $f_n \xrightarrow{\mu} f$, then $f_n \xrightarrow{L^1} f$.* □

Theorem 13. [26], [55], [106] *Let f_n, f be integrable functions. Suppose that $\mu(\Omega) < \infty$ and there exists $\alpha > 1$ such that*

$$\sup_n \int_\Omega |f_n|^\alpha d\mu < \infty.$$

If $f_n \xrightarrow{\mu} f$, then $f_n \xrightarrow{L^1} f$ and therefore $\lim_{n\to\infty} \int_\Omega f_n d\mu = \int_\Omega f d\mu$. □

Theorem 14. [43], [95] *If $f : [a, b] \to \mathbf{R}$ is an integrable function, then*

$$\lim_{h\to 0+} \frac{1}{h} \int_{\max\{t-h,a\}}^t f(s)ds = f(t) \quad a.e., \ t \in [a, b].$$ □

Definition 9. *Let μ_1 and μ_2 be two measures on the measurable space $(\Omega, \mathcal{F})$; we say that μ_1 is absolutely continuous with respect to μ_2 (and we write $\mu_1 \ll \mu_2$) if $\mu_2(A) = 0$ implies $\mu_1(A) = 0$.*

Theorem 15. *(Radon–Nicodym Theorem) If $\lambda \ll \mu$, $\lambda(\Omega) < \infty$, $\mu(\Omega) < \infty$, then there exists a unique (mod μ) integrable function f such that $\lambda(A) = \int_A f d\mu$ for all $A \in \mathcal{F}$.* □

Theorem 16. *(Fubini's Theorem) Let $(\Omega_1, \mathcal{F}_1, \mu_1)$, $(\Omega_2, \mathcal{F}_2, \mu_2)$ be two spaces with σ-finite measures μ_1 and μ_2, respectively. Then we have:*

(a) If $f : \Omega_1 \times \Omega_2 \to \mathbf{R}_+$ is a measurable function (with respect to $\mathcal{F}_1 \otimes \mathcal{F}_2$), then the function $\omega_2 \longmapsto \int_{\Omega_1} f(\omega_1, \omega_2)d\mu_1$ is $\mathcal{F}_2$ measurable, the function $\omega_1 \longmapsto \int_{\Omega_2} f(\omega_1, \omega_2)d\mu_2$ is $\mathcal{F}_1$ measurable, and

$$\int_{\Omega_1 \times \Omega_2} f d(\mu_1 \times \mu_2) = \int_{\Omega_1} \left(\int_{\Omega_2} f(\omega_1, \omega_2)d\mu_2 \right) d\mu_1$$

$$= \int_{\Omega_2} \left(\int_{\Omega_1} f(\omega_1, \omega_2)d\mu_1 \right) d\mu_2.$$

(b) A measurable function $f : \Omega_1 \times \Omega_2 \to \mathbf{R}$ is integrable (on the space $(\Omega_1 \times \Omega_2, \mathcal{F}_1 \otimes \mathcal{F}_2, \mu_1 \times \mu_2)$) if and only if

$$\int_{\Omega_1} \left(\int_{\Omega_2} |f(\omega_1, \omega_2)|d\mu_2 \right) d\mu_1 < \infty.$$

(c) If $f : \Omega_1 \times \Omega_2 \to \mathbf{R}$ is an integrable function, then:
(i) For a.a. $\omega_1 \in \Omega_1$ the function $\varphi_1(\omega_1) = \int_{\Omega_2} f(\omega_1, \omega_2)d\mu_2$ is well defined, finite, and measurable and integrable on the space $\{\Omega_1, \mathcal{F}_1, \mu_1\}$.
(ii) For a.a. $\omega_2 \in \Omega_2$ the function $\varphi_2(\omega_2) = \int_{\Omega_1} f(\omega_1, \omega_2)d\mu_1$ is well defined, finite, and measurable and integrable on the space $\{\Omega_2, \mathcal{F}_2, \mu_2\}$.
(iii) $\int_{\Omega_1 \times \Omega_2} f d(\mu_1 \times \mu_2) = \int_{\Omega_1} \varphi_1 d\mu_1 = \int_{\Omega_2} \varphi_2 d\mu_2$. □

1.3 Elements of probability theory

Throughout this section and throughout this monograph, $\{\Omega, \mathcal{F}, P\}$ is a given probability space (see Definition 3(c)).

In probability theory a measurable function is called a *random variable* and the integral of a random variable f is called the *expectation of* f and is denoted by Ef or $E(f)$, that is, $Ef = \int_\Omega f \, dP$.

A random vector is a vector whose components are random variables. All random vectors are considered column vectors. In probability theory the words *almost surely* (a.s.) and *with probability* 1 are often used instead of *almost everywhere*.

As usual, two random variables (random vectors) x, y are identified if $x = y$ a.s.

With this convention the space $L^2(\Omega, P)$ of all random variables x with $E|x^2| < \infty$ is a real Hilbert space with the *inner product* $\langle x, y \rangle = E(xy)$.

If $x_\alpha, \alpha \in \Delta$ is a family of random variables, by $\sigma(x_\alpha, \alpha \in \Delta)$ we denote the smallest σ-algebra $\mathcal{G} \subset \mathcal{F}$ with respect to which all functions $x_\alpha, \alpha \in \Delta$ are measurable.

1.3.1 Gaussian random vectors

Definition 10. *An n-dimensional random vector x is said to be* Gaussian *if there exist $m \in \mathbf{R}^n$ and K an $n \times n$ symmetric positive semidefinite matrix such that*

$$E e^{iu^*x} = e^{iu^*m - \frac{1}{2}u^*Ku}$$

for all $u \in \mathbf{R}^n$, where u^* denotes the transpose of u and $i := \sqrt{-1}$.

Remark 3. The above equality implies

$$m = Ex \text{ and } K = E(x - m)(x - m)^*. \tag{1.1}$$

Definition 11. *A Gaussian random vector x is said to be* nondegenerate *if K is a positive definite matrix. If x is a nondegenerate Gaussian random vector, then*

$$P(x \in A) = \frac{1}{((2\pi)^n \det K)^{\frac{1}{2}}} \int_A e^{-\frac{1}{2}(y-m)^*K^{-1}(y-m)} dy$$

for every $A \in \mathcal{B}(\mathbf{R}^n)$.

1.4 Independence

Definition 12. (i) *The σ-algebras $\mathcal{F}_1, \mathcal{F}_2, \ldots, \mathcal{F}_n, \mathcal{F}_i \subset \mathcal{F}$ are independent if*

$$P\left(\cap_{j=1}^n A_j\right) = \Pi_{j=1}^n P(A_j)$$

for all $A_j \in \mathcal{F}_j, 1 \le j \le n$.

(ii) *The random variables (random vectors) $x_1, x_2, \ldots, x_n$ are independent if the σ-algebras $\sigma(x_1), \sigma(x_2), \ldots, \sigma(x_n)$ are independent.*

(iii) *The set $\{x_1, x_2, \ldots, x_n\}$ of random variables (random vectors) is independent of the σ-algebra $\mathcal{G}$, $\mathcal{G} \subset \mathcal{F}$ if the σ-algebra $\sigma(x_i, 1 \le i \le n)$ is independent of $\mathcal{G}$.*

Theorem 17. (i) *If $x_1, x_2, \ldots, x_n$ are independent random variables and if x_i are integrable, $1 \leq i \leq n$, then the product $x_1 x_2 \ldots x_n$ is integrable and $E(x_1 x_2 \ldots x_n) = \Pi_{i=1}^{n} E(x_i)$.*

(ii) *If the random vectors $x_1, x_2, \ldots, x_n$, $n \geq 2$, are independent, then $\sigma(x_1, \ldots, x_k)$ is independent of $\sigma(x_{k+1}, \ldots, x_n)$ for every $1 \leq k \leq n - 1$.*   □

1.5 Conditional expectation

Let $\mathcal{G} \subset \mathcal{F}$ be a σ-algebra and x an integrable random variable. By the Radon–Nicodym Theorem (Theorem 15) it follows that there exists a unique (mod P) random variable y with the following properties:

(a) y is a measurable with respect to $\mathcal{G}$,

(b) $E|y| < \infty$, and

(c) $\int_A y \, dP = \int_A x \, dP$ for all $A \in \mathcal{G}$.

The random variable y with these properties is denoted by $E[x|\mathcal{G}]$ and is called the *conditional expectation of x with respect to the σ-algebra $\mathcal{G}$*.

By definition, for all $A \in \mathcal{F}$

$$P(A|\mathcal{G}) := E[\chi_A|\mathcal{G}] \text{ and}$$
$$E[x|y_1, \ldots, y_n] := E[x|\sigma(y_1, \ldots, y_n)],$$

where χ_A denotes the indicator function of A.

If x is an integrable random variable and $A \in \mathcal{F}$ with $P(A) > 0$, then by definition

$$E[x|A] := \int_{\Omega} x \, dP_A,$$

where

$$P_A : \mathcal{F} \to [0, \infty) \text{ by } P_A(B) = \frac{P(A \cap B)}{P(A)} \ \forall B \in \mathcal{F}.$$

$E[x|A]$ is called the *conditional expectation of x with respect to the event A*.

Since

$$P_A(B) = \frac{1}{P(A)} \int_B \chi_A \, dP,$$

we have

$$E[x|A] = \frac{1}{P(A)} \int_{\Omega} (x \chi_A) \, dP = \frac{1}{P(A)} \int_A x \, dP.$$

By definition,

$$P(B|A) := P_A(B), \ A \in \mathcal{F}, \ B \in \mathcal{F}, \ P(A) > 0.$$

Obviously, $P[B|A] = E(\chi_B|A)$.

Theorem 18. *Let x, y be integrable random variables and $\mathcal{G}$, $\mathcal{H} \subset \mathcal{F}$, σ-algebras. Then the following assertions hold:*

   (i) $E(E[x|\mathcal{G}]) = Ex$;

   (ii) $E[E[x|\mathcal{G}]|\mathcal{H}] = E[x|\mathcal{H}]$ *a.s. if $\mathcal{G} \supset \mathcal{H}$;*

   (iii) $E[(\alpha x + \beta y)|\mathcal{G}] = \alpha E[x|\mathcal{G}] + \beta E[x|\mathcal{G}]$ *a.s. if α, $\beta \in \mathbf{R}$;*

   (iv) $E[xy|\mathcal{G}] = y E[x|\mathcal{G}]$ *a.s. if y is measurable with respect to $\mathcal{G}$ and xy is integrable;*

   (v) *if x is independent of $\mathcal{G}$, then $E[x|\mathcal{G}] = Ex$;*

   (vi) $x \geq 0$ *implies $E[x|\mathcal{G}] \geq 0$ a.s.*   □

Remark 4. It is easy to verify that:

   (i) If x is an integrable random variable and y is a simple random variable with values $c_1, \ldots, c_n$, then

$$E[x|y] = \sum_{j \in M} \chi_{y=c_j} E[x|y = c_j],$$

where $M = \{j \in \{1, 2, \ldots, n\}; \; P(y = c_j) > 0\}$.

   (ii) If $A \in \mathcal{F}$, $\mathcal{G}_A = \{\Phi, \Omega, A, \Omega - A\}$, and x is an integrable random variable, then

$$E[x|\mathcal{G}_A] = \begin{cases} \chi_A E[x|A] + \chi_{\Omega-A} E[x|\Omega - A] \text{ if } 0 < P(A) < 1, \\ Ex \text{ if } P(A) = 0 \text{ or } P(A) = 1. \end{cases}$$

Therefore $E[x|\mathcal{G}_A]$ takes at most two values.

1.6 Stochastic processes

In this section $J \subseteq \mathbf{R}$ is an interval. Let us first introduce the following definition.

Definition 13. *An m-dimensional stochastic process is a function $x : J \times \Omega \to \mathbf{R}^m$ with the property that $x(t, \cdot)$ is a random vector for each $t \in J$.*

   Usually we denote a stochastic process by $\{x(t), t \in J\}$, $x = \{x(t)\}_{t \in J}$ or $x(t)$, $t \in J$, the dependence upon the second argument ω being omitted. The functions $t \to x(t, \omega)$ (with ω fixed) are called the *sample paths of the process*.

   If $m = 1$, we shall simply say that x is a *stochastic process*.

Definition 14. (i) *We say that the process $x = \{x(t)\}_{t \in J}$ is continuous if for a.a. ω the functions $x(\cdot, \omega)$ are continuous on J.*

   (ii) *x is called to be* right continuous *if for a.a. ω the functions $x(\cdot, \omega)$ are right continuous on J.*

   (iii) *The process $x = \{x(t)\}_{t \in J}$ is continuous in probability if $t_n \to t_0$ with $t_n, t_0 \in J$ implies $x(t_n) \xrightarrow{P} x(t_0)$.*

   (iv) *x is called to be a* measurable process *if it is measurable on the product space with respect to the σ-algebra $\mathcal{B}(J) \otimes \mathcal{F}$, $\mathcal{B}(J)$ being the σ-algebra of Borel sets in J.*

Remark 5. (i) Every right continuous stochastic process is a measurable process.

(ii) From the Fubini theorem it follows that if $x : J \times \Omega \to \mathbf{R}$ is a measurable process and $E \int_J |x(t)| dt < \infty$, then for a.a. ω, $\int_J x(t) dt$ is a random variable.

Definition 15. *Two stochastic processes $x_1 = \{x_1(t)\}_{t \in J}$, $x_2 = \{x_2(t)\}_{t \in J}$ are called stochastically equivalent if $P\{x_1(t) \neq x_2(t)\} = 0$ for all $t \in J$. We then say that x_2 is a version of x_1.*

Now let us consider a family $\mathcal{M} = \{\mathcal{M}_t\}_{t \in J}$ of σ-algebras $\mathcal{M}_t \subset \mathcal{F}$ with the property that $t_1 < t_2$ implies $\mathcal{M}_{t_1} \subset \mathcal{M}_{t_2}$.

Definition 16. *We say that the process $x = \{x(t)\}_{t \in J}$ is* nonanticipative *with respect to the family $\mathcal{M}$, if*
  (i) *x is a measurable process;*
  (ii) *for each $t \in J$, $x(t, \cdot)$ is measurable with respect to the σ-algebra $\mathcal{M}_t$.*
When (ii) *holds we say that $x(t)$ is $\mathcal{M}_t$-adapted.*

As usual by $L^p(J \times \Omega, \mathbf{R}^m)$, $p \geq 1$, we denote the space of all m-dimensional measurable stochastic processes $x : J \times \Omega \to \mathbf{R}^m$. By $L^p_{\mathcal{M}}(J)$ we denote the space of all $x \in L^p(J \times \Omega, \mathbf{R}^m)$ which are nonanticipative with respect to the family $\mathcal{M} = (\mathcal{M}_t)$, $t \in J$.

Theorem 19. *If for every $t \in J$, the σ-algebra $\mathcal{M}_t$ contains all sets $M \in \mathcal{F}$ with $P(M) = 0$, then $L^p_{\mathcal{M}}(J)$ is a closed subspace of $L^p(J \times \Omega, \mathbf{R}^m)$.*

Proof. Let $x_n \in L^p_{\mathcal{M}}(J)$, $n \geq 1$, be a sequence which converges to $x \in L^p(J \times \Omega, \mathbf{R}^m)$. We have to prove that there exists $\hat{x} \in L^p_{\mathcal{M}}(J)$ such that x_n converges to $\hat{x}$ in the space $L^p(J \times \Omega, \mathbf{R}^m)$. Indeed, since

$$\lim_{n \to \infty} \int_J E|x_n(t) - x(t)|^p dt = 0,$$

by Theorem 9 the sequence of functions $E|x_n(t) - x(t)|^p$ converges in measure to zero. Hence by virtue of Riesz's Theorem there exists a subsequence x_{n_k} and a set $N \subset J$ with $\mu(N) = 0$ (μ being the Lebesgue measure) such that

$$\lim_{n \to \infty} E|x_n(t) - x(t)|^p = 0$$

for all $t \in J - N$. Let $t \in J - N$ be fixed. Again applying Theorem 9 and Riesz's Theorem, one concludes that the sequence $x_{n_k}(t)$, $k \geq 1$, has a subsequence which converges a.e. to $x(t)$. But $x_{n_k}(t)$ are $\mathcal{M}_t$-adapted and $\mathcal{M}_t$ contains all sets $M \in \mathcal{F}$ with $P(M) = 0$. Therefore $x(t)$ is measurable with respect to $\mathcal{M}_t$ for each $t \in J - N$. Now, define $\hat{x} : J \times \Omega \to \mathbf{R}^m$ as follows:

$$\hat{x}(t, \omega) = \begin{cases} x(t, \omega) \text{ if } t \in J - N, \ \omega \in \Omega, \\ 0 \quad \text{if } t \in N \text{ and } \omega \in \Omega. \end{cases}$$

Obviously $\hat{x} \in L^p_{\mathcal{M}}(J)$ and $\lim_{n \to \infty} \int_J E|x_n(t) - x(t)|^p dt = 0$. The proof is complete. $\square$

The next result is proved in [81, Chap. 4, Section 2].

Theorem 20. *Let* $\mathcal{M} = \{\mathcal{M}_t\}_{t \in [a,b]}$ *be an increasing family of* σ*-algebras with the property that for each* t, $\mathcal{M}_t$ *contains all sets* $M \in \mathcal{F}$ *with* $P(M) = 0$. *If* $x = \{x(t)\}_{t \in [a,b]}$ *is a nonanticipative process with respect to the family* $\mathcal{M}$ *and if* $E \int_a^b |x(t)| dt < \infty$, *then the process*

$$y = \{y(t)\}_{t \in [a,b]}, \, y(t) = \int_a^t x(s)ds$$

is nonanticipative with respect to the family $\mathcal{M}$. $\square$

1.7 Stochastic processes with independent increments

Definition 17. *An* r*-dimensional stochastic process* $x(t)$, $t \in [0, \infty)$, *is said to have independent increments if for all* $0 \leq t_0 < t_1 < \cdots < t_k$, *the random vectors* $x(t_0), x(t_1) - x(t_0), \ldots, x(t_k) - x(t_{k-1})$ *are independent.*

Theorem 21. *If* $x(t)$, $t \geq 0$, *is an* r*-dimensional stochastic process with independent increments, then* $\sigma(x(t) - x(a), t \in [a, b])$ *is independent of* $\sigma(x(b + h) - x(b), h > 0)$ *for all* $0 \leq a < b$.

Proof. Let $\mathcal{M}$ be the family of all sets of the form $\cap_{i=1}^p (x(t_i) - x(a))^{-1}(A_i)$ where $a < t_i \leq b$ and $A_i \in \mathcal{B}(\mathbf{R}^r)$, $1 \leq i \leq p$, and let $\mathcal{N}$ be the family of all sets of the form $\cap_{i=1}^m (x(b + h_i) - x(b))^{-1}(B_i)$, where $0 < h_i$, $B_i \in \mathcal{B}(\mathbf{R}^r)$, $1 \leq i \leq m$. Obviously $\mathcal{M}$ and $\mathcal{N}$ are π-systems and

$$\sigma(\mathcal{M}) = \sigma(x(t) - x(a), t \in [a, b]), \, \sigma(\mathcal{N}) = \sigma(x(b + h) - x(b), h > 0).$$

First, we prove that $P(M \cap N) = P(M) \cdot P(N)$ if $M \in \mathcal{M}$ and $N \in \mathcal{N}$. Indeed, let $M = \cap_{i=1}^p (x(t_i) - x(a))^{-1}(A_i)$, $N = \cap_{i=1}^m (x(b + h_i) - x(b))^{-1}(B_i)$ with

$$a < t_1 < \cdots < t_p \leq b, \, 0 < h_1 < \cdots < h_n, \, A_i \in \mathcal{B}(\mathbf{R}^r), \, B_i \in \mathcal{B}(\mathbf{R}^r).$$

Since

$$\sigma(x(t_i) - x(a), 1 \leq i \leq p)$$
$$= \sigma(x(t_1) - x(a), x(t_2) - x(t_1), \ldots, x(t_p) - x(t_{p-1}))$$

and

$$\sigma(x(b + h_i) - x(b), 1 \leq i \leq m)$$
$$= \sigma(x(b + h_1) - x(b), x(b + h_2) - x(b + h_1), \ldots,$$
$$x(b + h_m) - x(b + h_{m-1}))$$

from Theorem 17(ii) it follows that $P(M \cap N) = P(M) \cdot P(N)$. Further, by using Theorem 1 and the equality $A - B = A - (A \cap B)$ one can prove that $P(M \cap B) = P(M) \cdot P(B)$ if $M \in \mathcal{M}$ and $B \in \sigma(x(b + h) - x(b), h > 0)$. Then, applying Theorem 1 again, we prove that $P(A \cap B) = P(A) \cdot P(B)$ if $A \in \sigma(\mathcal{M})$ and $B \in \sigma(\mathcal{N})$. The proof is complete. $\square$

Theorem 22. [106] *If $x(t)$, $t \geq 0$, is a continuous r-dimensional stochastic process with independent increments, then all increments $x(t_2) - x(t_1)$ are Gaussian random vectors.* □

1.8 Wiener process and Markov chain processes

In the following definitions, I is the interval $[0, \infty)$.

Definition 18. *A continuous stochastic process $\beta = \{\beta(t)\}_{t \in I}$ is called a* standard Brownian motion *or a* standard Wiener process *if:*
   (i) $\beta(0) = 0$;
   (ii) $\beta(t)$ *is a stochastic process with independent increments;*
   (iii) $E\beta(t) = 0, t \in I, E|\beta(t) - \beta(s)|^2 = |t - s|$ *with $t, s \in I$.*

Definition 19. *An r-dimensional stochastic process $w(t) = (w_1(t), \ldots, w_r(t))^*$, $t \in I$, is called an* r-dimensional standard Wiener process *if each process $w_i(t)$ is a* standard Brownian motion *and the σ-algebras $\sigma(w_i(t), t \in I), 1 \leq i \leq r$, are independent.*

For each $t \geq 0$, by $\mathcal{F}_t$ we denote the smallest σ-algebra which contains all sets $M \in \mathcal{F}$ with $P(M) = 0$ and with respect to which all random vectors $\{w(s)\}_{s \leq t}$ are measurable.
   For $t \geq 0, \mathcal{U}_t = \sigma(w(t + h) - w(t), h > 0)$.
   From Theorem 21 it follows that for each $t \in I$, $\mathcal{F}_t$ is independent of $\mathcal{U}_t$.

Remark 6. (i) Since $w(t) - w(s)$ is independent of $\mathcal{F}_s$ if $t > s$ (see Theorem 21), from Theorem 18(v) it follows that

$$E[(w(t) - w(s)) \mid \mathcal{F}_s] = 0, \tag{1.2}$$
$$E[(w(t) - w(s))(w(t) - w(s))^* \mid \mathcal{F}_s] = I_r(t - s), \ t > s, \ \text{a.e.}$$

(ii) The increments $w(t) - w(s)$, $t \neq s$ are nondegenerate Gaussian random vectors (see Theorem 22 and (1.1)).

The converse assertion in (i) is also valid.

Theorem 23. [52], [81] *Let $w(t), t \geq 0$, be a continuous r-dimensional stochastic process with $w(0) = 0$ and adapted to an increasing family of σ-algebras $\mathcal{F}_t, t \geq 0$, such that (1.2) hold. Then $w(t), t \geq 0$, is a standard r-dimensional Wiener process and all increments $w(t_2) - w(t_1), t_2 \neq t_1$, are nondegenerate Gaussian random vectors.* □

Theorems 22 and 23 will not be used in this book, but they are given because they are interesting by themselves and they give a more detailed image of the properties of these stochastic processes.

Definition 20. *A family* $P(t) = [p_{ij}(t)], t \in (0, \infty)$, *of* $d \times d$ *matrices is said to be a transition semigroup if the following two conditions are satisfied:*

(i) *For each* $t > 0$, $P(t)$ *is a stochastic matrix, that is,* $0 \leq p_{ij}(t) \leq 1$ *and* $\sum_{j=1}^{d} p_{ij}(t) = 1$, $1 \leq i \leq d$.

(ii) $P(t + s) = P(t)P(s)$ *for all* $t > 0$, $s > 0$.

The equality (ii) *is termed the* homogeneous Chapman–Kolmogorov relation.

Definition 21. *A stochastic process* $\eta(t), t \in [0, \infty)$, *is called a* standard homogeneous Markov chain *with state-space the set* $\mathcal{D} = \{1, 2, \ldots, d\}$ *and the transition semigroup* $P(t) = [p_{ij}(t)], t > 0$, *if:*

(i) $\eta(t, \omega) \in \mathcal{D}$ *for all* $t \geq 0$ *and* $\omega \in \Omega$;

(ii) $P[\eta(t + h = j)|\eta(s), s \leq t] = p_{\eta(t)j}(h)$ *a.s. for all* $t \geq 0$, $h > 0$, $j \in \mathcal{D}$;

(iii) $\lim_{h \to 0_+} P(t) = I_d$, I_d *is the identity matrix of dimension* $(d \times d)$;

(iv) $\eta(t), t \geq 0$ *is a right continuous stochastic process.*

In fact, the above definition says that a standard homogeneous Markov chain is a *triplet* $\{\eta(t), P(t), \mathcal{D}\}$ satisfying (i)–(iv), $P(t), t > 0$, being a transition semigroup.

The next result is proved in [26].

Theorem 24. *The standard homogeneous Markov chain has the following properties:*

(i) $P\{\eta(t + h) = j|\eta(t) = i\} = p_{ij}(h)$ *for all* $i, j \in \mathcal{D}, h > 0, t \geq 0$ *with* $P\{\eta(t) = i\} > 0$.

(ii) $P\{\eta(t + h) = j|\eta(s), s \leq t\} = P[\eta(t + h) = j|\eta(t)], t \geq 0, h > 0$, $j \in \mathcal{D}$, *a.s.*

(iii) *If* x *is a bounded random variable measurable with respect to the* σ-algebra $\sigma(\eta(s), s \geq t)$, *then* $E[x|\eta(u), u \leq t] = E[x|\eta(t)]$, *a.s.,* $t \geq 0$.

(iv) $\eta(t)$ *is continuous in probability.*

(v) $p_{ii}(t) > 0$ *for all* $i \in \mathcal{D}, t > 0$.

(vi) $\lim_{t \to \infty} P(t)$ *exists.*

(vii) *There exists a constant matrix* Q *such that* $P(t) = e^{Qt}, t > 0, Q = [q_{ij}]$ *is a matrix with* $q_{ij} \geq 0$ *if* $i \neq j$ *and* $\sum_{j=1}^{d} q_{ij} = 0$. $\square$

In fact (ii) follows from (iii) since $\chi_{\eta(t+h)=j}$ is measurable with respect to the σ-algebra $\sigma(\eta(u), u \geq t)$.

The assertion (iii) in Theorem 24 is termed *the Markov property of the process* $\eta(t)$.

The fact that a transition semigroup $P(t)t > 0$, with the property that $\lim_{t \to 0_+} P(t) = I_d$ admits an infinitesimal generator Q ($P(t) = e^{Qt}, t > 0$) follows from the general theory of semigroups in Banach algebras [63], but in the theory of Markov processes a probabilistic proof is given in [16], [26], [55].

We assume in the following that $\pi_i := P\{\eta(0) = i\} > 0$ for all $i \in \mathcal{D}$.

Remark 7. From the above assumption and from the equality

$$P\{\eta(t) = i\} = \sum_{j=1}^{d} \pi_j P\{\eta(t) = i|\eta(0) = j\},$$

we deduce that $P\{\eta(t) = i\} \geq \pi_i p_{ii}(t) > 0, t \geq 0, i \in \mathcal{D}$.

In the following developments $\mathcal{G}_t, t \geq 0$, denotes the family of σ-algebras $\mathcal{G}_t = \sigma(\eta(s); 0 \leq s \leq t)$ and $\mathcal{V}_t, t \geq 0$ is the family of σ-algebras $\mathcal{V}_t = \sigma(\eta(s), s \geq t)$.

1.9 Stochastic integral

Throughout this section and throughout the monograph we consider the pair $(w(t), \eta(t)), t \geq 0$, where $w(t)$ is an r-dimensional standard Wiener process and $\eta(t)$ is a standard homogeneous Markov chain (see Definitions 19, 21). Assume that the σ-algebra $\mathcal{F}_t$ is independent of $\mathcal{G}_t$ for every $t \geq 0$, where $\mathcal{F}_t$ and $\mathcal{G}_t$ have been defined in the preceding section.

Denote by $\mathcal{H}_t := \mathcal{F}_t \vee \mathcal{G}_t, t \geq 0$.

Let $\widetilde{\mathcal{G}} = \sigma(\eta(t), t \geq 0)$.

Theorem 25. *For every $t \geq 0$, $\mathcal{F}_t$ is independent of $\widetilde{\mathcal{G}}$ and $\mathcal{U}_t$ is independent of $\mathcal{F}_t \vee \widetilde{\mathcal{G}}$. Therefore, $\mathcal{U}_t$ and $\mathcal{H}_t$ are independent σ-algebras for every $t \geq 0$.*

Proof. First one proves that $\mathcal{F}_t$ is independent of $\mathcal{G}_s$ for all $t \geq 0$, $s \geq 0$. Indeed, if $t < s$ we have $\mathcal{F}_t \subset \mathcal{F}_s$ and since $\mathcal{F}_s$ is independent of $\mathcal{G}_s$ it follows that $\mathcal{F}_t$ and $\mathcal{F}_s$ are independent σ-algebras. Similarly one proves that $t > s$.

Now let $\mathcal{M}_0$ be the family of all sets of the form $\cap_{k=1}^m \{\eta(t_k) = i_k\}$, with $t_k \geq 0$, $t_k \neq t_\ell$, if $k \neq \ell$ and $i_k \in D$, $1 \leq k \leq m$,

$$\mathcal{M} = \{A; A \in M_0 \text{ or } A = \emptyset\}, \quad \mathcal{N}_t = \{G \cap F; G \in \widetilde{\mathcal{G}}, F \in \mathcal{F}_t\},$$

and S_t be the family of all sets of the form $\cap_{i=1}^p (w(t + h_i) - w(t))^{-1}(B_i)$ with $h_i > 0$, $B_i \in \mathcal{B}(\mathbf{R}^r)$, $1 \leq i \leq p$. Obviously $\mathcal{M}, \mathcal{N}_t$, and S_t are π-systems and $\sigma(\mathcal{M}) = \widetilde{\mathcal{G}}$, $\sigma(\mathcal{N}_t) = \mathcal{F}_t \vee \widetilde{\mathcal{G}}$, and $\sigma(S_t) = \mathcal{U}_t$.

Define $\mathcal{G}(F) = \{G \in \widetilde{\mathcal{G}}; P(G \cap F) = P(G)P(F)\}$ for each $F \in \mathcal{F}_t$. Since $\mathcal{F}_t$ is independent of $\mathcal{G}_s$ for all $s \geq 0$, it follows that $\mathcal{M} \subset \mathcal{G}(F)$. By using the equality $F - G = F - (F \cap G)$ one verifies easily that $\mathcal{G}(F)$ satisfies conditions (ii) and (iii) in Theorem 1. Thus, by virtue of Theorem 1, $\mathcal{G}(F) = \widetilde{\mathcal{G}}$ for all $F \in \mathcal{F}_t$ and thus the first assertion in the theorem is proved.

Further, if $S \in S_t$, $H \in \mathcal{N}_t$, $H = G \cap F$, $G \in \widetilde{\mathcal{G}}$, $F \in \mathcal{F}_t$, since $\mathcal{F}_u$ is independent of $\widetilde{\mathcal{G}}$ for every $u \geq 0$ and $\mathcal{U}_t$ is independent of $\mathcal{F}_t$ (see Theorem 21), we have

$$P(S \cap H) = P(S \cap G \cap F) = P(G)P(S \cap F)$$
$$= P(G)P(S)P(F) = P(S)P(H).$$

Therefore, by using Theorem 1, one gets $P(U \cap H) = P(U)P(H)$ for all $U \in \mathcal{U}_t$, $H \in \mathcal{N}_t$ and applying Theorem 1 again, one concludes that $P(U \cap V) = P(U)P(V)$ if $U \in \mathcal{U}_t$, $V \in \mathcal{F}_t \vee \widetilde{\mathcal{G}}$. The proof is complete. $\qquad\square$

If $[a, b] \subset [0, \infty)$ we denote by $L_{\eta,w}^{2p}[a, b]$ the space of all nonanticipative processes $f(t), t \in [a, b]$, with respect to the family $\mathcal{H} = (\mathcal{H}_t), t \in [a, b]$, with $E \int_a^b f^{2p}(t)dt < \infty$.

Let $k \in \{1, \ldots, r\}$ be fixed and let $\beta(t) = w_k(t), t \geq 0$.

Since the family of σ-algebras $\mathcal{H}_t, t \in [a, b]$, has the properties used in the theory of the *Itô stochastic integral*, namely:

(a) $\mathcal{H}_{t_1} \subset \mathcal{H}_{t_2}$ if $t_1 < t_2$;

(b) $\sigma(\beta(t + h) - \beta(t), h > 0)$ is independent of $\mathcal{H}_t$ (see Theorem 25);

(c) $\beta(t)$ is measurable with respect to $\mathcal{H}_t$;

(d) $\mathcal{H}_t$ contains all sets $M \in \mathcal{F}$ with $P(M) = 0$ for every $t \geq 0$,

we can define the Itô stochastic integral $\int_a^b f(t)d\beta(t)$ (see [52], [55], [81], [97], [98]) with $f \in L^2_{\eta,w}[a, b]$.

Definition 22. *A stochastic process $f(t)$, $t \in [a, b]$, is called a* step function *if there exists a partition $a = t_0 < t_1 \cdots < t_m = b$ of $[a, b]$ such that $f(t) = f(t_i)$ if $t \in [t_i, t_{i+1}), 1 \leq i \leq m - 1$.*

If f is a nonanticipative step function, by definition

$$\int_a^b f(t)d\beta(t) = \sum_{i=1}^{m-1} f(t_i)(\beta(t_{i+1}) - \beta(t_i)).$$

Further, let us remember some properties of the integral $\int_a^b f(t)d\beta(t)$ that are proved in [52].

Theorem 26. *If $f \in L^2_{\eta,w}[a, b]$ we have the following properties:*

(i) *There exists a sequence f_n of step functions in $L^2_{\eta,w}[a, b]$ such that $E \int_a^b |f_n(t) - f(t)|^2 dt \rightarrow 0$ and the sequence $\int_a^b f_n(t)d\beta(t)$ is convergent in probability; its limit is by definition $\int_a^b f(t)d\beta(t)$.*

(ii) $E\left[\int_a^b f(t)d\beta(t) | \mathcal{H}_a\right] = 0$ *and therefore* $E\left[\int_a^b f(t)d\beta(t) | \eta(a) = i\right] = 0$, $i \in \mathcal{D}$.

(iii) $E\left[(\int_a^b f(t)d\beta(t))^2 | \mathcal{H}_a\right] = E\left[\int_a^b f^2(t)dt | \mathcal{H}_a\right]$ *and therefore*

$$E\left[\left(\int_a^b f(t)d\beta(t)\right)^2 \Big| \eta(a) = i\right] = E\left[\int_a^b f^2(t)dt | \eta(a) = i\right], i \in \mathcal{D}.$$

(iv) *If ξ is a bounded random variable measurable with respect to $\mathcal{H}_a$, then*

$$\int_a^b \xi f(t)d\beta(t) = \xi \int_a^b f(t)d\beta(t).$$

(v) *The process $x(t) = \int_a^t f(t)d\beta(t), t \in [a, b]$, admits a continuous version and $x(t)$ is $\mathcal{H}_t$ adapted.* $\square$

Theorem 27. *Let $f \in L^{2p}_{\eta,w}[a, b]$ where p is a positive integer. Then*

$$E\left(\int_a^b f(t)d\beta(t)\right)^{2p} \leq p(2p - 1)^p(b - a)^{p-1}E\left(\int_a^b f^{2p}(t)dt\right). \qquad \square$$

Remark 8. (i) Since almost all sample paths of a Brownian motion have infinite variation on any finite interval (see [52]) the stochastic Itô integral cannot be defined in the usual Lebesgue–Stieljes sense, with ω fixed; therefore the assertion (iv) in Theorem 26 is not trivial and it must be proved.

(ii) The stochastic Itô integral can be defined for nonanticipative functions f with the property $\int_a^b |f(t)|dt < \infty$ a.s., but the equalities in (ii) and (iii) of Theorem 26 hold if $E \int_a^b |f(t)|^2 dt < \infty$.

Remark 9. The proof of assertion (i) in Theorem 26 shows (see also Lemma 6.2, Chapter 4, in [52]) that if $f \in L_{\mathcal{M}}^{2p}([a, b])$ where the increasing family $\mathcal{M}$ of σ-algebras has the property in Theorem 19, then there exists a sequence f_n of step functions $f_n \in L_{\mathcal{M}}^{2p}([a, b])$ such that $\lim_{n\to\infty} E \int_a^b |f_n - f|^{2p} dt = 0$.

The next result has been proved in [80].

Theorem 28. *If $f \in L_{\eta,w}^2[a, b]$ we have $E\left[\chi_{\eta(b)=i} \int_a^b f(t)d\beta(t)|\mathcal{H}_a \right] = 0$ for every $i \in \mathcal{D}$.*

Proof. We prove first that if $f \in L_{\eta,w}^2[a, b]$ is a step function, then

$$E\left(\chi_{\eta(b)=i} \int_a^b f(t)d\beta(t) \right) = 0.$$

Indeed, let $f(t) = \sum_{k=0}^{m-1} f(t_k)\chi_{[t_k,t_{k+1}]}$, $f(t_k)$ being measurable with respect to $\mathcal{H}_{t_k}$. Since $\mathcal{H}_{t_k} \vee \sigma(\eta(b)) \subset \mathcal{F}_{t_k} \vee \tilde{\mathcal{G}}$ by Theorem 25, it follows that $\beta(t_{k+1}) - \beta(t_k)$ is independent of the σ-algebra $\mathcal{H}_{t_k} \vee \sigma(\eta(b))$, and thus by Theorem 18(v) one gets

$$E[(\beta(t_{k+1}) - \beta(t_k))|\mathcal{H}_{t_k} \vee \sigma(\eta(b))] = E(\beta(t_{k+1}) - \beta(t_k)) = 0.$$

Hence, by using the properties of the conditional expectation (see Theorem 18), one can write

$$E\chi_{\eta(b)=i} \int_a^b f(t)d\beta(t) = \sum_{k=1}^{m-1} E\chi_{\eta(b)=i} f(t_k)(\beta(t_{k+1}) - \beta(t_k))$$

$$= \sum_{k=1}^{m-1} E(E[\chi_{\eta(b)=i} f(t_k)(\beta(t_{k+1}) - \beta(t_k))$$
$$|\mathcal{H}_{t_k} \vee \sigma(\eta(b))])$$

$$= \sum_{k=1}^{m-1} E(\chi_{\eta(b)=i} f(t_k)E[(\beta(t_{k+1}) - \beta(t_k))$$
$$|\mathcal{H}_k \vee \sigma(\eta(b))])$$

$$= 0.$$

Further, by Theorem 26, let f_n be a sequence of step functions in $L^2_{\eta,w}[a,b]$ with $E\int_a^b |f_n(t) - f(t)|^2 dt \to 0$. We have by virtue of Corollary 8 and Theorem 26

$$\left| E\left(\chi_{\eta(b)=i}\int_a^b f(t)d\beta(t)\right)\right| = \left| E\left[\chi_{\eta(b)=i}\left(\int_a^b f(t)d\beta(t) - \int_a^b f_n(t)d\beta(t)\right)\right]\right|$$

$$\leq E\left|\int_a^b (f_n(t) - f(t))d\beta(t)\right|$$

$$\leq \left[E\left(\int_a^b (f_n(t) - f(t))d\beta(t)\right)^2\right]^{1/2}$$

$$= \left(E\int_a^b (f_n(t) - f(t))^2 dt\right)^{1/2} \to 0 \text{ for } n \to \infty.$$

Hence

$$E\chi_{\eta(b)=i}\int_a^b f(t)d\beta(t) = 0. \tag{1.3}$$

Let ξ be a bounded random variable measurable with respect to $\mathcal{H}_a$.

Then it follows that $\xi f \in L^2_{\eta,w}[a,b]$ and hence (1.3) gives

$$E\chi_{\eta(b)=i}\int_a^b \xi f(t)d\beta(t) = 0.$$

But, according to Theorem 26(iv), we can write

$$E\chi_{\eta(b)=i}\xi\int_a^b f(t)d\beta(t) = E\chi_{\eta(b)=i}\int_a^b \xi f(t)d\beta(t) = 0.$$

Hence, by Theorem 18 we have

$$E\left(\xi E\left[\chi_{\eta(b)=i}\int_a^b f(t)d\beta(t)|\mathcal{H}_a\right]\right) = E\left(E\left[\xi\chi_{\eta(b)=i}\int_a^b f(t)d\beta(t)|\mathcal{H}_a\right]\right)$$

$$= E\left[\xi\chi_{\eta(b)=i}\int_a^b f(t)d\beta(t)\right] = 0.$$

Taking in the above equality $\xi = \chi_A$, $A \in \mathcal{H}_a$, we get that

$$E\left[\chi_{\eta(b)=i}\int_a^b f(t)d\beta(t)|\mathcal{H}_a\right] = 0 \text{ a.s.},$$

and the proof is complete. $\qquad\qquad\square$

Further, let $\sigma = (\sigma_{kl})$ be an $n \times r$ matrix whose elements are in $L^2_{\eta,w}[a,b]$. Then the stochastic integral $\int_a^b \sigma(t)dw(t)$ is an n-column vector whose k's component is given by

$$\sum_{\ell=1}^r \int_a^b \sigma_{k\ell}(t)dw_\ell(t), 1 \leq k \leq n,$$

where the integral $\int_a^b \sigma_{k\ell}(t)dw_\ell(t)$ is the Itô integral for $\beta = w_\ell$ with respect to the family of σ-algebra $\mathcal{H}_t$.

Here $w(t) = (w_1(t), \ldots, w_r(t))^*$.

Remark 10. From Theorem 26 it follows directly that if ξ is a bounded random variable measurable with respect to $\mathcal{H}_a$, then

$$\xi \int_a^b \sigma(t)dw(t) = \int_a^b \xi\sigma(t)dw(t) \text{ a.s.,}$$

the elements of $\sigma(t)$ being in $L_{\eta,w}^2[a,b]$.

The next result follows from Theorem 26 and it can be found in all books containing the theory of the stochastic Itô integral.

Theorem 29. *If the elements of $\sigma(t)$ are in $L_{\eta,w}^2[a,b]$, then*

$$E\int_a^b \sigma(t)dw(t) = 0 \ \text{and} \ E\left|\int_a^b \sigma(t)dw(t)\right|^2 = E\int_a^b \|\sigma(t)\|^2 dt,$$

where

$$\|\sigma(t)\|^2 = \sum_{k,\ell} \sigma_{k,\ell}^2(t) = Tr(\sigma^*(t)\sigma(t)). \qquad \square$$

Theorem 27 implies the following result directly.

Theorem 30. *If all elements of the matrix $\sigma(t)$ are in $L_{\eta,w}^{2p}[a,b]$, p being a positive integer, then*

$$E\left|\int_a^b \sigma(t)dw(t)\right|^{2p} \leq nr[p(2p-1)]^p(b-a)^{p-1}\sum_{k,\ell} E\int_a^b \sigma_{k,\ell}^{2p}(t)dt. \qquad \square$$

Applying Theorems 29 and 30 for $\chi_{\eta(a)=i} \cdot \sigma$ and taking into account Remark 10, one gets the following results.

Theorem 31. *Under the assumption of Theorem 29 we have*

$$E\left[\int_a^b \sigma(t)dw(t) \mid \eta(a) = i\right] = 0,$$

$$E\left[\left|\int_a^b \sigma(t)dw(t)\right|^2 \mid \eta(a) = i\right] = E\left[\int_a^b \|\sigma(t)\|^2 dt \mid \eta(a) = i\right]$$

for all $i \in \mathcal{D}$. $\qquad \square$

Theorem 32. *Under the assumption of Theorem 30 we have*

$$E\left[\left|\int_a^b \sigma(t)dw(t)\right|^{2p} \mid \eta(a) = i\right]$$

$$\leq nr[p(2p-1)]^p(b-a)^{p-1}\sum_{k,\ell} E\left[\int_a^b \sigma_{k,\ell}^{2p}(t)dt \mid \eta(a) = i\right]$$

for all $i \in \mathcal{D}$. $\qquad \square$

Definition 23. *Let $x(t), t \in [t_0, T]$, be an n-dimensional stochastic process verifying*

$$x(t) - x(t_0) = \int_{t_0}^{t} a(s)ds + \int_{t_0}^{t} \sigma(s)dw(s), \ a.s. \ if \ [t_0, T],$$

where $a = (a_1, \ldots, a_n)^*, \sigma = (\sigma_{k\ell})$ with $1 \le k \le n, 1 \le \ell \le r$, and $a_k, \sigma_{k\ell}$ being in $L^2_{\eta,w}[t_0, T]$ for all k and ℓ. Then we say that $x(t)$ has a stochastic differential $dx(t)$ given by

$$dx(t) = a(t)dt + \sigma(t)dw(t), \ t \in [t_0, T]. \tag{1.4}$$

Obviously if $x(t_0)$ is measurable with respect to $\mathcal{H}_{t_0}$ and $E|x(t_0)|^2 < \infty$, the above stochastic process $x = (x(t)), t \in [t_0, T]$, is a continuous process and $x \in L^2_{\eta,w}[t_0, T]$.

Theorem 33. *(Itô's formula) Let $v(t, x)$ be a continuous function in $(t, x) \in [0, T] \times \mathbf{R}^n$ together with its derivatives v_t, v_x, v_{xx}. If $x(t)$ verifies (1.4), then*

$$dv(t, x(t)) = \left[\frac{\partial v}{\partial t}(t, x(t)) + \left(\frac{\partial v}{\partial x}(t, x(t)) \right)^* a(t) \right.$$
$$\left. + \frac{1}{2} Tr\sigma^*(t) \frac{\partial^2 v}{\partial x \partial x}(t, x(t))\sigma(t) \right] dt$$
$$+ \left(\frac{\partial v}{\partial x}(t, x(t)) \right)^* \sigma(t)dw(t),$$

a.s., if $t \in [t_0, T]$. □

1.10 An Itô-type formula

We are interested in the following to obtain an Itô-type formula for (1.4) with functions $v(t, x, i), i \in \mathcal{D}$, rather than $v(t, x)$, namely for functions depending upon the states i of the Markov process $\eta(t)$.

Since $\mathcal{H}_t$ incorporates properties of $\eta(t)$, we would like to exploit the properties of both $w(t)$ and $\eta(t)$. This fact will be more clear in the following developments when stochastic differential equations with Markovian jumping will be investigated.

A strong argument for considering functions $v(t, x, i)$ instead of $v(t, x)$ is that the Itô formula for the function $v(t, x)$ (Theorem 33) does not retain the fundamental elements of the process $\eta(t)$ as $p_{ij}(t)$ and q_{ij}.

We must emphasize the fact that by contrast with the Itô formula given in Theorem 33, which is valid for a.a. $\omega \in \Omega$, when considering functions $v(t, x, i)$ we cannot expect to obtain a similar formula for $v(t, x(t), \eta(t))$ holding a.s. This is due to the fact that the coefficients q_{ij} are strongly related by considering the conditional expectation with respect to the events $\{\eta(t) = i\}$.

In order to prove an Itô-type formula for functions $v(t, x, i)$, we need the following result, which is interesting by itself.

Let us denote $\mathcal{R}_t = \mathcal{U}_t \vee \mathcal{V}_t, t \ge 0$, where the σ-algebras $\mathcal{U}_t$ and $\mathcal{V}_t$ are as defined in Section 1.8.

Theorem 34. *If ξ is an integrable random variable measurable with respect to $\mathcal{R}_t$, that is, $\xi \in L^1(\Omega, \mathcal{R}_t, P)$, then $E[\xi|\mathcal{H}_t] = E[\xi|\eta(t)]$ a.s.*

Proof. The proof is made in two steps. In the first step we show that the equality in the statement holds for $\xi = \chi_B$ for all $B \in \mathcal{R}_t$, and in the second step we consider the general situation when ξ is integrable.

Step 1 Define $z = E[\xi|\eta(t)]$. We must prove that

$$E(z\chi_A) = E(\xi\chi_A) \; \forall A \in \mathcal{H}_t. \tag{1.5}$$

First we shall prove that (1.5) holds in the particular case when $\xi = \chi_M \chi_N$, $M \in \mathcal{U}_t$, $N \in \mathcal{V}_t$.

Let $\mathcal{M}$ be the family of all sets $A \in \mathcal{F}$ verifying (1.5). It is obvious that $\mathcal{M}$ verifies (ii) and (iii) in Theorem 1.

Let $\mathcal{C} = \{F \cap G; \; F \in \mathcal{F}_t, \; G \in \mathcal{G}_t\}$; it is easy to check that $\mathcal{C}$ is a π system. We show now that $\mathcal{C} \subset \mathcal{M}$. Indeed, let $F \in \mathcal{F}_t$, $G \in \mathcal{G}_t$; we must prove that $E(z\chi_F\chi_G) = E(\xi\chi_F\chi_G)$. But since χ_M is independent of $\{\chi_N, \eta(t)\}$ (see Theorem 25) we can write

$$\int_{\{\eta(t)=i\}} E(\chi_M) E[\chi_N|\eta(t)] dP = E(\chi_M) \int_{\{\eta(t)=i\}} E[\chi_N|\eta(t)] dP$$
$$= E(\chi_M) E(\chi_N \chi_{\eta(t)=i}) = E(\chi_M \chi_N \chi_{\eta(t)=i})$$
$$= \int_{\{\eta(t)=i\}} \chi_M \chi_N dP.$$

Hence $z = E(\chi_M) E[\chi_N|\eta(t)]$ (in our case $z = E[\chi_M \chi_N|\eta(t)]$).

From Theorem 24(iii) we have $E[\chi_N|\eta(t)] = E[\chi_N|\mathcal{G}_t]$.

Further, since χ_M is independent of $\{\chi_F, \chi_G, \chi_N\}$ and χ_F is independent of $\{\chi_G, E[\chi_N|\eta(t)]\}$ (see Theorem 25), we can write, applying Theorems 17 and 18, that:

$$E(\xi\chi_F\chi_G) = E(\chi_M\chi_N\chi_F\chi_G) = E(\chi_M)E(\chi_N\chi_F\chi_G)$$
$$= E(\chi_M)E(\chi_F)E(\chi_N\chi_G),$$
$$E(z\chi_F\chi_G) = E(\chi_M)E(\chi_F\chi_G E[\chi_N|\eta(t)])$$
$$= E(\chi_M)E(\chi_F)E(\chi_G E[\chi_N|\eta(t)])$$
$$= E(\chi_M)E(\chi_F)E(\chi_G E[\chi_N|\mathcal{G}_t])$$
$$= (E\chi_M)(E\chi_F)E(E[\chi_G\chi_N|\mathcal{G}_t])$$
$$= E(\chi_M)E(\chi_F)E(\chi_N\chi_G).$$

Thus we proved that $\mathcal{C} \subset \mathcal{M}$. Hence by Theorem 1 $\sigma(\mathcal{C}) \subset \mathcal{M}$. But $\sigma(\mathcal{C}) = \mathcal{H}_t$, thus $E[\chi_M\chi_N|\mathcal{H}_t] = E[\chi_M\chi_N|\eta(t)]$ for all $M \in \mathcal{U}_t$, $N \in \mathcal{V}_t$.

Now let $\mathcal{N}$ be the family of $B \in \mathcal{F}$ with $E[\chi_B|\mathcal{H}_t] = E[\chi_B|\eta(t)]$.

We know that $\mathcal{N}$ contains $\widehat{\mathcal{C}} = \{M \cap N, \; M \in \mathcal{U}_t, N \in \mathcal{V}_t\}$. $\widehat{\mathcal{C}}$ is a π system and since $\mathcal{N}$ verifies (ii) and (iii) in Theorem 1 it follows that $\mathcal{N} \supset \sigma(\widehat{\mathcal{C}}) = \mathcal{R}_t$.

Step 2 First assume that $\xi \geq 0$; by Theorem 6 there exists a sequence of simple random variables $\xi_n(\omega)$ with the properties $0 \leq \xi_n \leq \xi_{n+1}$; $\lim_{n \to \infty} \xi_n(\omega) = \xi(\omega)$ and ξ_n are measurable with respect to $\mathcal{R}_t$. For each $n \geq 1$ we have $E[\xi_n | \mathcal{H}_t] = E[\xi_n | \eta(t)]$.

Applying Theorem 11, the equality in the statement is valid in the case when ξ is nonnegative, integrable, and measurable with respect to $\mathcal{R}_t$.

In the general case we can write $\xi = \xi^+ - \xi^-$, where $\xi^+ = \frac{1}{2}(|\xi| + \xi)$ and $\xi^- = \frac{1}{2}(|\xi| - \xi), \xi^+ \geq 0, \xi^- \geq 0$, and thus the equality in the statement takes place for ξ^+ and ξ^- and therefore, according to Theorem 18, the proof is complete. □

Theorem 35. *(Itô-type formula) Let us consider* $a = (a_1, \ldots, a_n)^*$ *with* $a_k \in L^2_{\eta,w} \times ([t_0, T]), 1 \leq k \leq n, \sigma = [\sigma_{ij}]_{1 \leq i \leq n, 1 \leq j \leq r}$ *with* $\sigma_{ij} \in L^2_{\eta,w}([t_0, T])$ *and* ξ *an n-dimensional random vector* $\mathcal{H}_{t_0}$ *measurable with* $E|\xi|^2 < \infty$ *and let the function*

$$v(t, x, i) = x^* K(t, i)x + 2k^*(t, i)x + k_0(t, i),$$

where $K : [t_0, T] \times \mathcal{D} \to \mathbf{R}^{n \times n}, K = K^*, k : [t_0, T] \times \mathcal{D} \to \mathbf{R}^n, k_0 : [t_0, T] \times \mathcal{D} \to \mathbf{R}$ *are* C^1*-functions with respect to* t. *Then the following equality is true:*

$$E\left[\left(v(t, x(t), \eta(t)) - v(t_0, \xi, i) \right) | \eta(t_0) = i \right]$$
$$= E\left[\int_{t_0}^{t} \left\{ \frac{\partial v}{\partial t}(s, x(s), \eta(s)) + a^*(s) \frac{\partial v}{\partial x}(s, x(s), \eta(s)) \right. \right.$$
$$\left. \left. + Tr(\sigma^*(s)K(s, \eta(s))\sigma(s)) + \sum_{j=1}^{d} v(s, x(s), j)q_{\eta(s),j} \right\} ds | \eta(t_0) = i \right]$$

(1.6)

for all $i \in \mathcal{D}$ *and for the stochastic process* $x(t), t \in [t_0, T]$, *verifying*

$$dx(t) = a(t)dt + \sigma(t)dw(t), t \in [t_0, T], \text{ and } x(t_0) = \xi.$$

Proof. The proof consists of three steps.

Step 1 Assume that ξ, a, σ satisfy the assumption in the statement and additionally ξ is a bounded random vector a, σ are bounded on $[t_0, T] \times \Omega$, and $a(t), \sigma(t)$ are, with probability 1, right continuous functions on $[t_0, T]$.

Under these assumptions, applying Theorem 30, we deduce that

$$\sup_{t \in [t_0, T]} E|x(t)|^{2k} < \infty$$

for all $k \in \mathbf{N}, k \geq 1$. We can write

$$v(t + h, x(t + h), \eta(t + h)) - v(t, x(t), \eta(t))$$
$$= v(t + h, x(t + h), \eta(t + h)) - v(t, x(t), \eta(t + h))$$
$$+ v(t, x(t), \eta(t + h)) - v(t, x(t), \eta(t))$$
$$= \sum_{j=1}^{d} \chi_{\eta(t+h)=j}(v(t + h, x(t + h), j) - v(t, x(t), j))$$
$$+ v(t, x(t), \eta(t + h)) - v(t, x(t), \eta(t)),$$

where χ_M is the indicator function of the set M.

For each fixed $j \in \mathcal{D}$, we can apply the Itô formula (Theorem 33) and obtain

$$v(t+h, x(t+h), j) - v(t, x(t), j)$$
$$= \int_t^{t+h} m_j(s)ds + 2 \int_t^{t+h} (x^*(s)K(s, j) + k^*(s, j))\sigma(s)dw(s),$$

where

$$m_j(s) = x^*(s)\dot{K}(s, j)x(s) + 2\dot{k}^*(s, j)x(s) + \dot{k}_0(s, j) + 2x^*(s)K(s, j)a(s)$$
$$+ 2k^*(s, j)a(s) + Tr(\sigma^*(s)K(s, j)\sigma(s)),$$

$j \in \mathcal{D}$. Using Theorem 28, we deduce that

$$E\left[\chi_{\eta(t+h)=j} \int_t^{t+h} [x^*(s)K(s, j) + k^*(s, j)]\sigma(s)dw(s)|\mathcal{H}_t\right] = 0.$$

Hence

$$E\left[\chi_{\eta(t+h)=j} \int_t^{t+h} (x^*(s)K(s, j) + k^*(s, j))\sigma(s)dw(s)|\eta(t_0) = i\right] = 0,$$

and finally we deduce

$$E[(v(t+h, x(t+h), \eta(t+h)) - v(t, x(t), \eta(t+h)))|\eta(t_0) = i]$$
$$= \sum_{j=1}^d E\left[\chi_{\eta(t+h)=j} \int_t^{t+h} m_j(s)ds|\eta(t_0) = i\right]. \tag{1.7}$$

It is obvious that $m_j(s)$ is, with probability 1, right continuous, and hence we have

$$\lim_{h\searrow 0} \frac{1}{h} \int_t^{t+h} m_j(s)ds = m_j(t), \quad t \in [t_0, T), j \in \mathcal{D}.$$

Since $\eta(t)$ is right continuous we can write

$$\lim_{h\searrow 0} \frac{1}{h} \chi_{\eta(t+h)=j} \int_t^{t+h} m_j(s)ds = \chi_{\eta(t)=j}m_j(t). \tag{1.8}$$

On the other hand, since $\sup_{t\in[t_0.T]} E|x(t)|^4 < \infty$ we obtain that there exists $\beta > 0$ (not depending upon t, h) such that:

$$E\left|\frac{1}{h}\chi_{\eta(t+h)=j} \int_t^{t+h} m_j(s)ds\right|^2 \leq \beta.$$

Thus, from (1.7) and (1.8) and Theorem 13, it follows that

$$\lim_{h\searrow 0} \frac{1}{h} E[(v(t+h, x(t+h), \eta(t+h)) - v(t, x(t), \eta(t+h)))|\eta(t_0) = i]$$
$$= \sum_{j=1}^r E[\chi_{\eta(t)=j}m_j(t)|\eta(t_0) = i] = E[\tilde{m}(t)|\eta(t_0) = i], \tag{1.9}$$

$t \in [t_0, T), i \in \mathcal{D}$, where

$$\tilde{m}(t) = x^*(t)\dot{K}(t, \eta(t))x(t) + 2\dot{K}(t, \eta(t))x(t) + \dot{k}_0(t, \eta(t))$$
$$+2[x^*(t)K(t, \eta(t)) + k^*(t, \eta(t))]a(t) + Tr(\sigma^*(t)K(t, \eta(t))\sigma(t)),$$

where $\dot{K}(t, \eta(t)) = \frac{\partial}{\partial t}K(t, \eta(t))$. Further, by using Theorem 18, we can write

$$E[(v(t, x(t), \eta(t + h)) - v(t, x(t), \eta(t)))|\eta(t_0) = i]$$

$$= E\left[\left(\sum_{j=1}^{d} \chi_{\eta(t+h)=j}v(t, x(t), j) - v(t, x(t), \eta(t))\right)|\eta(t_0) = i\right] \quad (1.10)$$

$$= \sum_{j=1}^{d} E[v(t, x(t), j)E[\chi_{\eta(t+h)=j}|\mathcal{H}_t]|\eta(t_0) = i]$$

$$-E[v(t, x(t), \eta(t))|\eta(t_0) = i].$$

By virtue of Theorem 34 we have

$$E[\chi_{\eta(t+h)=j}|\mathcal{H}_t] = E[\chi_{\eta(t+h)=j}|\eta(t)] = p_{\eta(t),j}(h). \quad (1.11)$$

Hence from (1.10) and (1.11) we have

$$E[(v(t, x(t), \eta(t + h)) - v(t, x(t), \eta(t)))|\eta(t_0) = i]$$

$$= E\left[\sum_{j \neq \eta(t)} (v(t, x(t), j) - v(t, x(t), \eta(t)))p_{\eta(t)j}(h)|\eta(t_0) = i\right].$$

Recall that $P(h) = [p_{ij}(h)] = e^{Qh}, h > 0$, with $\sum_{j=1}^{d} q_{ij} = 0$. Applying Lebesgue's Theorem we obtain that

$$\lim_{h \searrow 0} \frac{1}{h} E[(v(t, x(t), \eta(t + h)) - v(t, x(t), \eta(t)))|\eta(t_0) = i] \quad (1.12)$$

$$= \sum_{j=1}^{d} E[v(t, x(t), j)q_{\eta(t)j}|\eta(t_0) = i].$$

Combining (1.9) with (1.12) we conclude that

$$\lim_{h \searrow 0} \frac{1}{h} E[(v(t + h), x(t + h), \eta(t + h)) - v(t, x(t), \eta(t)))|\eta(t_0) = i]$$

$$= E\left[\left(\tilde{m}(t) + \sum_{j=1}^{d} v(t, x(t), j)q_{\eta(t)j}\right)|\eta(t_0) = i\right].$$

Denote

$$G_i(t) = E[v(t, x(t), \eta(t))|\eta(t_0) = i], i \in \mathcal{D},$$

and

$$h_i(t) = E\left[\left(\tilde{m}(t) + \sum_{j=1}^{d} v(t, x(t), j)q_{\eta(t)j}\right) \,|\, \eta(t_0) = i\right].$$

Since $\sup_{t \in [t_0, T]} E(\tilde{m}(t) + \sum_{j=1}^{r} v(t, x(t), j)q_{\eta(t)j})^2 < \infty$, it follows by Theorem 13 that $h_i(t)$ is right continuous and therefore

$$\lim_{h \searrow 0} \frac{1}{h} \int_t^{t+h} h_i(s)ds = h_i(t), t \in [t_0, T).$$

Hence

$$\lim_{h \searrow 0} \frac{1}{h}\left(G_i(t+h) - G_i(t) - \int_t^{t+h} h_i(s)ds\right) = 0, t \in [t_0, T), i \in \mathcal{D}. \qquad (1.13)$$

Since the process $\eta(t)$ is continuous in probability (see Theorem 24) it follows by using Corollary 4 that $v(t, x(t), \eta(t))$ is continuous in probability.

Having $\sup_{t \in [t_0, T]} E|v(t, x(t), \eta(t))|^2 < \infty$ it follows from Theorem 13 that $G_i(t), i \in \mathcal{D}$, is a continuous function, and thus from (1.13) we conclude that

$$G_i(t) - G_i(t_0) = \int_{t_0}^t h_i(t)dt, t \in [t_0, T], i \in \mathcal{D},$$

and so the equality (1.4) holds.

Step 2 Assume that ξ is $\mathcal{H}_{t_0}$-measurable; $E|\xi|^2 < \infty$; a, σ are bounded on $[t_0, T] \times \Omega$; and $a(t), \sigma(t)$ are $\mathcal{H}_t$-adapted. Let

$$\xi_k = \xi \chi_{|\xi| \le k},$$

$$a_k(t) = k \int_{\max\left\{t - \frac{1}{k}, t_0\right\}}^t a(s)ds,$$

$$\sigma_k(t) = \int_{\max\left\{t - \frac{1}{k}, t_0\right\}}^t \sigma(s)ds.$$

It is obvious that a_k and σ_k are continuous (with probability 1), bounded on $[t_0, T] \times \Omega$, and $\mathcal{H}_t$-adapted (see Theorem 20). From Theorem 14 and from Lebesgue's Theorem it follows that

$$\lim_{k \to \infty} \int_{t_0}^T \left(|a_k(t) - a(t)|^2 + \|\sigma_k(t) - \sigma(t)\|^2\right) dt = 0 \qquad (1.14)$$

and applying Lebesgue's Theorem again we have

$$\lim_{k \to \infty} E \int_{t_0}^T \left(|a_k(t) - a(t)|^2 + \|\sigma_k(t) - \sigma(t)\|^2\right) dt = 0.$$

From Lebesgue's Theorem it follows that

$$\lim_{k \to \infty} E|\xi_k - \xi|^2 = 0.$$

It is easy to verify by using Theorem 29 that $\sup_{t\in[t_0,T]} E|x(t)|^2 < \infty$ and

$$\sup_{t\in[t_0,T]} E|x_k(t) - x(t)|^2 \le 3E\left[|\xi_k - \xi|^2 + (T - t_0)\int_{t_0}^{T} |a_k(t) - a(t)|^2 dt \right.$$

$$\left. + \|\sigma_k(t) - \sigma(t)\|^2 dt\right], k \ge 1,$$

where

$$x_k = \xi_k + \int_{t_0}^{t} a_k(s)ds + \int_{t_0}^{t} \sigma_k(s)dw(s).$$

Applying the result of Step 1 for each $k \ge 1$ we obtain

$$E[(v(t, x_k(t), \eta(t)) - v(t_0, \xi_k, i))|\eta(t_0) = i] \tag{1.15}$$

$$= E\left\{\int_{t_0}^{t} \left[x_k^*(s)\dot{K}(s, \eta(s))x_k(s) + 2\dot{k}^*(s, \eta(s))x_k(s) + \dot{k}_0(s, \eta(s))\right.\right.$$

$$+ 2\left(x_k^*(s)K(s, \eta(s)) + k^*(s, \eta(s))\right)a_k(s) + Tr(\sigma_k^*(s)K(s, \eta(s))\sigma_k(s))$$

$$\left.\left. + \sum_{j=1}^{d} v(s, x_k(s), j)q_{\eta(s)j}\right] ds \middle| \eta(t_0) = i\right\}.$$

Taking the limit for $k \to \infty$ we conclude that (1.4) holds.

Step 3 Now consider that ξ, a, σ verify the general assumptions in the statement. Define

$$\bar{a}_k(t) = a(t)\chi_{|a(t)|\le k},$$

$$\bar{\sigma}_k(t) = \sigma(t)\chi_{|\sigma(t)|\le k}.$$

Applying Lebesgue's Theorem it follows that $\bar{a}_k$ and $\bar{\sigma}_k$ verify an equality of type (1.14). On the other hand it can be proved by using Theorem 29:

$$\sup_{t\in[t_0,T]} E|\bar{x}_k(t) - x(t)|^2 \le 2E\left[\int_{t_0}^{T} (T - t_0)|\bar{a}_k(t) - a(t)|^2 + \|\bar{\sigma}_k(t) - \sigma(t)\|^2 dt\right]$$

where

$$\bar{x}_k(t) = \xi + \int_{t_0}^{t} \bar{a}_k(s)ds + \int_{t_0}^{t} \bar{\sigma}_k(s)dw(s).$$

Now, applying the results from Step 2 for $\xi, \bar{a}_k, \bar{\sigma}_k, \bar{x}_k$ we obtain an equality of type (1.15) with $\xi_k, a_k, \sigma_k, x_k$ replaced by $\xi, \bar{a}_k, \bar{\sigma}_k, \bar{x}_k$.

Taking again the limit for $k \to \infty$ we conclude that (1.4) holds and the proof is complete. □

Remark 11. (i) The proof of Theorem 35 has been performed in several steps, since only poor information is available concerning a and σ, namely that their elements are in $L^2_{\eta,w}([t_0, T])$.

(ii) The particular form for $v(t, x, i)$ is essentially used when making $k \to \infty$ in Steps 2 and 3 of the proof.

(iii) The proof shows that the result is true for functions $v(t, x, i)$ in C^1 with respect to t and in C^2 with respect to x; the functions $v(t, x, i)$, $\frac{\partial v}{\partial t}(t, x, i)$, and $\frac{\partial v}{\partial x}(t, x, i)$ have increments with respect to x of the same type as the increments of the quadratic function used in the theorem. Moreover, $\frac{\partial^2 v}{\partial x \partial x}(t, x, i)$ must be bounded on $[t_0, T] \times \mathbf{R}^n \times \mathcal{D}$.

1.11 Stochastic differential equations

Stochastic differential equations depending on the pair $(w(t), \eta(t))$ with the above properties are considered in [60], [80], and [83], where stability and control problems are investigated.

In [117], Wonham emphasizes the importance of the differential equations subjected to the white noise perturbations $w(t)$ and Markovian jumping $\eta(t)$ for control problems.

Consider the system of stochastic differential equations

$$dx(t) = [f(t, x(t), \eta(t)) + a(t)]dt + [F(t, x(t), \eta(t)) + \sigma(t)] dw(t), \quad (1.16)$$

where the processes $w(t) = (w_1(t), \ldots, w_r(t))^*$ and $\eta(t)$, $t \geq 0$, have the properties in Section 1.9. Assume that a, σ, f, and F satisfy the following conditions:

(C1) $a : \mathbf{R}_+ \times \Omega \to \mathbf{R}^n$, $\sigma : \mathbf{R}_+ \times \Omega \to \mathbf{R}^{n \times r}$ and their elements are in $L^2_{\eta, w}[0, T]$, for all $T > 0$;

(C2) $f : \mathbf{R}_+ \times \mathbf{R}^n \times \mathcal{D} \to \mathbf{R}^n$, $F : \mathbf{R}_+ \times \mathbf{R}^n \times \mathcal{D} \to \mathbf{R}^{n \times r}$ and for each $i \in \mathcal{D}$, $f(\cdot, \cdot, i)$ and $F(\cdot, \cdot, i)$ are measurable with respect to $\mathcal{B}(\mathbf{R}_+ \times \mathbf{R}^n)$, where $\mathcal{B}(\mathbf{R}_+ \times \mathbf{R}^n)$ denotes the σ-algebra of Borel sets in $\mathbf{R}_+ \times \mathbf{R}^n$;

(C3) For each $T > 0$ there exists $\gamma(T) > 0$ such that

$$|f(t, x_1, i) - f(t, x_2, i)| + \|F(t, x_1, i) - F(t, x_2, i)\| \leq \gamma(T)|x_1 - x_2| \quad (1.17)$$

for all $t \in [0, T]$, $x_1, x_2 \in \mathbf{R}^n$, $i \in \mathcal{D}$, and

$$|f(t, x, i)| + \|F(t, x, i)\| \leq \gamma(T)(1 + |x|), \quad (1.18)$$

for all $t \in [0, T]$, $x \in \mathbf{R}^n$, $i \in \mathcal{D}$.

Using the same technique as in the proof of Theorem 1.1 from [52, Chap. 5], one can prove the following result.

Theorem 36. *Assume that a, σ, f, and F satisfy the conditions (C1) ÷ (C3). Then for all $t_0 \geq 0$ and ξ measurable with respect to $\mathcal{H}_{t_0}$ and $E|\xi|^2 < \infty$ there exists a unique continuous solution $x(t) = x(t, x_0, \xi)$, $t \geq t_0$, of (1.16), verifying $x(t_0) = \xi$*

and which components belong to $L^2_{\eta,w}[t_0, T]$ for all $T > t_0$. Moreover we have

$$\sup_{t_0 \le t \le T} E\left[|x(t)|^2 |\eta(t_0) = i\right]$$

$$\le K\left(1 + E\left[\left(|\xi|^2 + \int_{t_0}^T \left(|a(t)|^2 + \|\sigma(t)\|^2\right)dt\right)|\eta(t_0) = i\right]\right),$$

where K depends on T and $T - t_0$. The uniqueness must be understood in the sense that if $x_1(t)$ and $x_2(t)$ are two solutions of (1.16) satisfying $x_1(t_0) = x_2(t_0) = \xi$ and whose components are in $L^2_{\eta,w}[t_0, T]$, then $E|x_1(t) - x_2(t)| = 0, t \in [t_0, T]$. □

For the particular case when $a(t) = 0$ and $\sigma(t) = 0$, one obtains the following result.

Theorem 37. *Assume that f and F satisfy (C2), (C3), and $a(t) = 0, \sigma(t) = 0$, for all $t \ge 0$. Then for all $t_0 \ge 0$ and ξ measurable with respect to $\mathcal{H}_{t_0}$ with $E|\xi|^2 < \infty$, the system (1.16) has a unique continuous solution $x(t), t \ge t_0$, verifying $x(t_0) = \xi$ whose elements are in $L^2_{\eta,w}[t_0, T]$ for all $T > t_0$. Moreover, if $E|\xi|^{2p} < \infty$, then we have*

$$\sup_{t_0 \le t \le T} E[|x(t)|^{2p}|\eta(t_0) = i] \le K(1 + E[|\xi|^{2p}|\eta(t_0) = i]), \qquad (1.19)$$

$i \in \mathcal{D}$, *where K depends on T, $T - t_0$, and p.*

Proof. Consider the sequence of successive approximations defined by

$$x_0(t) = \xi, t \in [t_0, T],$$

$$x_{m+1}(t) = \xi + \int_{t_0}^t f(s, x_m(s), \eta(s))ds + \int_{t_0}^t F(s, x_m(s), \eta(s))dw(s), m \ge 0.$$

Using (1.17), (1.18), and Theorem 32 it is easy to verify by induction that

$$E\left[|x_{m+1}(t)|^{2p}|\eta(t_0) = i\right] \le \left[c + c^2(t - t_0) + \cdots + c^{m+2}\frac{(t - t_0)^{m+1}}{(m+1)!}\right]$$
$$\times \left(1 + E\left[|\xi|^{2p} \mid \eta(t_0) = i\right]\right),$$
$$t_0 \le t \le T, i \in \mathcal{D}, m \ge 0,$$

where $c > 0$ depends only on T, $T - t_0$, and p. Hence

$$E[|x_{m+1}(t)|^{2p}|\eta(t_0) = i] \le ce^{c(t-t_0)}(1 + E[|\xi|^{2p} \mid \eta(t_0) = i]).$$

Since $x_m(t) \to x(t)$ a.s. uniform on $[t_0, T]$ (see [52]) from Fatou's Lemma it follows that

$$E[|x(t)|^{2p}|\eta(t_0) = i] \le K(1 + E[|\xi|^{2p}|\eta(t_0) = i]), t \in [t_0, T], i \in \mathcal{D}$$

and the proof is complete.   □

With the same proof used for stochastic differential Itô systems (see [97], [111]) one can prove the following result.

Theorem 38. *Under the assumptions of Theorem 37, suppose that f and F are continuous functions for each $i \in D$. Then the function*

$$(t, x) \in [t_0, \infty) \times \mathbf{R}^n \to x(t, t_0, x)$$

is a.s. continuous for each $t_0 \geq 0$, hence $x(t, t_0, \cdot)$ defined on $\mathbf{R}^n \times \Omega$ is measurable with respect to $\mathcal{B}(\mathbf{R}^n) \otimes \mathcal{H}_{t_0.t}$, $t > t_0$, where

$$\mathcal{H}_{t_0.t} = \sigma(w(s) - w(t_0), \eta(s); \ s \in [t_0, t]).$$

Based on the inequality (1.19) one can obtain an Itô-type formula for the solution of the system (1.16) in case $a = 0, \sigma = 0$ and in more general assumptions for the functions $v(t, x, i)$ than the ones used in Theorem 35.

The result giving such a formula has been proved in [80].

Theorem 39. *Assume that the hypotheses of Theorem 37 are fulfilled and additionally $f(\cdot, \cdot, i)$, $F(\cdot, \cdot, i)$ are continuous on $\mathbf{R}_+ \times \mathbf{R}^n$, for all $i \in D$. Let $v : \mathbf{R}_+ \times \mathbf{R}^n \times D$ be a function which for each $i \in D$ is continuous together with its derivatives v_t, v_x and v_{xx}.*

Assume also that there exists $\gamma > 0$ such that

$$|v(t, x, i)| + \left| \frac{\partial v}{\partial t}(t, x, i) \right| + \left| \frac{\partial v}{\partial x}(t, x, i) \right| + \left\| \frac{\partial^2 v}{\partial x \partial x}(t, x, i) \right\|$$
$$\leq K_T(1 + |x|^\gamma), t \in [0, T], x \in \mathbf{R}^n, i \in D,$$

where $K_T > 0$ depends on T. Then we have:

$$E[v(s, x(s), \eta(s)) | \eta(t_0) = i] - v(t_0, x_0, i)$$

$$= E\left[\int_{t_0}^s \left\{ \frac{\partial v}{\partial t}(t, x(t), \eta(t)) + \left(\frac{\partial v}{\partial x}(t, x(t), \eta(t)) \right)^* f(t, x(t), \eta(t)) \right.\right.$$

$$+ \frac{1}{2} Tr F^*(t, x(t), \eta(t)) \frac{\partial^2 v}{\partial x \partial x}(t, x(t), \eta(t)) \qquad (1.20)$$

$$\left.\left. \times F(t, x(t), \eta(t)) + \sum_{j=1}^d v(t, x(t), \eta(t)) q_{\eta(t)j} \right\} dt | \eta(t_0) = i \right],$$

$$x(t) = x(t, t_0, x_0), \ x_0 \in \mathbf{R}^n, \ t \geq t_0 \geq 0,$$

for all $s \geq t_0, i \in D$.

Proof. From Theorem 37 it follows that for all positive integers p we have

$$\sup_{t_0 \leq t \leq T} E\left[|x(t)|^{2p}|\eta(t_0) = i\right] \leq K\left(1 + |x_0|^{2p}\right).$$

Therefore using Theorem 13 for $\alpha = 2$ it follows that it is possible to take the limits in the integrals from the first step in the proof of Theorem 35, obtaining that

$$\lim_{h \to 0_+} \frac{1}{h} E\left[\left\{v((t+h), x(t+h), \eta(t+h))\right.\right. \tag{1.21}$$

$$\left.\left. -v(t, x(t), \eta(t)) - \int_t^{t+h} m(s)ds\right\}|\eta(t_0) = i\right] = 0$$

where

$$m(t) = \frac{\partial v}{\partial t}(t, x(t), \eta(t)) + \left(\frac{\partial v}{\partial x}(t, x(t), \eta(t))\right)^* f(t, x(t), \eta(t))$$

$$+ \frac{1}{2} Tr F^*(t, x(t), \eta(t)) \frac{\partial^2 v}{\partial x \partial x}(t, x(t), \eta(t))$$

$$\times F(t, x(t), \eta(t)) + \sum_{j=1}^d v(t, x(t), \eta(t)) q_{\eta(t)j}.$$

Taking into account that $\eta(t)$ is continuous in probability and again using Theorems 37 and 13 for $\alpha = 2$, it follows immediately that

$$E\left[\left(v(t, x(t), \eta(t)) - \int_{t_0}^t m(s)ds\right)|\eta(t_0) = i\right]$$

is a continuous function, and therefore from (1.21) it results that (1.20) holds and the proof is complete. □

Remark 12. (i) The proof of the previous theorem shows that the result in the statement is also valid for random initial conditions ξ, $\mathcal{H}_{t_0}$-measurable and $E[|\xi|^{2p}] < \infty$ for $p \geq \gamma + 2$.

(ii) From Theorems 36 and 37, for the system (1.16), Theorem 39 is not applicable, while in the case when $a(t) \equiv 0$ and $\sigma(t) \equiv 0$ we can use Theorem 39 due to the estimate (1.19).

(iii) In many cases, in the following developments we shall consider the system (1.16) with $a(t) \neq 0$ and $\sigma(t) \neq 0$, being thus obliged to use Theorem 35.

1.12 Stochastic linear differential equations

Since the problems investigated in this book refer to stochastic linear controlled systems we recall here some facts concerning the solutions of stochastic linear differential equations.

Let us consider the system of linear differential equations

$$dx(t) = A_0(t, \eta(t))x(t)dt + \sum_{k=1}^{r} A_k(t, \eta(t))x(t)dw_k(t), \qquad (1.22)$$

where $t \to A_k(t, i) : \mathbf{R}_+ \to \mathbf{R}^{n \times n}$, $i \in \mathcal{D}$, are bounded and continuous matrix-valued functions.

The system (1.22) has two important particular forms:

(i) $A_k(t, i) = 0$, $k = 1, \ldots, r$, $t \geq 0$. In this case (1.22) becomes

$$\dot{x}(t) = A(t, \eta(t))x(t), \ t \geq 0, \qquad (1.23)$$

where $A(t, \eta(t))$ stands for $A_0(t, \eta(t))$ and it corresponds to the case when the system is subjected only to Markovian jumping.

(ii) $\mathcal{D} = \{1\}$; in this situation the system (1.22) becomes

$$dx(t) = A_0(t)x(t)dt + \sum_{k=1}^{r} A_k(t)x(t)dw_k(t), \ t \geq 0, \qquad (1.24)$$

where $A_k(t) := A_k(t, 1)$, $k = 0, \ldots, r$, $t \geq 0$, representing the case when the system is subjected only to white noise–type perturbations.

Definition 24. *We say that the system* (1.22) *is* time invariant *(or it is in the* stationary case*) if* $A_k(t, i) = A_k(i)$ *for all* $k = 0, \ldots, r$, $t \in \mathbf{R}_+$ *and* $i \in \mathcal{D}$. *In this case the system* (1.22) *becomes*

$$dx(t) = A_0(\eta(t))x(t)dt + \sum_{k=1}^{r} A_k(\eta(t))x(t)dw_k(t). \qquad (1.25)$$

Applying Theorem 37, it follows that for each $t_0 \geq 0$ and each random vector ξ, $\mathcal{H}_{t_0}$-measurable and $E|\xi|^2 < +\infty$, the system (1.22) has a unique solution $x(t; t_0, \xi)$ which verifies $x(t_0; t_0, \xi) = \xi$. Moreover, if $E|\xi|^{2p} < +\infty$, $p \geq 1$, then

$$\sup_{t \in [t_0, T]} E[|x(t, t_0, \xi)|^{2p} \mid \eta(t_0) = i] \leq cE[|\xi|^{2p} \mid \eta(t_0) = i],$$

$i \in \mathcal{D}$, $c > 0$ depending upon T, $T - t_0$, and p. For each $k \in \{1, 2, \ldots, n\}$ we denote $\Phi_k(t, t_0) = x(t, t_0, e_k)$ where $e_k = (0, 0, \ldots, 1, 0, \ldots, 0)^*$ and set

$$\Phi(t, t_0) = (\Phi_1(t, t_0) \ \Phi_2(t, t_0) \cdots \Phi_n(t, t_0)).$$

$\Phi(t, t_0)$ is the matrix-valued solution of the system (1.22), which verifies $\Phi(t_0, t_0) = I_n$. If ξ is a random vector $\mathcal{H}_t$-measurable with $E|\xi|^2 < \infty$, we denote $\tilde{x}(t) = \Phi(t, t_0)\xi$. By Remark 10 it is easy to verify that $\tilde{x}(t)$ is a solution of the system (1.22) verifying $\tilde{x}(t) = \xi$. Then, by uniqueness arguments, we conclude that $\tilde{x}(t) = x(t, t_0, \xi)$ a.s., $t \geq t_0$. Hence we have the representation formula

$$x(t, t_0, \xi) = \Phi(t, t_0)\xi \text{ a.s.}$$

The matrix $\Phi(t, t_0)$, $t \geq t_0 \geq 0$, will be termed the *fundamental matrix solution* of the system of stochastic linear differential equations (1.22). By the uniqueness argument it can be proved that

$$\Phi(t, s)\Phi(s, t_0) = \Phi(t, t_0) \text{ a.s., } t \geq s \geq t_0 \geq 0.$$

Proposition 40. *The matrix $\Phi(t, t_0)$ is invertible and its inverse is given by*

$$\Phi^{-1}(t, t_0) = \tilde{\Phi}^*(t, t_0) \text{ a.s., } t \geq t_0 \geq 0,$$

where $\tilde{\Phi}(t, t_0)$ is the fundamental matrix solution of the stochastic linear differential equation:

$$dy(t) = \left[-A_0^*(t, \eta(t)) + \sum_{k=1}^{r} \left(A_k^2(t, \eta(t)) \right)^* \right] y(t)dt \qquad (1.26)$$

$$- \sum_{k=1}^{r} A_k^*(t, \eta(t))y(t)dw_k(t).$$

Proof. Applying Itô's formula (Theorem 33) to the function

$$v(t, x, y) = y^*x, \quad t \geq t_0, \quad x, y \in \mathbf{R}^n$$

and to the systems (1.22) and (1.26), we obtain

$$y^*\tilde{\Phi}^*(t, t_0)\Phi(t, t_0)x - y^*x = 0 \text{ a.s., } t \geq t_0 \geq 0, \quad x, y \in \mathbf{R}^n;$$

hence $\tilde{\Phi}^*(t, t_0)\Phi(t, t_0) = I_n$ a.s., $t \geq t_0$, and the proof is complete. $\qquad \square$

Let us consider the affine system of stochastic differential equations:

$$dx(t) = [A_0(t, \eta(t))x(t) + f_0(t)]dt \qquad (1.27)$$

$$+ \sum_{k=1}^{r} [A_k(t, \eta(t))x(t) + f_k(t)]dw_k(t),$$

$t \geq 0$, where $f_k : \mathbf{R}_+ \times \Omega \to \mathbf{R}^n$ are stochastic processes with components in $L_{\eta, w}^2([0, T])$ for all $T > 0$. Using Theorem 36 we deduce that for all $t_0 \geq 0$ and for all random vectors ξ, $\mathcal{H}_{t_0}$-measurable with $E|\xi|^2 < \infty$, the system (1.27) has a unique solution $x_f(t, t_0, \xi)$, $f = (f_0, f_1, \ldots, f_r)$. Additionally, for all $T > t_0$, there exists a positive constant c depending on T, $T - t_0$ such that

$$\sup_{t \in [t_0, T]} E\left[\left| x_f(t, t_0, \xi) \right|^2 \mid \eta(t_0) = i \right] \qquad (1.28)$$

$$\leq c \left\{ E\left[\left(|\xi|^2 + \sum_{k=0}^{r} \int_{t_0}^{T} |f_k(s)|^2 ds \right) \mid \eta(t_0) = i \right] ds \right\}.$$

Let $\Phi(t, t_0)$, $t \geq t_0 \geq 0$, be the fundamental matrix solution of the linear system obtained by taking $f_k = 0$ in (1.27) and set $z(t) = \Phi^{-1}(t, t_0)x_f(t, t_0, \xi)$. Applying

Itô's formula (Theorem 33) to the function $v(t, x, y) = y^*x$, $x, y \in \mathbf{R}^n$, and to the systems (1.26) and (1.27), we obtain

$$
y^*z(t) = y^*z(t_0) + y^* \int_{t_0}^{t} \Phi^{-1}(s, t_0) \left[f_0(s) - \sum_{k=1}^{r} A_k(s, \eta(s)) f_k(s) \right] ds
$$

$$
+ \sum_{k=1}^{r} y^* \int_{t_0}^{t} \Phi^{-1}(s, t_0) f_k(s) dw_k(s) \text{ a.s.,}
$$

$t \geq t_0$, $y \in \mathbf{R}^n$. Since y is arbitrary in $\mathbf{R}^n$ we may conclude that

$$
z(t) = \xi + \int_{t_0}^{t} \Phi^{-1}(s, t_0) \left[f_0(s) - \sum_{k=1}^{r} A_k(s, \eta(s)) f_k(s) \right] ds
$$

$$
+ \sum_{k=1}^{r} \int_{t_0}^{t} \Phi^{-1}(s, t_0) f_k(s) dw_k(s) \text{ a.s.,}
$$

$t \geq t_0$. Thus we obtained the following representation formula:

$$
x_f(t, t_0, \xi) = \Phi(t, t_0) \tag{1.29}
$$

$$
+ \Phi(t, t_0) \int_{t_0}^{t} \Phi^{-1}(s, t_0) \left[f_0(s) - \sum_{k=1}^{r} A_k(s, \eta(s)) f_k(s) \right] ds
$$

$$
+ \sum_{k=1}^{r} \Phi(t, t_0) \int_{t_0}^{t} \Phi^{-1}(s, t_0) f_k(s) dw_k(s) \text{ a.s.,}
$$

$t \geq t_0$, which extends the well-known constant variation formula from the deterministic framework to the case of stochastic affine system (1.27). $\qquad \square$

2

Exponential Stability and Lyapunov-Type Linear Equations

In this chapter the problem of mean square exponential stability of the zero solution to the stochastic differential equations of type (1.22) is studied. The stabilization of a steady-state is one of the main tasks appearing in many design problems of controllers with prescribed performances.

In the case of stochastic systems there are several possibilities to define the concept of stability of a steady-state. Among them, one of the most popular is the so-called *exponential stability in mean square* (ESMS). The exponential stability in mean square has the advantage that it may be characterized by some conditions that are easy to check. Moreover, in some particular cases, such as the time-invariant case or the periodic case, the exponential stability in mean square is equivalent with other types of stability in mean square. From the representation formula proved in Theorem 4 in Section 2.2 one obtains that the ESMS to the zero solution of (1.22) is equivalent with the exponential stability of the zero solution of a deterministic linear differential equation on a finite-dimensional linear space adequately chosen. The deterministic differential equations are defined by the so-called *Lyapunov-type operators* acting on a space of symmetric matrices. Since criteria concerning the exponential stability of the zero solution of Lyapunov differential equations provide criteria for exponential stability in mean square of the zero solution to the stochastic equation of type (1.22), a great part of this chapter is devoted to studying the Lyapunov-type differential equations. In the first part of the chapter, we make a detailed investigation of the properties of the linear evolution operators and of the exponential stability for a class of Lyapunov-type differential equations. The results concerning the exponential stability in Section 2.4 are derived for a class of differential equations which contains as a particular case the Lyapunov-type equations arising in connection with the stochastic differential equation (1.22). A reason to consider the more general case when the Lyapunov operators (2.8) satisfy only condition (2.7) may be found later, in the following chapters. This allows us to simplify some proofs by using the so-called *dual systems*. In this case the matrix Q of the rates of the probability transition matrix will be replaced by its transpose Q^*, the entries of which verify only condition (2.7). In the last section of the chapter some useful estimates of the solutions of affine equations are derived. Some aspects concerning the exponential stability in mean square of the zero state equilibrium for nonlinear stochastic differential equations of type (1.16) will be discussed in Chapter 6.

2.1 Linear positive operators on the Hilbert space of symmetric matrices

Let $S_n \subset \mathbf{R}^{n \times n}$ be the subspace of $n \times n$ symmetric matrices, that is, $S \in S_n$, if and only if $S = S^*$. We denote by S_n^d the direct product

$$S_n^d = \underbrace{S_n \times \cdots \times S_n}_{d}.$$

Then $S \in S_n^d$ if and only if $S = (S(1), \ldots, S(d))$.

In the following we shall use either notations $S = (S(1), \ldots, S(d))$ or $S = (S_1, \ldots, S_d)$.

It is easy to prove that S_n^d is a finite-dimensional real Hilbert space with respect to the inner product:

$$\langle S, H \rangle = \sum_{i=1}^{d} Tr(S(i)H(i)), \ S, H \in S_n^d. \tag{2.1}$$

We introduce on S_n^d the following norm:

$$|S| = \max_{i \in \mathcal{D}} |S(i)|, \tag{2.2}$$

where $|S(i)|$ is the norm induced by the Euclidean norm on R^n, that is:

$$|S(i)| = \sup_{|x| \leq 1} |S(i)x| = \max_{\lambda \in \Lambda(S(i))} |\lambda| = \sup_{|x| \leq 1} |x^* S(i)x|,$$

where $\Lambda(A)$ is the spectrum of the matrix A. The norm defined by (2.2) differs from the norm provided by the inner product (2.1). The space S_n^d together with the norm (2.2) becomes a finite-dimensional Banach space.

It is not difficult to check that

$$|H| \leq \langle H, H \rangle^{\frac{1}{2}} \leq \sqrt{nd}|H| \tag{2.3}$$

for all $H \in S_n^d$.

If $T: S_n^d \rightarrow S_n^d$ is a linear operator, then $\|T\|$ stands for the operatorial norm induced by the norm (2.2), that is,

$$\|T\| = \sup_{|S| \leq 1} |TS|. \tag{2.4}$$

Remark 1. If $T^* : S_n^d \rightarrow S_n^d$ is the adjoint operator of T with respect to the inner product (2.1), then $\|T^*\|$ is not equal to $\|T\|$. However, based on (2.3), we obtain that there exist the positive constants c_1 and c_2 such that

$$c_1\|T^*\| \leq \|T\| \leq c_2\|T^*\|. \tag{2.5}$$

For $S \in S_n^d$, $S = (S(1), \ldots, S(d))$, we write $S \geq 0$ if $S(i) \geq 0$, for all $i \in \mathcal{D}$. Similarly, we write $S > 0$ if $S(i) > 0$, for all $i \in \mathcal{D}$.

We denote

$$S_{n,+}^d = \{S \in S_n^d; S \geq 0\}.$$

$S_{n,+}^d$ is a convex cone and it induces an order relation on S_n^d, namely $S \geq H$ if and only if $S - H \in S_{n,+}^d$.

By J^d we denote the element of S_n^d defined by

$$J^d = \underbrace{I_n \times \cdots \times I_n}_{d}.$$

Obviously, $J^d \in S_{n,+}^d$.

Definition 1. *We say that a function* $H : \mathcal{I} \subset \mathbf{R} \rightarrow S_n^d$ *is* uniform positive *and we write* $H \gg 0$ *if there exists a constant* $c > 0$ *such that* $H(t) \geq cJ^d$ *for all* $t \in \mathcal{I}$. *We shall also write* $H \ll 0$ *if and only if* $-H(t) \gg 0$.

Definition 2. *A linear operator* $T : S_n^d \rightarrow S_n^d$ *is said to be* positive *and we write* $T \geq 0$ *if* $TS_{n,+}^d \subseteq S_{n,+}^d$.

Lemma 1. *The inner product (2.1) has the following properties:*
    (i) *If* $\langle S, H \rangle \geq 0$ *for all* $H \in S_{n,+}^d$, *then* $S \in S_{n,+}^d$.
    (ii) *If* $H, S \in S_{n,+}^d$, *then* $\langle S, H \rangle \geq 0$.

Proof. (i) Let $x \in \mathbf{R}^n$ and $i_0 \in D$ be fixed. Set $H = (H(1), \ldots, H(d))$ by

$$H(i) = \begin{cases} xx^* & \text{if } i = i_0, \\ 0 & \text{if } i \neq i_0. \end{cases}$$

Obviously, $H \in S_{n,+}^d$. We have

$$0 \leq \langle S, H \rangle = Tr[S(i_0)H(i_0)] = x^* S(i_0)x.$$

Since x and i_0 are arbitrarily chosen in $\mathbf{R}^n$ and $\mathcal{D}$, respectively, we conclude that $S \geq 0$.

(ii) From (2.1), it is sufficient to show that if $S, M \in S_n$ with $S \geq 0$, $M \geq 0$, then $Tr[SM] \geq 0$. Since $S \geq 0$, there exist the orthogonal vectors $e_1, \ldots, e_n$ and the nonnegative numbers $\lambda_1, \ldots, \lambda_n$ such that

$$S = \sum_{i=1}^{n} \lambda_i e_i e_i^*$$

(see, e.g., [7]). Then we have

$$Tr[SM] = \sum_{i=1}^{n} \lambda_i Tr[e_i e_i^* M] = \sum_{i=1}^{n} \lambda_i e_i^* M e_i \geq 0$$

and the proof is complete.   □

Proposition 2. *If $T \in \mathcal{S}_n^d \rightarrow \mathcal{S}_n^d$ is a linear and positive operator then the adjoint operator $T^* : \mathcal{S}_n^d \rightarrow \mathcal{S}_n^d$ is positive too.*

Proof. Let $S \in \mathcal{S}_n^d$, $S \geq 0$. We show that $T^* S \geq 0$. Indeed, if $H \in \mathcal{S}_n^d$, $H \geq 0$, then $TH \geq 0$ and hence, according to Lemma 1(ii), we obtain $\langle S, TH \rangle \geq 0$. Therefore $\langle T^* S, H \rangle \geq 0$ for all $H \in \mathcal{S}_{n,+}^d$. Invoking part (i) in Lemma 1 we conclude that $T^* S \geq 0$ and the proof ends.   □

The result stated in the next theorem provides a method for determining $\|T\|$ for a positive operator T.

Theorem 3. *If $T : \mathcal{S}_n^d \rightarrow \mathcal{S}_n^d$ is a linear positive operator then $\|T\| = |TJ^d|$.*

Proof. From (2.4) one can see that $|TJ^d| \leq \|T\|$. Let $S \in \mathcal{S}_n^d$ with $|S| \leq 1$, that is, $|S(i)| \leq 1$ for all $i \in \mathcal{D}$. Hence $-I_n \leq S(i) \leq I_n$ for all $i \in \mathcal{D}$ and $-J^d \leq S \leq J^d$. Since T is a positive operator it follows that $-TJ^d \leq TS \leq TJ^d$ for all $S \in \mathcal{S}_n^d$ with $|S| \leq 1$. Further we have

$$-(TJ^d)(i) \leq (TS)(i) \leq (TJ^d)(i)$$

for all $i \in \mathcal{D}$, which leads to

$$|(TS)(i)| \leq |(TJ^d)(i)|$$

for all $i \in \mathcal{D}$ and

$$|TS| \leq |TJ^d|$$

for all $S \in \mathcal{S}_n^d$ with $|S| \leq 1$. Invoking (2.4) again, we conclude that $\|T\| \leq |TJ^d|$, which completes the proof.   □

Remark 2. If $T : \mathcal{S}_n^d \rightarrow \mathcal{S}_n^d$ is a linear and positive operator, then

$$(TJ^d)(i) \leq |(TJ^d)(i)| \cdot I_n$$

for all $i \in \mathcal{D}$, which leads to

$$TJ^d \leq \|T\| J^d. \tag{2.6}$$

Now we introduce another finite-dimensional Banach space which will be used in this book.

Let

$$\mathcal{M}_{n,m}^d = \underbrace{\mathbf{R}^{n \times m} \times \cdots \times \mathbf{R}^{n \times m}}_{d}.$$

Therefore

$$\mathcal{M}_{n,m}^d = \{M;\ M = (M(1), \ldots, M(d)),\ M(i) \in \mathbf{R}^{n \times m},\ i \in \mathcal{D}\}.$$

On $\mathcal{M}_{n,m}^d$ we introduce the norm

$$|M| = \max_{i \in \mathcal{D}} |M(i)|,$$

where

$$|M(i)| = \sup_{|x| \le 1} |M(i)x| = \lambda_{\max}^{\frac{1}{2}}(M^*(i)M(i)).$$

In the particular case when $m = n$ we shall write $\mathcal{M}_n^d$ instead of $\mathcal{M}_{n,m}^d$. It is obvious that $\mathcal{S}_n^d$ is a Banach subspace of the Banach space $\mathcal{M}_n^d$.

In this monograph $(\mathbf{R}^n)^d$ stands for the direct product

$$(\mathbf{R}^n)^d := \underbrace{\mathbf{R}^n \times \cdots \times \mathbf{R}^n}_{d},$$

that is, $y \in (\mathbf{R}^n)^d$ if and only if $y = (y(1), \ldots, y(d))$, $y(i) \in \mathbf{R}^n$, $i \in \mathcal{D}$.

On $(\mathbf{R}^n)^d$ we consider the inner product

$$\langle y, z \rangle = \sum_{i=1}^{d} y^*(i)z(i)$$

for all $y = (y(1), \ldots, y(d))$ and $z = (z(1), \ldots, z(d))$ in $(\mathbf{R}^n)^d$.

By $\|y\|$ we denote the norm defined by

$$\|y\|^2 = \langle y, y \rangle = \sum_{i=1}^{d} |y(i)|^2.$$

If $T : (\mathbf{R}^n)^d \to (\mathbf{R}^n)^d$ is a linear operator, then $\|T\|$ stands for the operational norm induced by the considered norm in $(\mathbf{R}^n)^d$.

2.2 Lyapunov-type differential equations on the space $\mathcal{S}_n^d$

Let $\mathcal{I} \subseteq \mathbf{R}$ be an interval and $A_k : \mathcal{I} \to \mathcal{M}_n^d$, $k = 0, \ldots, r$, be continuous functions

$$A_k(t) = (A_k(t, 1), \ldots, A_k(t, d)),\ k \in \{0, \ldots, r\},\ t \in \mathcal{I}.$$

Denote by $Q \in \mathbf{R}^{d \times d}$ a matrix whose elements q_{ij} verify the condition

$$q_{ij} \ge 0 \quad \text{if } i \ne j. \tag{2.7}$$

For each $t \in \mathcal{I}$ we define the linear operator $\mathcal{L}(t) : \mathcal{S}_n^d \to \mathcal{S}_n^d$ by

$$(\mathcal{L}(t)S)(i) = A_0(t, i)S(i) + S(i)A_0^*(t, i) \tag{2.8}$$

$$+ \sum_{k=1}^{r} A_k(t, i)S(i)A_k^*(t, i) + \sum_{j=1}^{d} q_{ji}S(j),$$

$i \in \mathcal{D}$, $S \in \mathcal{S}_n^d$. It is easy to see that $t \longmapsto L(t)$ is a continuous operator-valued function.

Definition 3. *The operator $\mathcal{L}(t)$ defined by* (2.8) *is called the* Lyapunov operator associated with $A_0, \ldots, A_r$ and Q.

The Lyapunov operator $\mathcal{L}(t)$ defines the following linear differential equation on $\mathcal{S}_n^d$:

$$\frac{d}{dt}S(t) = \mathcal{L}(t)S(t), \ t \in \mathcal{I}. \tag{2.9}$$

For each $t_0 \in \mathcal{I}$ and $H \in \mathcal{S}_n^d$, $S(t, t_0, H)$ stands for the solution of the differential equation (2.9) which verifies the initial condition $S(t_0, t_0, H) = H$.

Let us denote by $T(t, t_0)$ the linear evolution operator on $\mathcal{S}_n^d$ defined by the differential equation (2.9), that is

$$T(t, t_0)H = S(t, t_0, H); \ t, t_0 \in \mathcal{I}, \ H \in \mathcal{S}_n^d.$$

It is said that $T(t, t_0)$ is the *evolution operator associated with the system* $(A_0, \ldots, A_r; Q)$.

We have

$$\frac{d}{dt}T(t, t_0) = \mathcal{L}(t)T(t, t_0),$$

$$T(t, t_0) = \tilde{J}^d,$$

where $\tilde{J}^d : \mathcal{S}_n^d \to \mathcal{S}_n^d$ is the identity operator.

It is easy to check that $T(t, s)T(s, \tau) = T(t, \tau)$ for all $t, s, \tau \in \mathcal{I}$. For all pairs $t, \tau \in \mathcal{I}$, the operator $T(t, \tau)$ is invertible and its inverse is $T^{-1}(t, \tau) = T(\tau, t)$.

If $T^*(t, \tau)$ denotes the adjoint operator of $T(t, \tau)$, the following hold:

$$T^*(t, t_0) = T^*(s, t_0)T^*(t, s), \tag{2.10}$$

$$T^*(t, s) = (T^*(\tau, s))T^*(t, \tau), \tag{2.11}$$

$$\frac{d}{dt}T^*(t, s) = T^*(t, s)L^*(t), \tag{2.12}$$

$$\frac{d}{dt}T^*(s, t) = -\mathcal{L}^*(t)T^*(s, t). \tag{2.13}$$

It is not difficult to see that the adjoint operator $\mathcal{L}^*(t) : \mathcal{S}_n^d \to \mathcal{S}_n^d$ is given by

$$(\mathcal{L}^*(t)S)(i) = A_0^*(t, i)S(i) + S(i)A_0(t, i) \tag{2.14}$$

$$+ \sum_{k=1}^{r} A_k^*(t, i)S(i)A_k(t, i) + \sum_{j=1}^{d} q_{ij}S(j),$$

$i \in \mathcal{D}$, $S \in \mathcal{S}_n^d$.

Remark 3. (i) If $A_k(t,i), k = 1, \ldots, r$, do not depend on t, then the operator $\mathcal{L}$ defined by (2.8) is independent on t. More precisely, if $A_k = (A_k(1), \ldots, A_k(d))$, then

$$(\mathcal{L}S)(i) = A_0(i)S(i) + S(i)A_0^*(i) + \sum_{k=1}^{r} A_k(i)S(i)A_k^*(i) \qquad (2.15)$$

$$+ \sum_{j=1}^{d} q_{ji}S(j),$$

$i \in \mathcal{D}$, $S \in \mathcal{S}_n^d$. In this situation the evolution operator defined by the differential equation

$$\frac{d}{dt}S(t) = \mathcal{L}S(t)$$

is given by

$$T(t,t_0) = e^{\mathcal{L}(t-t_0)}, \qquad (2.16)$$

where

$$e^{\mathcal{L}t} := \sum_{k=0}^{\infty} \frac{\mathcal{L}^k t^k}{k!}$$

(the above series being uniform convergent on every compact subset of the real axis). $\mathcal{L}^k$ stands for *the k-iteration of the operator* $\mathcal{L}$ and $\mathcal{L}^0 = \tilde{J}^d$.

(ii) If $A_k : \mathcal{I} \to \mathcal{M}_n^d$ are θ-periodic functions, then $T(t + \theta, t_0 + \theta) = T(t, t_0)$ for all $t, t_0 \in \mathcal{I}$ such that $t + \theta, t_0 + \theta \in \mathcal{I}$.

In order to motivate the definition of the Lyapunov operator $\mathcal{L}(t)$ and its corresponding evolution operator $T(t, t_0)$, we shall prove the following result which establishes the relationship between the evolution operator $T(t, t_0)$ and the fundamental matrix solution of a system of stochastic linear differential equations of type (1.22).

Theorem 4. *Assume that $\mathcal{I} = \mathbf{R}_+$ and that the elements of Q satisfy (2.7) and the additional condition $\sum_{j=1}^{d} q_{ij} = 0$, $i \in \mathcal{D}$. Under these assumptions we have*

$$(T^*(t, t_0)H)(i) = E[\Phi^*(t, t_0)H(\eta(t))\Phi(t, t_0)|\eta(t_0) = i]$$

for all $t \geq t_0 \geq 0$, $H \in \mathcal{S}_n^d$, $i \in \mathcal{D}$, where $\Phi(t, t_0)$ is the fundamental matrix solution of the system (1.22).

Proof. Let $\mathcal{U}(t, t_0) : \mathcal{S}_n^d \to \mathcal{S}_n^d$ be defined by

$$(\mathcal{U}(t, t_0)(H))(i) = E[\Phi^*(t, t_0)H(\eta(t))\Phi(t, t_0)|\eta(t_0) = i],$$

$H \in \mathcal{S}_n^d$, $i \in \mathcal{D}$, $t \geq t_0$.

Taking $H \in \mathcal{S}_n^d$, we define $v(t, x, i) = x^*H(i)x$, $x \in \mathbf{R}^n$, $i \in \mathcal{D}$, $t \geq 0$.

Applying Theorem 35 of Chapter 1 to the function $v(t, x, i)$ and to the equation (1.22), we obtain

$$x^*(\mathcal{U}(t, t_0)(H))(i)x - x^* H(i)x = x^* \int_{t_0}^{t} (\mathcal{U}(s, t_0)(\mathcal{L}^*(s)H))(i) \, ds \, x.$$

Hence

$$\frac{d}{dt}\mathcal{U}(t, t_0) = \mathcal{U}(t, t_0)\mathcal{L}^*(t).$$

Since $\mathcal{U}(t_0, t_0) = T^*(t_0, t_0)$ and using (2.12) it follows that

$$\mathcal{U}(t, s) = T^*(t, s),$$

$t \geq s$, and the proof is complete. $\quad\square$

As we shall see in Section 2.5, the above result allows us to reduce the study of the exponential stability for the linear stochastic system (1.22) to the problem of the exponential stability for a deterministic system of type (2.9).

Remark 4. (i) If in the system (1.22) we have $A_k(t + \theta) = A_k(i)$, $t \geq 0$, $i \in \mathcal{D}$, then from Theorem 4 and Remark 3(ii) we deduce that

$$E\big[|\Phi(t + \theta, t_0 + \theta)x_0|^2 \mid \eta(t_0 + \theta) = i\big]$$
$$= E\big[|\Phi(t, t_0)x_0|^2 \mid \eta(t_0) = i\big]$$

for all $t \geq t_0 \geq 0$, $i \in \mathcal{D}$, $x_0 \in \mathbf{R}^n$.

(ii) If the system (1.22) is time invariant, then according to Theorem 4 and Remark 3(i), we have

$$E\big[|\Phi(t, t_0)x_0|^2 \mid \eta(t_0) = i\big]$$
$$= E\big[|\Phi(t - t_0, 0)x_0|^2 \mid \eta(0) = i\big]$$

for all $t \geq t_0 \geq 0$, $i \in \mathcal{D}$, $x_0 \in \mathbf{R}^n$.

Theorem 5. *If $T(t, t_0)$ are linear evolution operators on $\mathcal{S}_n^d$ defined by the linear differential equation (2.9), then the following hold:*

(i) *$T(t, t_0) \geq 0$, $T^*(t, t_0) \geq 0$ for all $t \geq t_0$, $t, t_0 \in \mathcal{I}$;*

(ii) *if $t \to A_k(t)$ are bounded functions, then there exist $\delta > 0$, $\gamma > 0$ such that*

$$T(t, t_0)J^d \geq \delta e^{-\gamma(t-t_0)} J^d, \quad T^*(t, t_0)J^d \geq \delta e^{-\gamma(t-t_0)} J^d$$

for all $t \geq t_0$, $t, t_0 \in \mathcal{I}$.

Proof. To prove (i) we consider the linear operators $\mathcal{L}_1(t) : \mathcal{S}_n^d \to \mathcal{S}_n^d$, $\tilde{\mathcal{L}}(t) : \mathcal{S}_n^d \to \mathcal{S}_n^d$ defined by

$$(\mathcal{L}_1(t)H)(i) = \left(A_0(t, i) + \frac{1}{2}q_{ii} I_n\right) H(i) + H(i) \left(A_0(t, i) + \frac{1}{2}q_{ii} I_n\right)^*,$$

$$(\tilde{\mathcal{L}}(t)H)(i) = \sum_{k=1}^{r} A_k(t, i)H(i)A_k^*(t, i) + \sum_{j=1, j \neq i}^{d} q_{ji} H(j), i \in \mathcal{D},$$

$$H = (H(1), \ H(2), \ldots, \ H(d)) \in \mathcal{S}_n^d, t \in \mathcal{I}.$$

It is easy to see that for each $t \in \mathcal{I}$, the operator $\tilde{\mathcal{L}}(t)$ is a positive operator on S_n^d. Let us consider the linear differential equation

$$\frac{d}{dt} S(t) = \mathcal{L}_1(t) S(t) \tag{2.17}$$

and denote $T_1(t, t_0)$ the linear evolution operator on S_n^d defined by (2.17). By direct calculation, we obtain that

$$(T_1(t, t_0) H)(i) = \Phi_i(t, t_0) H(i) \Phi_i^*(t, t_0)$$

for all $t \geq t_0, i \in \mathcal{D}, H \in S_n^d$, where $\Phi_i(t, t_0)$ is a fundamental matrix solution of the deterministic differential equation on $\mathbf{R}^n$,

$$\frac{d}{dt} x(t) = \left[A_0(t, i) + \frac{1}{2} q_{ii} I_n \right] x(t).$$

It is clear that for each $t \geq t_0, T_1(t, t_0) \geq 0$. Since the linear differential equation (2.9) is written as

$$\frac{d}{dt} S(t) = \mathcal{L}_1(t) S(t) + \tilde{\mathcal{L}}(t) S(t),$$

we may write the following representation formula:

$$T(t, t_0) H = T_1(t, t_0) H + \int_{t_0}^{t} T_1(t, s) \tilde{\mathcal{L}}(s) T(s, t_0) H \, ds$$

for all $H \in S_n^d, t \geq t_0, t, t_0 \in \mathcal{I}$.

Let $H \in S_n^d, H \geq 0$ be fixed. We define the sequence of Volterra approximations $S_k(t), k \geq 0, t \geq t_0$, by

$$S_0(t) = T_1(t, t_0) H,$$

$$S_{k+1}(t) = T_1(t, t_0) H + \int_{t_0}^{t} T_1(t, s) \tilde{\mathcal{L}}(s) S_k(s) \, ds, \quad k = 1, 2, \ldots.$$

Since $T_1(t, t_0)$ is a positive operator on S_n^d, we get inductively that $S_k(s) \geq 0$ for all $s \geq t_0, k = 1, 2, \ldots$. Taking into account that $\lim_{k \to \infty} S_k(t) = T(t, t_0) H$ we conclude that $T(t, t_0) H \geq 0$, hence $T(t, t_0) \geq 0$. By using Proposition 2 we get that the adjoint operator $T^*(t, t_0)$ is positive.

(ii) First, we show that there exist $\delta > 0, \gamma > 0$, such that

$$|T(t, t_0) H| \geq \delta e^{-\gamma(t-t_0)} |H|, \tag{2.18}$$

$$|T^*(t, t_0) H| \geq \delta e^{-\gamma(t-t_0)} |H|$$

for all $H \in S_n^d, t \geq t_0, t, t_0 \in \mathcal{I}$.

Let us denote

$$v(t) = \frac{1}{2}|||T(t, t_0)H|||^2 = \frac{1}{2}\langle T(t, t_0)H, T(t, t_0)H\rangle,$$

where $||| \cdot |||$ denotes the norm induced by the inner product, that is, $||| \cdot ||| := \langle \cdot, \cdot \rangle^{\frac{1}{2}}$. By direct calculation, we obtain

$$\frac{d}{dt}v(t) = \langle \mathcal{L}(t)T(t, t_0)H, T(t, t_0)H\rangle, \quad t \geq t_0.$$

Under the considered assumptions there exists $\gamma > 0$ such that

$$\left|\frac{d}{dt}v(t)\right| \leq \gamma |||T(t, t_0)H|||^2,$$

$$\left|\frac{d}{dt}v(t)\right| \leq 2\gamma v(t), \quad t \geq t_0.$$

Further, we have

$$\frac{d}{dt}v(t) \geq -2\gamma v(t), \quad t \geq t_0,$$

or equivalently

$$\frac{d}{dt}\left[v(t)e^{2\gamma(t-t_0)}\right] \geq 0$$

for all $t \geq t_0$. Hence the function $t \to v(t)e^{2\gamma(t-t_0)}$ is not decreasing and $v(t) \geq e^{-2\gamma(t-t_0)}v(t_0)$. Considering the definition of $v(t)$ and using (2.3), we conclude that there exists $\delta > 0$ such that

$$|T(t, t_0)H| \geq \delta e^{-\gamma(t-t_0)}|H|,$$

which is the first inequality in (2.18).

To prove the second inequality in (2.18), we consider the function

$$\hat{v}(s) = \frac{1}{2}|||T^*(t, s)H|||^2, \quad H \in \mathcal{S}_n^d, s \leq t, s, t \in \mathcal{I}.$$

By direct computation we obtain

$$\frac{d}{ds}\hat{v}(s) = -\langle \mathcal{L}^*(s)T^*(t, s)H, T^*(t, s)H\rangle.$$

Further, we have

$$\left|\frac{d}{ds}\hat{v}(s)\right| \leq 2\gamma \hat{v}(s)$$

and

$$\frac{d}{ds}\left[\hat{v}(s)e^{2\gamma(t-s)}\right] \leq 0.$$

Thus we obtain that the function $s \to \hat{v}(s)e^{2\gamma(t-s)}$ is not increasing and therefore $\hat{v}(s)e^{2\gamma(t-s)} \geq \hat{v}(t)$ for all $s \leq t$, and hence

$$|||T^*(t, s)H||| \geq e^{-\gamma(t-s)}|||H|||.$$

Using again (2.3) we obtain the second inequality in (2.18).

Let $x \in \mathbf{R}^n$, $i \in \mathcal{D}$, be fixed; consider $\widetilde{H} \in \mathcal{S}_n^d$ defined by

$$\widetilde{H}(j) = \begin{cases} 0 & \text{if } j \neq i, \\ xx^* & \text{if } j = i. \end{cases}$$

We may write successively

$$x^*(T(t, t_0)J^d)(i)x = Tr\big[xx^*(T(t, t_0)J^d)(i)\big] = \big\langle \widetilde{H}, T(t, t_0)J^d \big\rangle$$

$$= \big\langle T^*(t, t_0)\widetilde{H}, J^d \big\rangle = \sum_{i=1}^d Tr\big[T^*(t, t_0)\widetilde{H}\big](i)$$

$$\geq \sum_{i=1}^d \big|(T^*(t, t_0)\widetilde{H})(i)\big| \geq \max_{i \in \mathcal{D}} \big|(T^*(t, t_0)\widetilde{H})(i)\big|$$

$$= \big|T^*(t, t_0)\widetilde{H}\big| \geq \delta e^{-\gamma(t-t_0)}|x|^2.$$

Since $x \in \mathbf{R}^n$ is arbitrary we get

$$\big(T(t, t_0)J^d\big)(i) \geq \delta e^{-\gamma(t-t_0)} I_n, \ (\forall) i \in \mathcal{D}, t \geq t_0 \geq 0,$$

or equivalently $\big|T(t, t_0)J^d\big| \geq \delta e^{-\gamma(t-t_0)} J \ \forall t \geq t_0$. The second inequality in (ii) may be proved in the same way.   □

Remark 5. Combining the result in Theorem 5 with Remark 1 we obtain that

$$T(t, t_0)J^d \leq \|T(t, t_0)\| J^d, \tag{2.19}$$

$$T^*(t, t_0)J^d \leq \|T^*(t, t_0)\| J^d$$

for all $t, t_0 \in \mathcal{I}$. If the dependence $t \longmapsto \|\mathcal{L}(t)\|$ is a bounded function, we deduce easily that there exists $\hat{\gamma} > 0$ such that

$$\|T(t, t_0)\| \leq e^{\hat{\gamma}(t-t_0)},$$

$$\|T^*(t, t_0)\| \leq e^{\hat{\gamma}(t-t_0)}$$

for all $t \geq t_0, t, t_0 \in \mathcal{I}$.

Corollary 6. *Suppose that* $A_k, 0 \leq k \leq r$, *are continuous and bounded functions. Then there exist* $\delta > 0$ *and* $\gamma > 0$ *such that*

$$\delta e^{-\gamma(t-t_0)} J^d \leq T(t, t_0)J^d \leq e^{\gamma(t-t_0)} J^d, \tag{2.20}$$

$$\delta e^{-\gamma(t-t_0)} J^d \leq T^*(t, t_0)J^d \leq e^{\gamma(t-t_0)} J^d$$

for all $t \geq t_0, t, t_0 \in \mathcal{I}$.   □

Let us close this section with two important particular cases:

(a) $A_k(t) = 0$, $k = 1, \ldots, r$; in this case the linear operator (2.8) becomes

$$(\widehat{\mathcal{L}}(t)S)(i) = A_0(t, i)S(i) + S(i)A_0^*(t, i) \qquad (2.21)$$

$$+ \sum_{j=1}^{d} q_{ji} S(j),$$

$i \in \mathcal{D}$, $S \in \mathcal{S}_n^d$. It is easy to check that the evolution operator $T(t, t_0)$ defined by (2.9) has the representation

$$T(t, t_0) = \widehat{T}(t, t_0) + \int_{t_0}^{t} \widehat{T}(t, s)\mathcal{L}_2(s)T(s, t_0)\, ds, \qquad (2.22)$$

$t \geq t_0, t, t_0 \in \mathcal{I}$, where $\widehat{T}(t, t_0)$ is the evolution operator on $\mathcal{S}_n^d$ defined by the differential equation

$$\frac{d}{dt}S(t) = \widehat{\mathcal{L}}(t)S(t)$$

and $\mathcal{L}_2(t) : \mathcal{S}_n^d \to \mathcal{S}_n^d$ is defined by

$$(\mathcal{L}_2(t)H)(i) = \sum_{k=1}^{r} A_k(t, i)H(i)A_k^*(t, i),$$

$t \in \mathcal{I}$, $H \in \mathcal{S}_n^d$, $i \in \mathcal{D}$.

Remark 6. Since $T(t, t_0) \geq 0$, $\widehat{T}(t, t_0) \geq 0$, $t \geq t_0$, and $\mathcal{L}_2(t) \geq 0$, $t \in \mathcal{I}$, from (2.22) it follows that $T(t, t_0) \geq \widehat{T}(t, t_0)$ for all $t \geq t_0$, $t, t_0 \in \mathcal{I}$, and hence, using Theorem 3, we get

$$\|T(t, t_0)\| \geq \|\widehat{T}(t, t_0)\|,\ t \geq t_0,\ t, t_0 \in \mathcal{I}.$$

The evolution operator $\widehat{T}(t, t_0)$ will be called the *evolution operator on the space $\mathcal{S}_n^d$ defined by the pair* (A_0, Q). If additionally Q verifies the assumptions in Theorem 4, then (2.21) is the Lyapunov-type operator associated with the system (1.23).

(b) $\mathcal{D} = \{1\}$ and $q_{11} = 0$. In this case $\mathcal{S}_n^d$ reduces to $\mathcal{S}_n$ and the operator $\mathcal{L}(t)$ is defined by

$$\mathcal{L}(t)S = A_0(t)S + SA_0^*(t) + \sum_{k=1}^{r} A_k(t)SA_k^*(t), \qquad (2.23)$$

$t \in \mathcal{I}$, $S \in \mathcal{S}_n$, where we denoted $A_k(t) := A_k(t, 1)$. The evolution operator $T(t, t_0)$ will be called the *evolution operator on $\mathcal{S}_n$ defined by the system* $(A_0, \ldots, A_r)$. The operator (2.23) corresponds to the stochastic linear system (1.24).

Proposition 7. *If $\mathcal{I} = \mathbf{R}_+$ and $T(t, t_0)$ is the linear evolution operator on $\mathcal{S}_n$ defined by the Lyapunov operator* (2.23), *then we have the following representation formulae:*

$$T(t, t_0) = E[\Phi(t, t_0)\Phi^*(t, t_0)],$$

$$T^*(t, t_0) = E[\Phi^*(t, t_0)\Phi(t, t_0)]$$

for all $t \geq t_0 \geq 0$, $\Phi(t, t_0)$ denoting the fundamental matrix solution of the system (1.24).

Proof. The second equality follows directly from Theorem 4 and the first follows from the second one and the definition of the adjoint operator. □

Remark 7. Although in Theorem 4 we established a representation formula for the adjoint operator $T^*(t, t_0)$, a representation formula for $T(t, t_0)$ can be also be given, namely,

$$(T(t, t_0)H)(j) = \sum_{i=1}^{d} E\big[\Phi(t, t_0)H_i\Phi^*(t, t_0)\chi_{\eta(t)=j} \mid \eta(t_0) = i\big], \qquad (2.24)$$

$t \geq t_0 \geq 0$, $j \in \mathcal{D}$, $H \in \mathcal{S}_n^d$. Indeed, we have for $T = T(t, t_0)$,

$$\langle TH, G \rangle = \langle H, T^*G \rangle$$

$$= \sum_{i=1}^{d} Tr H_i E\big[\Phi^*(t, t_0)G(\eta(t))\Phi(t, t_0) \mid \eta(t_0) = i\big]$$

$$= \sum_{i=1}^{d}\sum_{j=1}^{d} Tr H_i E\big[\Phi^*(t, t_0)G(j)\Phi(t, t_0)\chi_{\eta(t)=j} \mid \eta(t_0) = i\big]$$

$$= \sum_{i=1}^{d}\sum_{j=1}^{d} E\big[Tr(H_i\Phi^*(t, t_0)G(j)\Phi(t, t_0))\chi_{\eta(t)=j} \mid \eta(t_0) = i\big]$$

$$= \sum_{i=1}^{d}\sum_{j=1}^{d} E\big[Tr(G(j)\Phi(t, t_0)H_i\Phi^*(t, t_0))\chi_{\eta(t)=j} \mid \eta(t_0) = i\big]$$

$$= \sum_{j=1}^{d} Tr G_j \left(\sum_{i=1}^{d} E\big[\Phi(t, t_0)H_i\Phi^*(t, t_0)\chi_{\eta(t)=j} \mid \eta(t_0) = i\big]\right),$$

from which (2.24) directly follows.

2.3 A class of linear differential equations on the space $(\mathbf{R}^n)^d$

Let $A: \mathbf{R}_+ \to \mathcal{M}_n^d$ be a bounded and continuous function, that is, $A(t) = (A(t, 1), \ldots, A(t, d))$, $t \in \mathbf{R}_+$. For each $t \geq 0$ we define the linear operator $M(t) : (\mathbf{R}^n)^d \to (\mathbf{R}^n)^d$ by

$$(M(t)y)(i) = A(t, i)y(i) + \sum_{j=1}^{d} q_{ji}y(j), \quad i \in \mathcal{D}, \qquad (2.25)$$

$y = (y(1), \ldots, y(d)) \in (\mathbf{R}^n)^d$, $Q = (q_{ij}) \in \mathbf{R}^{d \times d}$ satisfies the conditions $q_{ij} \geq 0$ for $i \neq j$ and $\sum_{j=1}^{d} q_{ij} = 0$. It is easy to check that for each $t \geq 0$, $M(t)$ is a linear and bounded operator on the Hilbert space $(\mathbf{R}^n)^d$ and $t \longmapsto \|M(t)\|$ is a bounded function, $\| \cdot \|$ denoting the operatorial norm induced by the norm in $(\mathbf{R}^n)^d$.

Let us consider the linear differential equation on $(\mathbf{R}^n)^d$:

$$\frac{d}{dt} y(t) = M(t) y(t). \tag{2.26}$$

Let $R(t, t_0)$ be the linear evolution operator associated with the equation (2.26), that is,

$$\frac{d}{dt} R(t, t_0) = M(t) R(t, t_0), \quad R(t_0, t_0) y = y$$

for all $t, t_0 \geq 0$, $y \in (\mathbf{R}^n)^d$.

By $M^*(t)$ and $R^*(t, t_0)$ we denote the adjoint operators of $M(t)$ and $R(t, t_0)$, respectively, on $(\mathbf{R}^n)^d$. One can easily see that

$$(M^*(t) y)(i) = A^*(t, i) y(i) + \sum_{j=1}^{r} q_{ij} y(j), \ i \in \mathcal{D}, \ y \in (\mathbf{R}^n)^d,$$

$$\frac{d}{dt} R^*(t, t_0) = R^*(t, t_0) M^*(t), \tag{2.27}$$

$$\frac{d}{dt} R^*(s, t) = -M^*(t) R^*(s, t)$$

for all $t, s \in R_+$. The operator $R(t, t_0)$ will be termed the *evolution operator on* $(\mathbf{R}^n)^d$ *defined by the pair* (A, Q).

The next result provides the relationship between the evolution operator $R(t, t_0)$ and the fundamental matrix solution $\Phi(t, t_0)$ of the stochastic system (1.23).

Proposition 8. *Under the assumptions given at the beginning of the section, the following equality holds:*

$$(R^*(t, t_0) y)(i) = E[\Phi^*(t, t_0) y(\eta(t)) \mid \eta(t_0) = i], \ t \geq t_0 \geq 0,$$

$i \in \mathcal{D}$, $y = (y(1), \ldots, y(d)) \in (\mathbf{R}^n)^d$.

Proof. Let $t \geq t_0 \geq 0$ and the operator $V(t, t_0) : (\mathbf{R}^n)^d \to (\mathbf{R}^n)^d$ be defined by

$$(V(t, t_0) y)(i) = E[\Phi^*(t, t_0) y(\eta(t)) \mid \eta(t_0) = i],$$

$i \in \mathcal{D}$, $y = (y(1), \ldots, y(d)) \in (\mathbf{R}^n)^d$. Let y be fixed and consider the function $v : \mathbf{R}^n \times \mathcal{D} \to \mathbf{R}$ by

$$v(x, i) = x^* y(i).$$

Applying the Itô-type formula (Theorem 35 of Chapter 1) to the function v and to the system (1.23), we obtain:

$$E[v(x(t), \eta(t)) \mid \eta(t_0) = i] - x_0^* y(i)$$

$$= E\left[\int_{t_0}^{t} x^*(s) \left(A_0^*(s, \eta(s)) y(\eta(s)) + \sum_{j=1}^{d} q_{\eta(s)j} y(j) \right) ds \mid \eta(t_0) = i \right],$$

where $x(s) = \Phi(s, t_0)x_0$. Further, we write

$$x_0^*(V(t, t_0)y)(i) - x_0^* y(i) = x_0^* \int_{t_0}^t (V(s, t_0)M^*(s)y)(i)\, ds$$

for all $t \geq t_0 \geq 0$, $x_0 \in \mathbf{R}^n$, $i \in \mathcal{D}$. Therefore, we may conclude that

$$V(t, t_0)y - y = \int_{t_0}^t V(s, t_0)M^*(s)y\, ds$$

for all $t \geq t_0$ and $y \in (\mathbf{R}^n)^d$.

By differentiation, we deduce that

$$\frac{d}{dt}V(t, t_0)y = V(t, t_0)M^*(t)y$$

for all $y \in (\mathbf{R}^n)^d$, and hence

$$\frac{d}{dt}V(t, t_0) = V(t, t_0)M^*(t), \; t \geq t_0.$$

Since $V(t_0, t_0) = R^*(t_0, t_0)$, from (2.27), $V(t, t_0) = R^*(t, t_0)$ for all $t \geq t_0 \geq 0$, and the proof ends. $\square$

2.4 Exponential stability for Lyapunov-type equations on $\mathcal{S}_n^d$

In this section $\mathcal{I} \subset \mathbf{R}$ denotes a right-unbounded interval. Consider the Lyapunov operator (2.8) on $\mathcal{S}_n^d$, where Q satisfies (2.7) and A_k are continuous and bounded functions. Let $T(t, t_0)$ be the linear evolution operator on $\mathcal{S}_n^d$ defined by (2.9).

Definition 4. *We say that the Lyapunov-type operator $\mathcal{L}(t)$ generates an* exponentially stable evolution, *or equivalently, the system $(A_0, \ldots, A_r; Q)$ is* stable *if there exist the constants $\beta \geq 1$, $\alpha > 0$ such that*

$$\|T(t, t_0)\| \leq \beta e^{-\alpha(t-t_0)}, \; t \geq t_0, \, t_0 \in \mathcal{I}. \tag{2.28}$$

Remark 8. From Remark 6 immediately follows that if $(A_0, \ldots, A_r; Q)$ is stable, then there exists $\beta \geq 1$ and $\alpha > 0$ such that

$$\|\widehat{T}(t, t_0)\| \leq \beta e^{-\alpha(t-t_0)}$$

for all $t \geq t_0, t, t_0 \in \mathcal{I}$, where $\widehat{T}(t, t_0)$ is the evolution operator on $\mathcal{S}_n^d$ defined by the pair (A_0, Q).

As usual we denote

$$\int_t^\infty T^*(s, t)H(s)\, ds := \lim_{\tau \to \infty} \int_t^\tau T^*(s, t)H(s)\, ds$$

each time when the limit in the right-hand side exists. In this case we say that the integral in the left-hand side is convergent.

The result stated in the next lemma will be used several times in this section.

Lemma 9. *Let* $H : \mathcal{I} \to \mathcal{S}_n^d$ *be a continuous function. Assume that the integral* $\int_t^\infty T^*(s,t)H(s)\,ds$ *is convergent for all* $t \in \mathcal{I}$. *Set*

$$K(t) := \int_t^\infty T^*(s,t)H(s)\,ds.$$

Then $K(t)$ *is a solution of the affine differential equation*

$$\frac{d}{dt}K(t) + \mathcal{L}^*(t)K(t) + H(t) = 0.$$

Proof. Let $\tau > t$ be fixed. Then we have

$$K(t) = \int_t^\tau T^*(s,t)H(s)\,ds + \int_\tau^\infty T^*(s,t)H(s)\,ds.$$

Based on (2.11) we get

$$K(t) = T^*(\tau,t)K(\tau) + T^*(\tau,t)\int_t^\tau T^*(s,\tau)H(s)\,ds.$$

Using (2.12) we obtain that $K(t)$ is differentiable and

$$\frac{d}{dt}K(t) = -\mathcal{L}^*(t)K(t) - H(t),$$

and the proof ends. □

The next lemma shows that the integrals used in this section are absolutely convergent.

Lemma 10. *Let* $H : \mathcal{I} \to \mathcal{S}_n^d$ *be a continuous function such that* $H(t) \geq 0$ *for all* $t \in I$. *Then the following are equivalent:*
(i) *The integral* $\int_t^\infty |T^*(s,t)H(s)|\,ds$ *is convergent for all* $t \in \mathcal{I}$.
(ii) *The integral* $\int_t^\infty T^*(s,t)H(s)\,ds$ *is convergent for all* $t \in \mathcal{I}$.

Proof. (i) $\Rightarrow$ (ii) follows immediately.
(ii) $\Rightarrow$ (i) Let

$$\gamma(t) = \left| \int_t^\infty T^*(s,t)H(s)\,ds \right|, \ t \in \mathcal{I}.$$

We have

$$\int_t^\infty T^*(s,t)H(s)\,ds \leq \gamma(t)J^d, \ t \in \mathcal{I},$$

which leads to

$$\int_t^\infty (T^*(s,t)H(s))(i)\,ds \leq \gamma(t)I_n, \ i \in \mathcal{D}, \ t \in \mathcal{I}.$$

Hence

$$\int_t^\infty Tr(T^*(s,t)H(s))(i)\,ds \leq n\gamma(t), \ i \in \mathcal{D}, \ t \in \mathcal{I},$$

from which we deduce that

$$\int_t^\tau Tr(T^*(s, t)H(s))(i)\, ds \le n\gamma(t), \ \tau \ge t.$$

The above inequality gives

$$\int_t^\tau |(T^*(s, t)H(s))(i)|\, ds \le n\gamma(t),$$

which leads to

$$\sum_{i=1}^d \int_t^\tau |(T^*(s, t)H(s))(i)|\, ds \le dn\gamma(t).$$

Since

$$|T^*(s, t)H(s)| \le \sum_{i=1}^d |(T^*(s, t)H(s))(i)|,$$

we get

$$\int_t^\tau |(T^*(s, t)H(s))|\, ds \le nd\gamma(t)$$

for all $\tau \ge t$ and the proof is complete. $\qquad\square$

The following result provides necessary and sufficient conditions ensuring exponential stability of the considered class of differential equations.

Theorem 11. *The following are equivalent:*
(i) *The system* $(A_0, \ldots, A_r; Q)$ *is stable.*
(ii) *There exists* $\delta > 0$ *such that*

$$\int_{t_0}^t \|T(t, s)\|\, ds < \delta$$

for all $t \ge t_0, t, t_0 \in \mathcal{I}$.
(iii) *There exists a constant* $\delta > 0$ *such that*

$$\int_{t_0}^t T(t, s)J^d\, ds < \delta J^d$$

for all $t \ge t_0, t, t_0 \in \mathcal{I}$.

Proof. (i) $\Rightarrow$ (ii) From (2.28) it follows that

$$\int_{t_0}^t \|T(t, s)\|\, ds \le \frac{\beta}{\alpha}$$

for all $t \ge t_0$.
(ii) $\Rightarrow$ (iii) immediately follows from (2.6) and Theorem 5.

(iii) $\Rightarrow$ (i) Let $H : \mathcal{I} \to \mathcal{S}_n^d$ be a continuous and bounded function. It follows that the real constants δ_1, δ_2 exist such that $\delta_1 J^d \leq H(s) \leq \delta_2 J^d$ for all $s \in \mathcal{I}$.

Since $T(t, s)$ is a positive operator defined on $\mathcal{S}_n^d$, we deduce $\delta_1 T(t, s) J^d \leq T(t, s)H(s) \leq \delta_2 T(t, s)J^d$ for all $t \geq s \geq t_0, t_0 \in \mathcal{I}$. Hence

$$\delta_1 \int_{t_0}^{t} T(t, s)J^d ds \leq \int_{t_0}^{t} T(t, s)H(s)ds \leq \delta_2 \int_{t_0}^{t} T(t, s)J^d ds$$

for all $t \geq t_0, t_0 \in \mathcal{I}$. Thus, if (iii) holds we deduce that the real constants $\tilde{\delta}_1, \tilde{\delta}_2$ exist such that

$$\tilde{\delta}_1 J^d \leq \int_{t_0}^{t} T(t, s)H(s)ds \leq \tilde{\delta}_2 J^d$$

for all $t \geq t_0, t_0 \in \mathcal{I}$, which shows that $t \to \int_{t_0}^{t} T(t, s)H(s)ds$ is bounded on $[t_0, \infty)$ uniformly with respect to $t_0 \in \mathcal{I}$ for all continuous and bounded functions $H(s)$.

Applying Perron's theorem (see [58]) we deduce that the constants $\beta \geq 1, \alpha > 0$ exist such that

$$\|T(t, s)\| \leq \beta e^{-\alpha(t-s)}, \forall t \geq s \geq t_0, t_0 \in \mathcal{I},$$

that is, the system $(A_0, \ldots, A_r; Q)$ is stable and thus the proof is complete. $\square$

Theorem 12. *The following are equivalent:*
  (i) *The system* $(A_0, \ldots, A_r; Q)$ *is stable.*
  (ii) *There exist the constants* $\beta_1 \geq 1, \alpha > 0$ *such that*

$$\|T^*(t, t_0)\| \leq \beta_1 e^{-\alpha(t-t_0)}$$

for all $t \geq t_0, t, t_0 \in \mathcal{I}$.
  (iii) *There exists a constant* $\delta > 0$ *such that*

$$\int_{t}^{\infty} \|T^*(s, t)\| ds \leq \delta$$

for all $t \in \mathcal{I}$.
  (iv) *There exists* $\delta > 0$ *such that*

$$\int_{t}^{\infty} T^*(s, t)J^d ds \leq \delta J^d$$

for all $t \in \mathcal{I}$.
  (v) *The affine differential equation*

$$\frac{d}{dt}K(t) + \mathcal{L}^*(t)K(t) + J^d = 0 \tag{2.29}$$

has a bounded and uniform positive solution on $\mathcal{I}$.
  (vi) *For each* $H : \mathcal{I} \to \mathcal{S}_n^d$ *continuous, bounded, and uniform positive function, the affine differential equation on* $\mathcal{S}_n^d$,

$$\frac{d}{dt}K(t) + \mathcal{L}^*(t)K(t) + H(t) = 0, \tag{2.30}$$

has a bounded and uniform positive solution defined on $\mathcal{I}$.

(vii) *For each* $H : \mathcal{I} \to \mathcal{S}_n^d$ *continuous, bounded, and uniform positive function, there exists a C^1-function* $\widetilde{K} : \mathcal{I} \to \mathcal{S}_n^d$, $\widetilde{K} \gg 0$, *bounded with bounded derivative, solving the following differential inequality on $\mathcal{S}_n^d$:*

$$\frac{d}{dt}K(t) + \mathcal{L}^*(t)K(t) + H(t) \ll 0, \ t \in \mathcal{I}. \tag{2.31}$$

(viii) *There exists a C^1-function $K : \mathcal{I} \to \mathcal{S}_n^d$, bounded with bounded derivative, $K \gg 0$ solving the differential inequality*

$$\frac{d}{dt}K(t) + \mathcal{L}^*(t)K(t) \ll 0, \ t \in \mathcal{I}.$$

Proof. (i) $\Longleftrightarrow$ (ii) immediately follows from (2.5)
(ii) $\Rightarrow$ (iii) From (ii),

$$\int_t^\infty \|T^*(s,t)\| \, ds \le \frac{\beta}{\alpha}$$

for all $t \in \mathcal{I}$.
(iii) $\Rightarrow$ (iv) immediately follows from (2.6) and Theorem 5.
(iv) $\Rightarrow$ (v) Define

$$\widetilde{K}(t) = \int_t^\infty T^*(s,t) J^d \, ds, t \in \mathcal{I}.$$

From Theorem 5(ii), there exists $\delta_1 > 0$ such that $\widetilde{K}(t) \ge \delta_1 J^d$ for all $t \in \mathcal{I}$, hence $\widetilde{K}(t) \gg 0$. On the other hand the function $t \longmapsto \widetilde{K}(t)$ is differentiable, and based on (2.13) (see Lemma 9), we get that $\widetilde{K}(t)$ is a solution of the equation (2.29).
(v) $\Rightarrow$ (iv) Let $\widehat{K} : \mathcal{I} \to \mathcal{S}_n^d$ be the bounded and uniform positive solution of the equation (2.29). Therefore there exist the constants $\mu_1 > 0$, $\mu_2 > 0$ such that

$$\mu_1 J^d \le \widehat{K}(t) \le \mu_2 J^d, \ t \in \mathcal{I}.$$

Using (2.13) and the constant variation formula we deduce that

$$\widehat{K}(t) = T^*(\tau, t)\widehat{K}(\tau) + \int_t^\tau T^*(s,t) J^d ds$$

for all $t \le \tau; t, \tau \in \mathcal{I}$. Since the operator $T^*(\tau, t)$ is positive, we can write

$$0 \le \int_t^\tau T^*(s,t) J^d ds \le \widehat{K}(t) < \mu_2 J^d.$$

Therefore the integral

$$\int_t^\infty T^*(s,t) J^d \, ds = \lim_{\tau \to \infty} \int_t^\tau T^*(s,t) J^d \, ds$$

is well defined and

$$\int_t^\infty T^*(s,t) J^d \, ds \le \mu_2 J^d, \ t \in \mathcal{I}.$$

(iv) $\Rightarrow$ (vi) Let $H: \mathcal{I} \to \mathcal{S}_n^d$ be a function with the properties in the statement; that is, there exist the constants $\nu_1 > 0$ and $\nu_2 > 0$ such that

$$\nu_1 J^d \leq H(t) \leq \nu_2 J^d, \ t \in \mathcal{I}.$$

Since the operator $T^*(s, t)$ is positive, we have

$$\nu_1 T^*(s, t) J^d \leq T^*(s, t) H(t) \leq \nu_2 T^*(s, t) J^d \tag{2.32}$$

for all $s \geq t$, $s, t \in \mathcal{I}$, which leads to

$$\int_t^\tau T^*(s, t) H(s)\, ds \leq \nu_2 \int_t^\tau T^*(s, t) J^d\, ds$$

for all $t \leq \tau$. Further, we obtain

$$\int_t^\tau T^*(s, t) H(s)\, ds \leq \nu_2 \delta J^d$$

for all $t \leq \tau, t, \tau \in \mathcal{I}$, which gives

$$\int_t^\infty T^*(s, t) H(s)\, ds \leq \nu_2 \delta J^d, \ t \in \mathcal{I}.$$

On the other hand, from (2.32) together with (2.20) we deduce that there exists $\tilde{\delta} > 0$ such that

$$\tilde{\delta} J^d \leq \int_t^\infty T^*(s, t) H(s)\, ds \leq \nu_2 \delta J^d$$

for all $t \in \mathcal{I}$.

We define

$$K(t) = \int_t^\infty T^*(s, t) H(s)\, ds.$$

Based on (2.13) we obtain that $K(t)$ defined above is a solution of (2.30).

(vi) $\Rightarrow$ (vii) From vi) it follows that the affine differential equation

$$\frac{d}{dt} K(t) + \mathcal{L}^*(t) K(t) + H(t) + J^d = 0$$

has a uniform positive and bounded solution which also solves (2.31).

(vii) $\Rightarrow$ (viii) It is obvious that any solution of (2.31) is a solution of

$$\frac{d}{dt} K(t) + \mathcal{L}^*(t) K(t) \ll 0. \tag{2.33}$$

(viii) $\Rightarrow$ (iv) Let $\overline{K} : \mathcal{I} \to \mathcal{S}_n^d$ be a bounded and uniform positive solution of (2.33) with bounded derivative. We define $M(t) = (M(t, 1), \ldots, M(t, d))$ by

$$M(t) = -\frac{d}{dt}\overline{K} - \mathcal{L}^*(t)\overline{K}.$$

Therefore, there exists the constants $\tilde{\mu}_1 > 0$ and $\tilde{\mu}_2 > 0$ such that

$$\tilde{\mu}_1 J^d \leq M(t) \leq \tilde{\mu}_2 J^d, \tag{2.34}$$

$t \in \mathcal{I}$. Based on (2.13) and the constant variation formula we obtain that

$$\overline{K}(t) = T^*(\tau, t)\overline{K}(\tau) + \int_t^\tau T^*(s, t)M(s)\,ds$$

for all $t \leq \tau$, $t, \tau \in \mathcal{I}$.

Since the operator $T^*(s, t)$ is positive, we deduce that

$$\int_t^\tau T^*(s, t)M(s)\,ds \leq \overline{K}(t) \leq \sup_{t \in \mathcal{I}} |\overline{K}(t)| J^d.$$

Therefore, there exists $\tilde{\delta} > 0$ such that

$$\int_t^\infty T^*(s, t)M(s)\,ds \leq \tilde{\delta} J^d, \tag{2.35}$$

$t \in \mathcal{I}$. From (2.34) and (2.35) we deduce that

$$\int_t^\infty T^*(s, t)J^d\,ds \leq \frac{1}{\tilde{\mu}_1}\int_t^\infty T^*(s, t)M(s)\,ds \leq \frac{\tilde{\delta}}{\tilde{\mu}_1}$$

for all $t \in \mathcal{I}$.

(iv) $\Rightarrow$ (ii) Let

$$\widetilde{K}(t) = \int_t^\infty T^*(s, t)J^d\,ds.$$

Then we have $\widetilde{K}(t) \leq \delta J^d$, and as in the proof of (iv) $\Rightarrow$ (v) we have

$$\delta_1 J^d \leq \widetilde{K}(t) \leq \delta J^d. \tag{2.36}$$

For $t \geq t_0$, $t_0 \in \mathcal{I}$, we define $G(t) := T^*(t, t_0)\widetilde{K}(t)$. Using (2.10) we get

$$G(t) = \int_t^\infty T^*(s, t)J^d\,ds.$$

Therefore

$$\frac{d}{dt}G(t) = -T^*(t, t_0)J^d$$

for all $t \geq t_0$. Since $T^*(t, t_0)$ is a positive operator, from (2.36) we obtain that

$$T^*(t, t_0)J^d \geq \frac{1}{\delta}G(t) \geq \frac{\delta_1}{\delta}T^*(t, t_0)J^d, \tag{2.37}$$

which leads to

$$\frac{d}{dt}G(t) \leq -\frac{1}{\delta}G(t), \ t \in \mathcal{I},$$

from which it follows that

$$\frac{d}{dt}G(t, i) \leq -\frac{1}{\delta}G(t, i), \ i \in \mathcal{D}.$$

Let $x \in \mathbf{R}^n$ be arbitrary and set $g_i(t) = x^*G(t, i)x$. Then we have

$$\frac{d}{dt}g_i(t) + \frac{1}{\delta}g_i(t) \leq 0$$

for all $t \geq t_0$, or equivalently

$$\frac{d}{dt}\left(g_i(t)e^{\frac{1}{\delta}(t-t_0)}\right) \leq 0,$$

which shows that the function

$$t \longmapsto g_i(t)e^{\frac{1}{\delta}(t-t_0)}$$

is not increasing. Hence we obtain

$$G(t, i) \leq e^{-\alpha(t-t_0)}G(t_0, i)$$

for all $t \geq t_0$, $i \in \mathcal{D}$, where $\alpha = \frac{1}{\delta}$, and with (2.36),

$$G(t) \leq \delta e^{-\alpha(t-t_0)}J^d.$$

From (2.37) we get

$$T^*(t, t_0)J^d \leq \frac{\delta}{\delta_1}e^{-\alpha(t-t_0)}$$

for all $t \geq t_0$, hence

$$(T^*(t, t_0)J^d)(i) \leq \frac{\delta}{\delta_1}e^{-\alpha(t-t_0)}I_n.$$

The above inequality leads to

$$|(T^*(t, t_0)J^d)(i)| \leq \frac{\delta}{\delta_1}e^{-\alpha(t-t_0)}$$

and therefore

$$|T^*(t, t_0)J^d| \leq \frac{\delta}{\delta_1}e^{-\alpha(t-t_0)}.$$

Using Theorem 3 we obtain that

$$\|T^*(t, t_0)\| \leq \frac{\delta}{\delta_1}e^{-\alpha(t-t_0)}, \ t \geq t_0, t_0 \in \mathcal{I},$$

and the proof is complete.   □

Proposition 13. *Assume that there exists a bounded uniform positive and continuous function $H: \mathcal{I} \rightarrow \mathcal{S}_n^d$, for which the affine Lyapunov-type function (2.30) has a bounded solution $K_0(t) = (K_0(t, 1), \ldots, K_0(t, d))$ with $K_0(t, i) \geq 0$, $t \in \mathcal{I}$. Then the system $(A_0, \ldots, A_r; Q)$ is stable.*

Proof. If $K_0: \mathcal{I} \rightarrow \mathcal{S}_n^d$ is a bounded solution of (2.30) and $K_0(t) \geq 0$, then

$$K_0(t) = T^*(\tau, t) K_0(\tau) + \int_t^\tau T^*(s, t) H(s) \, ds$$

for all $t \leq \tau$, $t, \tau \in \mathcal{I}$. Since $T^*(\tau, t) K_0(\tau) \geq 0$, we get

$$\int_t^\tau T^*(s, t) H(s) \, ds \leq K_0(t) \leq c J^d$$

for all $t \leq \tau$, $t, \tau \in \mathcal{I}$ and for some positive constant c.

On the other hand, $H \gg 0$ implies that there exists a positive constant $\tilde{c}$ such that $\tilde{c} J^d \leq H(s)$ for all $s \in \mathcal{I}$, which leads to

$$T^*(s, t) J^d \leq \frac{1}{\tilde{c}} T^*(s, t) H(s)$$

and therefore

$$\int_t^\infty T^*(s, t) J^d ds \leq \frac{1}{\tilde{c}} \int_t^\infty T^*(s, t) H(s) ds \leq \frac{c}{\tilde{c}} J^d,$$

and from Theorem 12 we conclude that the system $(A_0, \ldots, A_r; Q)$ is stable and the proof is complete. □

Remark 9. From the proof of Theorem 12 and of Proposition 13, we remark that if $H: \mathcal{I} \rightarrow \mathcal{S}_n^d$ is a bounded and continuous function $H(t) \geq 0$, then the differential equation

$$\frac{d}{dt} K(t) + \mathcal{L}^*(t) K(t) + H(t) = 0 \tag{2.38}$$

has a bounded solution $K(t) \geq 0$ if and only if there exists $\gamma > 0$ such that

$$\int_t^\infty T^*(s, t) H(s) \, ds \leq \gamma J^d \tag{2.39}$$

for all $t \in \mathcal{I}$. Moreover, if (2.39) is accomplished, then

$$K(t) = \int_t^\infty T^*(s, t) H(s) \, ds$$

is a bounded and semipositive solution of (2.38).

Proposition 14. *If the system* $(A_0, \ldots, A_r; Q)$ *is stable, then for all bounded and continuous functions* $H : \mathcal{I} \to \mathcal{S}_n^d$, *the corresponding Lyapunov-type equation* (2.30) *has a unique bounded solution given by*

$$\widetilde{K}(t) = \int_t^\infty T^*(s, t) H(s) \, ds.$$

Moreover, if $t \longmapsto A_k(t)$, $k = 0, \ldots, r$, $t \longmapsto H(t)$ *are* θ-*periodic functions, then the unique bounded solution of* (2.30) *is a* θ-*periodic function too.*

If $A_k(t) = A_k$, $k = 0, \ldots, r$, *and* $H(t) = H$, $t \in \mathcal{I}$, *then the unique bounded solution of* (2.30) *is constant and it solves the algebraic equation*

$$\mathcal{L}^* K + H = 0.$$

Proof. From Theorem 12 and Lemma 9 it follows directly that $\widetilde{K}(t) = \int_t^\infty T^*(s, t) H(s) \, ds, t \in \mathcal{I}$, is a bounded solution of (2.30). Further, let $K : \mathcal{I} \to \mathcal{S}_n^d$ be a bounded solution of (2.30). By the constant variation formula we obtain

$$K(t) = T^*(\tau, t) K(\tau) + \int_t^\tau T^*(s, t) H(s) \, ds \qquad (2.40)$$

for all $t \leq \tau, t, \tau \in \mathcal{I}$. Since the system $(A_0, \ldots, A_r; Q)$ is stable and $K(\tau)$ is bounded, it follows that

$$\lim_{\tau \to \infty} T^*(\tau, t) K(\tau) = 0,$$

$$\lim_{\tau \to \infty} \int_t^\tau T^*(s, t) H(s) \, ds = \int_t^\infty T^*(s, t) H(s) \, ds.$$

Hence, if in (2.40) we take the limit for $\tau \to \infty$, then we obtain

$$K(t) = \int_t^\infty T^*(s, t) H(s) \, ds,$$

which shows that $K(t) = \widetilde{K}(t)$. Assume now that $t \longmapsto A_k(t)$, $k = 0, \ldots, r$, $t \longmapsto H(t)$ are θ-periodic functions. In this case we have

$$\widetilde{K}(t + \theta) = \int_{t+\theta}^\infty T^*(s, t + \theta) H(s) \, ds.$$

Invoking Remark 3(ii) we may write

$$\widetilde{K}(t + \theta) = \int_t^\infty T^*(s, t) H(s + \theta) \, ds = \int_t^\infty T^*(s, t) H(s) \, ds.$$

Thus we proved that $\widetilde{K}(t + \theta) = \widetilde{K}(t)$ for all $t \in \mathcal{I}$, which shows that the unique bounded solution of equation (2.30) is a θ-periodic function.

If the functions $A_k, k \in \{0, \ldots, r\}$, and H are constant functions, based on Remark 3(i) we obtain that

$$\widetilde{K}(t) = \int_t^\infty e^{L^*(s-t)} H \, ds = \int_0^\infty e^{L^*s} H \, ds,$$

which shows that $\widetilde{K}(t) = \widetilde{K}(0)$ for all $t \in \mathcal{I}$; that is, in the time-invariant case, the unique bounded solution of the equation (2.30) is constant. It is obvious that it solves the algebraic equation $\mathcal{L}^* K + H = 0$ and the proof is complete. □

In the time-invariant case we have the following theorem.

Theorem 15. *Assume that the system* (1.22) *is in the stationary case. Then the following are equivalent:*

(i) *The system* $(A_0, \ldots, A_r; Q)$ *is stable.*

(ii) *For all* $H = (H(1), \ldots, H(d)) \in \mathcal{S}_n^d$, $H(i) > 0, i \in \mathcal{D}$, *the algebraic linear equation on* $\mathcal{S}_n^d$,

$$\mathcal{L}^* K + H = 0, \tag{2.41}$$

has a unique solution $K = (K(1), \ldots, K(d)) \in \mathcal{S}_n^d$, $K(i) > 0, i \in \mathcal{D}$.

(iii) *For each* $H = (H(1), \ldots, H(d)) \in \mathcal{S}_n^d$, $H(i) \geq 0, i \in \mathcal{D}$, *the linear inequality*

$$\mathcal{L}^* K + H < 0 \tag{2.42}$$

has a solution $K = (K(1), \ldots, K(d))$, $K(i) > 0, i \in \mathcal{D}$.

(iv) *There exists* $K \geq 0$ *satisfying* $\mathcal{L}^* K < 0$.

(v) *For each* $H \in \mathcal{S}_n^d$, $H > 0$, *the linear equation on* $\mathcal{S}_n^d$,

$$\mathcal{L} K + H = 0, \tag{2.43}$$

has a unique positive solution $K = (K(1), \ldots, K(d))$.

(vi) *For each* $H \in \mathcal{S}_n^d$, $H \geq 0$, *the linear inequality*

$$\mathcal{L} K + H < 0 \tag{2.44}$$

has a solution $K > 0$.

(vii) *There exists* $K \geq 0$ *satisfying* $\mathcal{L} K < 0$.

Proof. (i) $\Rightarrow$ (ii). From the equivalence (i) $\Longleftrightarrow$ (vi) in Theorem 12 we get that the equation

$$\frac{d}{dt} K(t) + \mathcal{L}^* K(t) + H = 0$$

has a unique bounded and uniform positive solution $\widetilde{K}(t)$. Moreover, $\widetilde{K}(t)$ is given by

$$\widetilde{K}(t) = \int_t^\infty e^{L^*(s-t)} H \, ds.$$

We have $\widetilde{K}(t) = \int_0^\infty e^{\mathcal{L}^* s} H ds = \widetilde{K}(0)$ for all $t \geq 0$. Hence $\widetilde{K}(t)$ is constant and it verifies the equation (2.41).

(ii) $\Rightarrow$ (iii). Indeed (ii) implies that the equation $\mathcal{L}^* K + H + J^d = 0$ has a solution $\hat{K} > 0$. Hence $\hat{K}$ verifies (2.42).

(iii) $\Rightarrow$ (iv) follows immediately (taking $H = J^d$).

(iv) $\Rightarrow$ (i) follows from Proposition 13.

(i) $\Rightarrow$ (v) Let $H > 0$. Therefore $\beta_2 J^d \leq H \leq \beta_1 J^d$ and with $\beta_1 \geq \beta_2 > 0$. Since $\|e^{\mathcal{L}t}\| \leq \beta e^{-\alpha t}, t \geq 0$, for some $\beta \geq 1, \alpha > 0$ the integral $\hat{K} = \int_0^\infty e^{\mathcal{L}t} H dt$ is convergent, and since $e^{\mathcal{L}t}$ is a positive operator we have according to (2.20)

$$\beta_3 J^d \leq \beta_2 \int_0^\infty e^{\mathcal{L}t} J^d dt \leq \hat{K} \leq \frac{\beta}{\alpha} \beta_1 J^d.$$

Further, we can write

$$\mathcal{L}\hat{K} = \int_0^\infty \frac{d}{dt}(e^{\mathcal{L}t} H) dt = -H,$$

and thus $\widehat{K}$ is a solution of (2.43). To prove the uniqueness, one observes that if K verifies (2.43), then K is a constant solution of the equation

$$\frac{d}{dt} K(t) = \mathcal{L}K(t) + H,$$

hence

$$K = e^{\mathcal{L}t} K + \int_0^t e^{\mathcal{L}(t-s)} H ds = e^{\mathcal{L}t} + \int_0^t e^{\mathcal{L}u} H du.$$

Since $\lim_{t\to\infty} e^{\mathcal{L}t} = 0$, taking $t \to \infty$ in the above inequality, one gets $K = \int_0^\infty e^{\mathcal{L}s} H ds = \widehat{K}$ and thus the proof of (i)$\Rightarrow$(v) is complete.

(v) $\Rightarrow$ (vi) follows by using the same reasoning as in the proof (ii) $\Rightarrow$ (iii).

(vi) $\Rightarrow$ (vii) follows immediately (taking $H = J^d$).

(vii) $\Rightarrow$ (i) Let $H = -\mathcal{L}K$. Thus $\mathcal{L}K + H = 0$ with $H > 0$ and $K \geq 0$. Since K is a constant solution of the equation $\frac{d}{dt} K(t) = \mathcal{L}K(t) + H$ we have

$$K = e^{\mathcal{L}(t-t_0)} K + \int_{t_0}^t e^{\mathcal{L}(t-s)} H ds, t \geq t_0.$$

Since $e^{\mathcal{L}t}$ is a positive operator and $H \geq \gamma J^d$ with some $\gamma > 0$ we can write

$$\gamma \int_{t_0}^t e^{\mathcal{L}(t-s)} J^d ds \leq \int_{t_0}^t e^{\mathcal{L}(t-s)} H ds \leq K \leq \delta J^d.$$

Thus, by Theorem 11 the proof is complete.   □

Remark 10. The affine differential equation (2.30) is the compact version of the following system of matrix linear differential equations:

$$\frac{d}{dt}K(t,i) + \left(A_0(t,i) + \frac{1}{2}q_{ii}I_n\right)^* K(t,i) + K(t,i)\left(A_0(t,i) + \frac{1}{2}q_{ii}I_n\right)$$

$$+ \sum_{k=1}^{r} A_k^*(t,i)K(t,i)A_k(t,i) + \sum_{\substack{j=1 \\ j\neq i}}^{d} q_{ij}K(t,j) + H(t,i) \tag{2.45}$$

$$= 0, \; i \in \mathcal{D}.$$

In the time-invariant case the algebraic equation $\mathcal{L}^*K + H = 0$ is the compact form of the following system of linear equations:

$$\left(A_0(i) + \frac{1}{2}q_{ii}I_n\right)^* K(i) + K(i)\left(A_0(i) + \frac{1}{2}q_{ii}I_n\right)$$

$$+ \sum_{k=1}^{r} A_k^*(i)K(i)A_k(i) + \sum_{\substack{j=1 \\ j\neq i}}^{d} q_{ij}K(j) + H(i) = 0. \tag{2.46}$$

A consequence of Theorem 12 and Proposition 14 is the following corollary.

Corollary 16. *If the system* $(A_0, \ldots, A_r; Q)$ *is stable, then for all* $i \in \mathcal{D}$ *the system of linear differential equations on* $\mathbf{R}^n$,

$$\frac{d}{dt}y_i(t) = \left(A_0(t,i) + \frac{1}{2}q_{ii}I_n\right)y_i(t), \; t \in \mathcal{I}, \tag{2.47}$$

defines an exponentially stable evolution.

In the invariant case, if the system $(A_0, \ldots, A_r; Q)$ *is stable, then for all* $i \in \mathcal{D}$, *the eigenvalues of the matrices* $A_0(i) + \frac{1}{2}q_{ii}I_n$ *are located in the half plane* $\mathbf{C}^- = \{z \in \mathbf{C} \mid \operatorname{Re}(z) < 0\}$.

Proof. Since the system $(A_0, \ldots, A_r; Q)$ is stable, from Theorem 12 it follows that (2.45) has a uniform positive and bounded solution $\widetilde{K}(t) = (\widetilde{K}(t,1), \ldots, \widetilde{K}(t,d))$. For each $i \in \mathcal{D}$ we can write

$$\frac{d}{dt}\widetilde{K}(t,i) + \left(A_0(t,i) + \frac{1}{2}q_{ii}I_n\right)^* \widetilde{K}(t,i)$$

$$+ \widetilde{K}(t,i)\left(A_0(t,i) + \frac{1}{2}q_{ii}I_n\right) + \widetilde{H}(t,i) = 0,$$

where

$$\widetilde{H}(t,i) := H(t,i) + \sum_{k=1}^{r} A_k^*(t,i)\widetilde{K}(t,i)A_k(t,i) + \sum_{\substack{j=1 \\ j\neq i}}^{d} q_{ij}\widetilde{K}(t,j).$$

It is obvious that $\widetilde{H}(t,i) \gg 0$ for all $t \in \mathcal{I}$. By standard Lyapunov function arguments we conclude that the system (2.47) is exponentially stable and the proof ends.   □

The next result shows that the bounded solution of (2.30) can be obtained as a limit of a sequence of bounded solutions of some Lyapunov equations.

Proposition 17. *Assume that the system $(A_0, \ldots, A_r; Q)$ is stable. Let $H : \mathcal{I} \to \mathcal{S}_n^d$ be a bounded and positive semidefinite continuous function, $H(t) = (H(t, 1), \ldots, H(t, d))$. For each $i \in \mathcal{D}$ we define the sequence $\{K_i^p(t)\}_{p \in \mathbb{N}}$, where $t \longmapsto K_i^p(t)$ is the unique bounded solution of the differential equation:*

$$\frac{d}{dt} K_i^p(t) + \left(A_0(t, i) + \frac{1}{2} q_{ii} I_n \right)^* K_i^p(t)$$

$$+ K_i^p(t) \left(A_0(t, i) + \frac{1}{2} q_{ii} I_n \right) + H_i^p(t) = 0, \ i \in \mathcal{D}, \qquad (2.48)$$

with

$$H_i^p(t) := H(t, i) + \sum_{k=1}^r A_k^*(t, i) K_i^{p-1}(t) A_k(t, i) + \sum_{\substack{j=1 \\ j \neq i}}^d q_{ij} K_j^{p-1}(t),$$

$$p = 1, \ldots, t \in \mathcal{I} \quad and \quad K_i^0(t) = 0.$$

The sequences $\{K_i^p(t)\}_{p \in \mathbb{N}}$, $i \in \mathcal{D}$, are increasing and bounded. If we denote

$$K^\infty(t, i) = \lim_{p \to \infty} K_i^p(t), \ i \in \mathcal{D}, t \in \mathcal{I},$$

then $K^\infty(t) = (K^\infty(t, 1), \ldots, K^\infty(t, d))$ is the unique bounded solution of the equation (2.30).

Proof. Let $\widetilde{K}(t) = (\widetilde{K}(t, 1), \ldots, \widetilde{K}(t, d))$ be the unique bounded solution of equation (2.30). From Proposition 14 it follows that $\widetilde{K} \geq 0$; then we have

$$\frac{d}{dt} \widetilde{K}(t, i) + \left(A_0(t, i) + \frac{1}{2} q_{ii} I_n \right)^* \widetilde{K}(t, i) + \widetilde{K}(t, i) \left(A_0(t, i) + \frac{1}{2} q_{ii} I_n \right)$$

$$+ \sum_{k=1}^r A_k^*(t, i) \widetilde{K}(t, i) A_k(t, i) + \sum_{\substack{j=1 \\ j \neq i}}^d q_{ij} \widetilde{K}(t, j)$$

$$+ H(t, i) = 0, \ i \in \mathcal{D}, t \in \mathcal{I}.$$

By direct calculations we obtain

$$\frac{d}{dt} \left(\widetilde{K}(t, i) - K_i^p(t) \right) + \left(A_0(t, i) + \frac{1}{2} q_{ii} I_n \right)^* \left(\widetilde{K}(t, i) - K_i^p(t) \right)$$

$$+ \left(\widetilde{K}(t, i) - K_i^p(t) \right) \left(A_0(t, i) + \frac{1}{2} q_{ii} I_n \right) + \Delta_i^p(t) = 0, \ i \in \mathcal{D}, \quad (2.49)$$

where

$$\Delta_i^p(t) = \sum_{k=1}^r A_k^*(t, i) \left(\widetilde{K}(t, i) - K_i^{p-1}(t) \right) A_k(t, i) + \sum_{\substack{j=1 \\ j \neq i}}^d q_{ij} \left(\widetilde{K}(t, j) - K_j^{p-1}(t) \right),$$

$i \in \mathcal{D}$, $p \geq 2$, and for $p = 1$ we have

$$\Delta_i^1(t) = \sum_{k=1}^{r} A_k^*(t, i)\, \tilde{K}(t, i) A_k(t, i) + \sum_{\substack{j=1 \\ j \neq i}}^{d} q_{ij} \tilde{K}(t, j) \geq 0, \ i \in \mathcal{D}, t \in \mathcal{I}.$$

Since for each $i \in \mathcal{D}$, $A_0(t, i) + \frac{1}{2} q_{ii} I_n$ defines an exponentially stable evolution, from (2.49) for $p = 1$ we deduce that $\tilde{K}(t, i) - K_i^1(t) \geq 0$, $i \in \mathcal{D}, t \in \mathcal{I}$. Further, by induction with respect to p we obtain that $\Delta_i^{p-1}(t) \geq 0$, which shows together with (2.49) that $\tilde{K}(t, i) - K_i^p(t) \geq 0$ for all $p \geq 1, i \in \mathcal{D}, t \in \mathcal{I}$; that is, the sequence $\{K_i^p(t)\}_{p \in \mathbb{N}}$ is bounded. On the other hand, for each $p \geq 1$, (2.48) gives:

$$\frac{d}{dt}\left(K_i^{p+1}(t) - K_i^p(t)\right) + \left(A_0(t, i) + \frac{1}{2} q_{ii} I_n\right)^* \left(K_i^{p+1}(t) - K_i^p(t)\right)$$

$$+ \left(K_i^{p+1}(t) - K_i^p(t)\right)\left(A_0(t, i) + \frac{1}{2} q_{ii} I_n\right) + \tilde{\Delta}_i^p(t) = 0, \ i \in \mathcal{D}, \quad (2.50)$$

where

$$\tilde{\Delta}_i^p(t) = \sum_{k=1}^{r} A_k^*(t, i)\left(K_i^p(t) - K_i^{p-1}(t)\right) A_k(t, i)$$

$$+ \sum_{\substack{j=1 \\ j \neq i}}^{d} q_{ij}\left(K_i^p(t) - K_i^{p-1}(t)\right),$$

$i \in \mathcal{D}$, $p \geq 2$, and for $p = 1$

$$\tilde{\Delta}_i^1(t) = \sum_{k=1}^{r} A_k^*(t, i)\, K_i^1(t) A_k(t, i) + \sum_{\substack{j=1 \\ j \neq i}}^{d} q_{ij} K_i(t) \geq 0.$$

By induction with respect to p, one can easily show that $\tilde{\Delta}_i^p(t) \geq 0$, which implies that $K_i^{p+1}(t) - K_i^p(t) \geq 0$, $i \in \mathcal{D}$, $p \geq 0$; that is, the sequence $\{K_i^p(t)\}_{p \in \mathbb{N}}$ is increasing and therefore the sequence is convergent. Let $K_i^\infty(t, i) = \lim_{p \to \infty} K_i^p(t)$. By standard arguments based on the Lesbegue Theorem (Chapter 1) we deduce that $t \longmapsto K^\infty(t, i)$, $i \in \mathcal{D}$, is a solution of the system (2.45). Since $K^\infty(t, i)$ is bounded with respect to t, it follows that $K^\infty(t, i) = \tilde{K}(t, i)$ and the proof ends. $\quad \square$

Remark 11. (i) In the time-invariant case the unique bounded solution of (2.48) is constant and it solves the standard Lyapunov equation

$$\left(A_0(i) + \frac{1}{2} q_{ii} I_n\right)^* K_i^p + K_i^p \left(A_0(i) + \frac{1}{2} q_{ii} I_n\right) + H_i^p = 0,$$

where

$$H_i^p := \sum_{k=1}^{r} A_k^*(i) \, K_i^{p-1} A_k(i) + \sum_{\substack{j=1 \\ j \neq i}}^{d} q_{ij} K_j^{p-1} + H(i), \quad i \in \mathcal{D}.$$

(ii) If $t \longmapsto A_k(t)$, $t \longmapsto H(t)$ are θ-periodic functions, then for each p and $i \in \mathcal{D}$, the unique bounded solution on $\mathcal{I}$ of the Lyapunov differential equation (2.48) is a θ-periodic function. Therefore, it is sufficient to compute only the values of $K_i^p(t)$ on the interval $[t_0, t_0 + \theta]$. We have

$$K_i^p(t) = \Phi_i^*(t_0 + \theta, t) K_i^p(t_0 + \theta) \Phi_i(t_0 + \theta, t)$$
$$+ \int_t^{t_0+\theta} \Phi_i^*(s, t) H_i^p(s) \Phi_i(s, t) \, ds, \quad t \le t_0 + \theta,$$

$\Phi_i(s, t)$ denoting the fundamental matrix solution of the equation (2.47). The periodicity condition $K_i^p(t) = K_i^p(t + \theta)$ shows that $K_i^p(t_0 + \theta)$ is a solution of the following algebraic discrete-time Lyapunov equation:

$$X_i = \Phi_i^*(t_0 + \theta, t_0) X_i \Phi_i(t_0 + \theta, t_0)$$
$$+ \int_{t_0}^{t_0+\theta} \Phi_i^*(s, t_0) H_i^p(s) \Phi_i(s, t_0) \, ds, \quad i \in \mathcal{D}. \tag{2.51}$$

The eigenvalues of the matrices $\Phi_i(t_0 + \theta, t_0)$ which are the Floquet multipliers [58] of the system (2.47) are inside the unit disk $|\lambda| < 1$, $\lambda \in \mathbf{C}$ and therefore (2.51) has a unique positive semidefinite solution.

2.5 Mean square exponential stability

In this section we introduce the concept of mean square exponential stability of the zero solution of the stochastic linear differential equations of type (1.22) and we also give necessary and sufficient conditions ensuring this kind of stability. The results proved in this section extend to a more general case, the existing results corresponding to the particular cases referring to the system (1.23) and (1.24), respectively.

Definition 5. *We say that the zero solution of the linear system* (1.22) *is* exponentially stable in mean square (ESMS), *or equivalently, that the system* (1.22) *defines an ESMS evolution if there exist $\beta \ge 1$ and $\alpha > 0$ such that*

$$E\left[|\Phi(t, t_0)x_0|^2 \mid \eta(t_0) = i \right] \le \beta e^{-\alpha(t-t_0)} |x_0|^2 \tag{2.52}$$

for all $t \ge t_0 \ge 0$, $i \in \mathcal{D}$, $x_0 \in \mathbf{R}^n$, where $\Phi(t, t_0)$ is the fundamental matrix solution of (1.22).

Proposition 18. *The following are equivalent:*

   (i) *The system* (1.22) *defines an ESMS evolution.*

   (ii) *There exist* $\beta \geq 1$, $\alpha > 0$ *such that*

$$E\left[|\Phi(t, t_0)|^2 \mid \eta(t_0) = i\right] \leq \beta e^{-\alpha(t-t_0)}, \; t \geq t_0 \geq 0, \; i \in \mathcal{D}.$$

   (iii) *There exist* $\beta_1 \geq 1$, $\alpha_1 > 0$ *such that*

$$E\left[|\Phi(t, t_0)|^2 \mid \eta(t_0)\right] \leq \beta_1 e^{-\alpha_1(t-t_0)}, \; a.s., \; t \geq t_0 \geq 0.$$

   (iv) *There exist* $\tilde{\beta} \geq 1$, $\tilde{\alpha} > 0$ *such that*

$$E\left[|\Phi(t, t_0)\xi|^2 \mid \eta(t_0)\right] \leq \tilde{\beta} e^{-\tilde{\alpha}(t-t_0)} E\left[|\xi|^2 \mid \eta(t_0)\right],$$

$t \geq t_0 \geq 0, i \in \mathcal{D}$, *and* ξ *is any random vector* $\mathcal{H}_{t_0}$-*measurable and* $E[|\xi|^2] < \infty$.

   Proof. (i) $\Longleftrightarrow$ (ii), (iii) $\Rightarrow$ (ii), and (iv) $\Rightarrow$ (i) are obvious.

   We now prove the implication (i) $\Rightarrow$ (iii) Let $e_1, \ldots, e_n$ be the canonical basis in $\mathbf{R}^n$, that is, $e_k = (0, \ldots, 0, 1, 0, \ldots, 0)^*$, with 1 being the kth element. From the inequality

$$|\Phi(t, t_0)|^2 \leq \sum_{k=1}^{n} |\Phi(t, t_0)e_k|^2,$$

we deduce that

$$E\left[|\Phi(t, t_0)|^2 \mid \eta(t_0)\right] \leq \sum_{k=1}^{n} \left[|\Phi(t, t_0)e_k|^2 \mid \eta(t_0)\right].$$

Since $\eta(t_0)$ takes a finite number of values we have

$$E\left[|\Phi(t, t_0)|^2 \mid \eta(t_0)\right] \leq \sum_{k=1}^{n} \sum_{j=1}^{d} \chi_{\eta(t_0)=j} E\left[|\Phi(t, t_0)e_k|^2 \mid \eta(t_0) = j\right] \text{ a.s.}$$

Using (2.52) we can write

$$E\left[|\Phi(t, t_0)|^2 \mid \eta(t_0)\right] \leq \beta \sum_{k=1}^{n} \sum_{j=1}^{d} \chi_{\eta(t_0)=j} e^{-\alpha(t-t_0)} |e_k|^2$$

$$= \beta n d e^{-\alpha(t-t_0)} \text{ a.s.}$$

   (iii) $\Rightarrow$ (iv) Let ξ be an arbitrary random vector $\mathcal{H}_{t_0}$-measurable and $E[|\xi|^2] < \infty$. From the inequality

$$|\Phi(t, t_0)\xi|^2 \leq |\Phi(t, t_0)|^2 |\xi|^2$$

we deduce that

$$E\left[|\Phi(t, t_0)\xi|^2 \mid \mathcal{H}_{t_0}\right] \leq E\left[|\Phi(t, t_0)|^2 |\xi|^2 \mid \mathcal{H}_{t_0}\right]$$

$$= |\xi|^2 E\left[|\Phi(t, t_0)|^2 \mid \mathcal{H}_{t_0}\right].$$

Since the components of $\Phi(t, t_0)$ are measurable with respect to $\eta(s)$, $w_j(s)$, $t_0 \leq s \leq t$, $j = 1, \ldots, r$, it follows that we may apply Theorem 34 from Chapter 1 and get

$$E\left[|\Phi(t, t_0)\xi|^2 \mid \mathcal{H}_{t_0}\right] \leq |\xi|^2 E\left[|\Phi(t, t_0)|^2 \mid \eta(t_0)\right] \text{ a.s.}$$

Using (iii) we deduce that

$$E\left[|\Phi(t, t_0)\xi|^2 \mid \mathcal{H}_{t_0}\right] \leq \beta_1 e^{-\alpha_1(t-t_0)}|\xi|^2, \text{ a.s., } t \geq t_0 \geq 0,$$

from which one easily deduces that

$$E\left[|\Phi(t, t_0)\xi|^2 \mid \eta(t_0) = i\right] \leq \beta_1 e^{-\alpha_1(t-t_0)} E\left[|\xi|^2 \mid \eta(t_0) = i\right]$$

for all $t \geq t_0 \geq 0, i \in \mathcal{D}$, and the proof is complete. $\qquad\square$

Remark 12. (i) In the particular case of the considered system of stochastic differential equations of type (1.24), the definition of the mean square exponential stability reduces to

$$E\left[|\Phi(t, t_0)x_0|^2\right] \leq \beta e^{-\alpha(t-t_0)}|x_0|^2 \qquad (2.53)$$

for all $t \geq t_0, x_0 \in \mathbf{R}^n$. Let us remark that it is possible to define the mean square exponential stability for systems subjected to Markovian jumping of type (1.22) and (1.23), using (2.53) instead of (2.52). However, we can notice that in the presence of Markovian perturbations in the system, if (2.52) is fulfilled, then (2.53) also holds, but the reverse implication is not true.

(ii) In the time-invariant case, based on Remark 4(ii) we obtain that the system (1.25) defines an ESMS evolution if and only if there exist $\beta \geq 1, \alpha > 0$ such that

$$E\left[|\Phi(t, 0)x_0|^2 \mid \eta(0) = i\right] \leq \beta e^{-\alpha t}|x_0|^2$$

for all $t \geq 0, i \in \mathcal{D}, x_0 \in \mathbf{R}^n$. Since $P(\eta(0) = i) > 0, i \in \mathcal{D}$, we obtain that the system (1.25) defines an ESMS evolution if and only if there exist $\beta \geq 1, \alpha > 0$ such that

$$E\left[|\Phi(t, 0)x_0|^2\right] \leq \beta e^{-\alpha t}|x_0|^2, t \geq 0, x_0 \in \mathbf{R}^n.$$

Based on Theorems 4 and 12 and Proposition 13 we get the following theorem.

Theorem 19. *The following are equivalent:*
  (i) *The system (1.22) defines an ESMS evolution.*
  (ii) *There exists $\delta > 0$ such that*

$$E\left[\int_t^\infty |\Phi(s, t)x_0|^2 \, ds \mid \eta(t) = i\right] \leq \delta|x_0|^2$$

for all $t \geq 0$ and $x_0 \in \mathbf{R}^n$.

(iii) *The system of linear differential equations*

$$\frac{d}{dt} K(t, i) + A_0^*(t, i) K(t, i) + K(t, i) A_0(t, i)$$

$$+ \sum_{k=1}^{r} A_k^*(t, i) K(t, i) A_k(t, i)$$

$$+ \sum_{j=1}^{d} q_{ij} K(t, j) + I_n = 0,$$

$i \in \mathcal{D}, t \geq 0$, *has a bounded solution* $K \gg 0$:

$$K(t) = (K(t, 1), \ldots, K(t, d)).$$

(iv) *There exists a bounded uniform positive and continuous function* $H : \mathbf{R}_+ \to S_n^d$, $H(t) = (H(t, 1), \ldots, H(t, d))$, *such that the system of linear differential equations*

$$\frac{d}{dt} K(t, i) + A_0^*(t, i) K(t, i) + K(t, i) A_0(t, i)$$

$$+ \sum_{k=1}^{r} A_k^*(t, i) K(t, i) A_k(t, i)$$

$$+ \sum_{j=1}^{d} q_{ij} K(t, j) + H(t, i) = 0 \qquad (2.54)$$

has a bounded and uniform positive solution $K(t) = (K(t, 1), \ldots, K(t, d))$.

(v) *For every bounded uniform positive and continuous function* $H : \mathbf{R}_+ \to S_n^d$, *the system (2.54) has a bounded and uniform positive solution.*

(vi) *For each* $H(t)$ *as above, there exists a* C^1 *function* $K : \mathbf{R}_+ \to S_n^d$, *bounded with bounded derivative* $K \gg 0$, *which solves the following system of linear differential inequalities:*

$$\frac{d}{dt} K(t, i) + A_0^*(t, i) K(t, i) + K(t, i) A_0(t, i)$$

$$+ \sum_{k=1}^{r} A_k^*(t, i) K(t, i) A_k(t, i)$$

$$+ \sum_{j=1}^{d} q_{ij} K(t, j) + H(t, i) < 0$$

$i \in \mathcal{D}$, *uniformly with respect to t, with* $t \geq 0$.

(vii) *There exists a C^1 function $K : \mathbf{R}_+ \to \mathcal{S}_n^d$, bounded with bounded derivative $K \gg 0$, which solves the following system of linear differential inequalities:*

$$\frac{d}{dt}K(t, i) + A_0^*(t, i)K(t, i) + K(t, i)A_0(t, i)$$

$$+ \sum_{k=1}^{r} A_k^*(t, i)K(t, i)A_k(t, i)$$

$$+ \sum_{j=1}^{d} q_{ij}K(t, j) < 0,$$

$i \in \mathcal{D}$, *uniformly with respect to t, with $t \geq 0$.*   □

Combining the results of Theorems 4 and 15 we obtain the following result for the time-invariant case.

Theorem 20. *The following are equivalent:*
  (i) *The system (1.25) defines an ESMS evolution.*
  (ii) *The system of linear matrix equalities (LMEs)*

$$A_0^*(i)X(i) + X(i)A_0(i) + \sum_{k=1}^{r} A_k^*(i)X(i)A_k(i) + \sum_{j=1}^{d} q_{ij}X(j) + I_n = 0,$$

$i \in \mathcal{D}$, *has a solution $X = (X(1), \ldots, X(d))$ with $X(i) > 0, i \in \mathcal{D}$.*
  (iii) *There exists $H = (H(1), \ldots, H(d)) \in \mathcal{S}_n^d$ with $H(i) > 0$ such that the system of LMEs*

$$A_0^*(i)X(i) + X(i)A_0(i) + \sum_{k=1}^{r} A_k^*(i)X(i)A_k(i) + \sum_{j=1}^{d} q_{ij}X(j) + H(i) = 0,$$

$$(2.55)$$

$i \in \mathcal{D}$, *has a positive solution $X = (X(1), \ldots, X(d))$.*
  (iv) *For every $H = (H(1), \ldots, H(d)) \in \mathcal{S}_n^d$ with $H > 0$, the system of LMEs (2.55) has a positive solution $X = (X(1), \ldots, X(d))$.*
  (v) *For each $H = (H(1), \ldots, H(d)) \in \mathcal{S}_n^d$ with $H > 0$, the system of linear matrix inequalities (LMIs)*

$$A_0^*(i)X(i) + X(i)A_0(i) + \sum_{k=1}^{r} A_k^*(i)X(i)A_k(i) + \sum_{j=1}^{d} q_{ij}X(j) + H(i) < 0$$

has a positive solution $X = (X(1), \ldots, X(d))$.
  The system of LMIs

$$A_0^*(i)X(i) + X(i)A_0(i) + \sum_{k=1}^{r} A_k^*(i)X(i)A_k(i) + \sum_{j=1}^{d} q_{ij}X(j) < 0$$

has a positive solution $X = (X(1), \ldots, X(d))$.   □

Similarly we have the following theorem.

Theorem 21. *The following are equivalent:*
(i) *The system (1.25) defines an ESMS evolution.*
(ii) *The system of LMEs*

$$A_0(i)Y(i) + Y(i)A_0^*(i) + \sum_{k=1}^{r} A_k(i)Y(i)A_k^*(i) + \sum_{j=1}^{d} q_{ji}Y(j) + I_n = 0,$$

$i \in \mathcal{D}$, *has a solution* $Y = (Y(1), \ldots, Y(d))$ *with* $Y(i) > 0$, $i \in \mathcal{D}$.
(iii) *There exists* $H = (H(1), \ldots, H(d)) \in \mathcal{S}_n^d$ *with* $H(i) > 0$ *such that the system of LMEs*

$$A_0(i)Y(i) + Y(i)A_0^*(i) + \sum_{k=1}^{r} A_k(i)Y(i)A_k^*(i) + \sum_{j=1}^{d} q_{ji}Y(j) + H(i) = 0,$$

$$(2.56)$$

$i \in \mathcal{D}$, *has a positive solution* $Y = (Y(1), \ldots, Y(d))$.
(iv) *For every* $H = (H(1), \ldots, H(d)) \in \mathcal{S}_n^d$ *with* $H > 0$, *the system of LMEs* (2.56) *has a positive solution* $Y = (Y(1), \ldots, Y(d))$.
(v) *For each* $H = (H(1), \ldots, H(d)) \in \mathcal{S}_n^d$ *with* $H > 0$, *the system of LMIs*

$$A_0(i)Y(i) + Y(i)A_0^*(i) + \sum_{k=1}^{r} A_k(i)Y(i)A_k^*(i) + \sum_{j=1}^{d} q_{ji}Y(j) + H(i) < 0$$

has a positive solution $Y = (Y(1), \ldots, Y(d))$.
The system of LMIs

$$A_0(i)Y(i) + Y(i)A_0^*(i) + \sum_{k=1}^{r} A_k(i)Y(i)A_k^*(i) + \sum_{j=1}^{d} q_{ji}Y(j) < 0$$

has a positive solution $Y = (Y(1), \ldots, Y(d))$. □

The following result shows that in the time-invariant case the ESMS is equivalent to a type of attractivity of the zero solution.

Theorem 22. *The following assertions are equivalent:*
(i) *The system (1.25) defines an ESMS evolution*
(ii)

$$\lim_{t \to \infty} E[|x(t)|^2] = 0$$

for any solution $x(t)$ *of the system (1.25) with* $x(0) = x_0$, $x_0 \in \mathbf{R}^n$.
(iii)

$$\lim_{t \to \infty} E[x(t)x^*(t)] = 0$$

for all solutions $x(t)$ *of (1.25) as above.*
(iv)

$$\lim_{t \to \infty} E[\Phi^*(t, 0)\Phi(t, 0)] = 0.$$

Proof. (i) $\Rightarrow$ (ii) directly follows from Remark 12(ii).
(ii) $\Rightarrow$ (iii) follows from the inequality

$$0 \le x(t)x^*(t) \le |x(t)|^2 I_n.$$

(iii) $\Rightarrow$ (ii) follows from

$$|x(t)|^2 = Tr[x(t)x^*(t)].$$

(ii) $\Rightarrow$ (iv) easily follows using the identity

$$E[x^* \Phi^*(t, 0)\Phi(t, 0)y] = \frac{1}{2}\{E[|\Phi(t, 0)(x + y)|^2] - E[|\Phi(t, 0)(x - y)|^2]\} \tag{2.57}$$

for all $x, y \in \mathbf{R}^n$.
(iv) $\Rightarrow$ (i) Since $P(\eta(0) = i) > 0$, $i \in \mathcal{D}$, then from (iv) we have

$$\lim_{t \to \infty} E[\Phi^*(t, 0)\Phi(t, 0) \mid \eta(0) = i] = 0, \ i \in \mathcal{D}.$$

Based on Theorem 4 and Remark 3(i), the above equality gives

$$\lim_{t \to \infty} \left(e^{\mathcal{L}^* t} J^d\right)(i) = 0, \ i \in \mathcal{D},$$

and therefore $\lim_{t \to \infty} |e^{\mathcal{L}^* t} J^d| = 0$. Applying Theorem 3 we conclude that $\lim_{t \to \infty} \|e^{\mathcal{L}^* t}\| = 0$, and from (2.5) we obtain that

$$\lim_{t \to \infty} \|e^{\mathcal{L} t}\| = 0. \tag{2.58}$$

Since $\mathcal{L}$ is a linear operator on a finite-dimensional Hilbert space, from (2.58) we deduce that the eigenvalues of the operator $\mathcal{L}$ are located in the half-plane $\mathbf{C}^-$, and hence there exists $\beta \ge 1$, $\alpha > 0$ such that $\|e^{\mathcal{L} t}\| \le \beta e^{-\alpha t}$. Combining Theorems 15 and 21 we deduce that the system (1.25) defines an ESMS evolution and the proof is complete.   □

In the case of periodic coefficients we obtain the following analogous result.

Theorem 23. *Assume that* $t \longmapsto A_k(t, i)$, $k = 0, \ldots, r$, *are* θ-*periodic and continuous functions. Then the following are equivalent:*
   (i) *The system* (1.22) *defines an ESMS evolution.*
   (ii)

$$\lim_{p \to \infty} E[|x(p\theta)|^2] = 0$$

for all solution $x(t)$ *of* (1.22) *with* $x(0) = x_0$, $x_0 \in \mathbf{R}^n$.
   (iii)

$$\lim_{p \to \infty} E[x(p\theta)x^*(p\theta)] = 0$$

for any solution $x(t)$ *of* (1.22) *as above.*
   (iv)

$$\lim_{p \to \infty} E[\Phi^*(p\theta, 0)\Phi(p\theta, 0)] = 0.$$

Proof. (i) $\Rightarrow$ (ii) and (ii) $\Longleftrightarrow$ (iii) are similar to the proof of Theorem 22.
(ii) $\Rightarrow$ (iv) immediately follows from (2.57) and Remark 4(i).
(iv) $\Rightarrow$ (i) If (ii) is fulfilled, then

$$\lim_{p \to \infty} E\big[\Phi^*(p\theta, 0)\Phi(p\theta, 0) \mid \eta(0) = i\big] = 0, \ i \in \mathcal{D}.$$

Using Theorem 4 we obtain

$$\lim_{p \to \infty} \big(T^*(p\theta, 0)J^d\big)(i) = 0, \ i \in \mathcal{D},$$

and therefore

$$\lim_{p \to \infty} \big|(T^*(p\theta, 0)J^d)(i)\big| = 0, \ i \in \mathcal{D},$$

which leads to

$$\lim_{p \to \infty} \big|T^*(p\theta, 0)J^d\big| = 0.$$

Based on Theorem 3 we deduce that

$$\lim_{p \to \infty} \|T^*(p\theta, 0)\| = 0.$$

Using (2.5) we get

$$\lim_{p \to \infty} \|T(p\theta, 0)\| = 0,$$

which is equivalent to

$$\lim_{p \to \infty} \|(T(\theta, 0))^p\| = 0, \tag{2.59}$$

$T(\theta, 0)$ being the monodromy operator associated with the differential equation (2.9).
From (2.59) we deduce that the eigenvalues of $T(\theta, 0)$ are inside the unit disk $|\lambda| < 1$.
Applying a result in [58], we may conclude that the zero solution of (2.9) is exponentially stable, which implies via Theorem 4 that (1.22) defines an ESMS evolution, and therefore the proof is complete. $\qquad\square$

In the following we consider the cases when the stochastic system (1.22) is subjected only to either Markov jumping or multiplicative white noise. Thus, in the case of system (1.23), Theorem 19 becomes the following.

Theorem 24. *The following assertions are equivalent:*
   (i) *The system* (1.23) *defines an ESMS evolution.*
   (ii) *The system of linear differential equations*

$$\frac{d}{dt}K(t, i) + A^*(t, i)K(t, i) + K(t, i)A(t, i)$$

$$+ \sum_{j=1}^{d} q_{ij}K(t, j) + I_n = 0,$$

$i \in \mathcal{D}, t \geq 0,$ *has a bounded and uniform positive solution*

$$K(t) = (K(t, 1), \ldots, K(t, d)).$$

   (iii) *There exists a bounded uniform positive and continuous function* $H : \mathbf{R}_+ \to$
$\mathcal{S}_n^d, H(t) = (H(t, 1), \ldots, H(t, d))$ *such that the system of linear differential*

equations

$$\frac{d}{dt}K(t,i) + A^*(t,i)K(t,i) + K(t,i)A(t,i)$$

$$+ \sum_{j=1}^{d} q_{ij} K(t,j) + H(t,i) = 0 \tag{2.60}$$

has a bounded and uniform positive solution $K(t) = (K(t,1), \ldots, K(t,d))$.

(iv) *For every bounded uniform positive and continuous function* $H : \mathbf{R}_+ \to \mathcal{S}_n^d$, *the system (2.60) has a bounded and uniform positive solution.*

(v) *For each* $H(t)$ *as above, there exists a* C^1 *function* $K : \mathbf{R}_+ \to \mathcal{S}_n^d$, *bounded with bounded derivative* $K \gg 0$, *which solves the following system of linear differential inequalities:*

$$\frac{d}{dt}K(t,i) + A^*(t,i)K(t,i) + K(t,i)A(t,i)$$

$$+ \sum_{j=1}^{d} q_{ij} K(t,j) + H(t,i) < 0,$$

$i \in \mathcal{D}$, *uniformly with respect to* t, *with* $t \geq 0$.

(vi) *There exists a* C^1 *function* $K : \mathbf{R}_+ \to \mathcal{S}_n^d$ *bounded with bounded derivative* $K \gg 0$, *which solves the following system of linear differential inequalities:*

$$\frac{d}{dt}K(t,i) + A^*(t,i)K(t,i) + K(t,i)A(t,i) + \sum_{j=1}^{d} q_{ij} K(t,j) < 0,$$

$i \in \mathcal{D}$, *uniformly with respect to* t, *with* $t \geq 0$. $\qquad\qquad \square$

Remark 13. If the system (1.23) is in the time-invariant case, that is $A(t,i) = A(i)$ for all $t \geq 0, i \in \mathcal{D}$, similar results in Theorems 20 and 21 can also be formulated. In this case one obtains the well-known results concerning the ESMS of linear systems with jump Markov perturbations.

Theorem 25. *Assume that the system* (1.23) *defines an ESMS evolution; then there exist* $\beta \geq 1$ *and* $\alpha > 0$ *such that* $\|R(t,t_0)\| \leq \beta e^{-\alpha(t-t_0)}$ *for all* $t \geq t_0 > 0$, $R(t,t_0)$ *being the linear evolution operator on* $(\mathbf{R}^n)^d$ *defined by the differential equation* (2.26).

Proof. Let $y = (y(1), \ldots, y(d)) \in (\mathbf{R}^n)^d$; then we have

$$|E[\Phi^*(t,t_0)y(\eta(t)) \mid \eta(t_0) = i]|^2 \tag{2.61}$$

$$\leq E\big[|\Phi^*(t,t_0)|^2 \mid \eta(t_0) = i\big]E\big[|y(\eta(t))|^2 \mid \eta(t_0) = i\big],$$

$t \geq t_0 \geq 0$. On the other hand,

$$E\big[|y(\eta(t))|^2 \mid \eta(t_0) = i\big] = \sum_{j=1}^{d} E\big[\chi_{\eta(t)=j} \mid \eta(t_0) = i\big]|y(j)|^2$$

$$= \sum_{j=1}^{d} p_{ij}(t - t_0)|y(j)|^2 \leq \sum_{j=1}^{d} |y(j)|^2 = \|y\|^2.$$

Thus (2.61) leads to

$$|E[\Phi^*(t, t_0)y(\eta(t)) \mid \eta(t_0) = i]|^2 \leq E[|\Phi^*(t, t_0)|^2 \mid \eta(t_0) = i]^2 \|y\|^2.$$

Because the system (1.23) defines an ESMS evolution and $|\Phi^*(t, t_0)| = |\Phi(t, t_0)|$, there exist $\beta \geq 1$, $\alpha > 0$ such that

$$E\big[|\Phi^*(t, t_0)|^2 \mid \eta(t_0) = i\big]^2 \leq \beta e^{-\alpha(t-t_0)}.$$

Therefore

$$|E[\Phi^*(t, t_0)y(\eta(t)) \mid \eta(t_0) = i]|^2 \leq \beta e^{-\alpha(t-t_0)} \|y\|^2$$

for all $t \geq t_0 \geq 0$. Based on Proposition 8 we deduce that

$$|(R^*(t, t_0)y)(i)|^2 \leq \beta e^{-\alpha(t-t_0)} \|y\|^2$$

and hence

$$\|R^*(t, t_0)y\|^2 = \sum_{i=1}^{d} |(R^*(t, t_0)y)(i)|^2 \leq d\beta e^{-\alpha(t-t_0)} \|y\|^2,$$

which gives

$$\|R^*(t, t_0)\| \leq \sqrt{d\beta}\, e^{-\frac{\alpha}{2}(t-t_0)}$$

for all $t \geq t_0 \geq 0$. Since $\|R^*(t, t_0)\| = \|R(t, t_0)\|$ we conclude that

$$\|R(t, t_0)\| \leq \sqrt{d\beta}\, e^{-\frac{\alpha}{2}(t-t_0)}$$

and the proof is complete.   □

Corollary 26. *If the system (1.23) defines an ESMS evolution, then for all $h : \mathbf{R}_+ \to (\mathbf{R}^n)^d$ continuous and bounded, the affine differential equation*

$$\frac{d}{dt}y(t) + M^*(t)y(t) + h(t) = 0$$

has a unique bounded-on-$\mathbf{R}_+$ solution, $M(t)$ being defined by (2.25).

Combining the results in Theorems 19 and 24, we obtain the following corollary.

Corollary 27. *If the system* (1.22) *defines an ESMS evolution, then the linear system*

$$\dot{x}(t) = A_0(t, \eta(t))x(t),$$

obtained by ignoring the white noise perturbations in (1.22), *defines an ESMS evolution, too.* □

Let us now consider the case when the system (1.22) is subjected only to white noise perturbations, that is, when the system under consideration is of form (1.24). In this case, from Theorem 19 one obtains some known results concerning the exponential stability of linear systems described by Itô differential equations [74].

Theorem 28. *The following assertions are equivalent:*

(i) *The system* (1.24) *defines an ESMS evolution.*

(ii) *The affine differential equation over the space of symmetric matrices*

$$\frac{d}{dt}X(t) + A_0^*(t)X(t) + X(t)A_0(t) + \sum_{k=1}^{r} A_k^*(t)X(t)A_k(t) + I_n = 0$$

has a bounded and uniform positive solution $X(t)$.

(iii) *There exists an* $H : \mathbf{R}_+ \to \mathcal{S}_n$ *bounded and continuous function,* $H(t) \gg 0$, *such that the affine differential equation*

$$\frac{d}{dt}X(t) + A_0^*(t)X(t) + X(t)A_0(t) + \sum_{k=1}^{r} A_k^*(t)X(t)A_k(t) + H(t) = 0 \quad (2.62)$$

has a bounded and uniform positive solution $X(t)$.

(iv) *For each* $H : \mathbf{R}_+ \to \mathcal{S}_n$ *bounded, continuous and* $H \gg 0$, *the affine differential equation* (2.62) *has a bounded solution* $X \gg 0$.

(v) *For each* $H : \mathbf{R}_+ \to \mathcal{S}_n$ *bounded, continuous function,* $H \gg 0$, *the linear differential inequality*

$$\frac{d}{dt}X(t) + A_0^*(t)X(t) + X(t)A_0(t) + \sum_{k=1}^{r} A_k^*(t)X(t)A_k(t) + H(t) < 0,$$

uniformly with respect to $t \geq 0$, *has a solution* $X(t)$ *bounded with bounded derivative* $X \gg 0$.

(vi) *The linear differential inequality*

$$\frac{d}{dt}X(t) + A_0^*(t)X(t) + X(t)A_0(t) + \sum_{k=1}^{r} A_k^*(t)X(t)A_k(t) < 0,$$

uniformly with respect to $t \geq 0$, *has a* C^1 *solution* $X : \mathbf{R}_+ \to \mathcal{S}_n$, *which is bounded with bounded derivative and* $X(t) \gg 0$. □

Remark 14. If the system (1.24) is in the time-invariant case, similar results to those in Theorems 20 and 21 can also be stated.

The next result is proved in a more general situation in [79].

Theorem 29. *The linear system of stochastic differential equations*

$$dx(t) = Ax(t)dt + bc^*xdw_1(t), \ b, c \in \mathbf{R}^n, \tag{2.63}$$

has an ESMS evolution if and only if A is stable and $\int_0^\infty |c^*e^{At}b|^2 dt < 1$.

Proof. From Theorem 28 and Remark 14 it follows that (2.63) has an ESMS evolution if and only if there exists $X > 0$ such that

$$A^*X + XA + cb^*Xbc^* = -I_n,$$

or equivalently,

$$A^*X + XA + cb^*Xbc^* + I_n = 0. \tag{2.64}$$

Assume that (2.64) is fulfilled for $X > 0$. Then it follows that A is stable, and therefore we can define the linear operator $\mathcal{G} : S_n \to S_n$ by

$$\mathcal{G}(G) = \int_0^\infty e^{A^*t} G e^{At},$$

and $H = \mathcal{G}(G)$ is the unique solution of the Lyapunov equation

$$A^*H + HA = -G. \tag{2.65}$$

If $G > 0$ then $\mathcal{G}(G) > 0$; applying the operator $\mathcal{G}$ to the matrix from the left side of (2.64) and using (2.65), we obtain that

$$-X + b^*Xb\mathcal{G}(cc^*) + \mathcal{G}(I_n) = 0.$$

Hence

$$-b^*Xb + (b^*Xb)b^*\mathcal{G}(cc^*)b + b^*\mathcal{G}(I_n)b = 0$$

and therefore

$$b^*Xb(1 - b^*\mathcal{G}(cc^*)b) = b^*\mathcal{G}(I_n)b,$$

which implies that $1 - b^*\mathcal{G}(cc^*)b > 0$, since if $b = 0$ the inequality is obvious, and if $b \neq 0$ we have $b^*Xb > 0$, $b^*\mathcal{G}(I_n)b > 0$. Taking into account that

$$b^*\mathcal{G}(cc^*)b = \int_0^\infty |c^*e^{At}b|^2 dt,$$

the inequality in the statement directly follows.

The condition in the statement is sufficient. Indeed, assume that A is stable and that $\int_0^\infty |c^*e^{At}b|^2 dt < 1$, namely $b^*\mathcal{G}(cc^*)b < 1$. Let

$$X = \mathcal{G}(I_n) + \frac{b^*\mathcal{G}(I_n)b}{1 - b^*\mathcal{G}(cc^*)b}\mathcal{G}(cc^*).$$

It is obvious that $X > 0$ and a direct calculation using (2.65) shows that X verifies (2.64) and the proof is complete.   □

Remark 15. From Parseval's formula one easily obtains that

$$\int_0^\infty |c^* e^{At} b|^2 dt = \frac{1}{2\pi} \int_{-\infty}^\infty |c^*(A - i\lambda I_n)^{-1} b|^2 d\lambda.$$

For each $i \in \mathcal{D}$ we can consider the following system subjected only to white noise perturbations:

$$dx_i(t) = \left(A_0(t, i) + \frac{1}{2} q_{ii} I_n \right) x_i(t) dt + \sum_{k=1}^r A_k(t, i) x_i(t) dw_k(t), \qquad (2.66)$$

$t \geq 0, i \in \mathcal{D}$. In this case one obtains the following corollary.

Corollary 30. *If the system* (1.22) *defines an ESMS evolution, then*
    (i) *The system* (2.66) *defines an ESMS evolution for each $i \in \mathcal{D}$.*
    (ii) *For each $i \in \mathcal{D}$ the deterministic system*

$$\dot{x}_i(t) = \left(A_0(t, i) + \frac{1}{2} q_{ii} I_n \right) x_i(t)$$

defines an exponentially stable evolution.   □

At the end of this section we prove the following result.

Theorem 31. *Assume that there exists a bounded and uniform positive function $K :$ $\mathbf{R}_+ \to \mathcal{S}_n^d, K(t) = (K(t, 1), \ldots, K(t, d))$, and the constants $\tau > 0, \delta \in (0, 1)$ such that*

$$(T^*(t + \tau, t) K(t + \tau))(i) \leq \delta K(t, i), \ t \geq 0, \ i \in \mathcal{D}$$

for all $t \geq 0, i \in \mathcal{D}$. Then the system $(A_0, A_1, \ldots, A_r; Q)$ is stable.

Proof. From the statement of the theorem it follows that

$$T^*(t + \tau, t) K(t + \tau) \leq \delta K(t), \ t \geq 0.$$

Let $t_0 \geq 0$ be fixed; since $T^*(t, t_0)$ is a positive operator, we obtain by induction that

$$T^*(t_0 + m\tau, t_0) K(t_0 + m\tau) \leq \delta^m K(t_0)$$

for all $m \geq 1$. Taking into account that

$$T^*(t_0 + m\tau, t_0) J^d \leq \beta \delta^m J^d$$

leads to

$$|T^*(t_0 + m\tau, t_0) J^d| \leq \beta \delta^m, m \geq 1.$$

Based on Theorem 3 we obtain

$$\|T^*(t_0 + m\tau, t_0) J^d\| \leq \beta \delta^m.$$

Since $\sup_{t \geq 0} \|L^*(t)\| < \infty$, we easily deduce (using (2.13)) that $\|T^*(t, s)\| \leq \beta_1$ for all $0 \leq t - s \leq \tau$. Using (2.11) we deduce that $\|T^*(t, t_0)\| \leq \beta_2 e^{-\alpha(t-t_0)}$ for all $t \geq t_0 \geq 0$ for some $\beta_2 > 0$ and $\alpha = -\frac{1}{\tau} \ln \delta$, and by virtue of Theorem 4 the proof is complete.   □

2.6 Numerical examples

Example 1. Let us consider the particular case $n = 1$ in which situation the system (1.24) reduces to the linear differential equation

$$dx(t) = a(\eta(t))x(t)dt + \sum_{k=1}^{r} g_k(\eta(t))x(t)dw_k(t), \ t \geq 0. \qquad (2.67)$$

We shall prove that if

$$2a(i) + \sum_{k=1}^{r} g_k^2(i) < 0, \ i \in \mathcal{D}, \qquad (2.68)$$

then (2.67) defines an ESMS evolution.

Indeed, taking $K = (1, \ldots, 1)$ and using the fact that $\sum_{j=1}^{d} q_{ij} = 0$, we get

$$2a(i)K(i) + \sum_{k=1}^{r} g_k^2(i)K(i) + \sum_{j=1}^{d} q_{ij}K(j) = 2a(i) + \sum_{j=1}^{d} g_k^2(i),$$

$i \in \mathcal{D}$. Since the left side in the above equation coincides with L^*K and $K > 0$, from Theorem 20 it follows that if (2.68) is fulfilled then the system (2.67) defines an ESMS evolution.

Remark 16. (i) The above example shows that (2.68) are sufficient conditions under which (2.67) defines an ESMS evolution. As we shall see in the next example, these conditions are not necessary.

(ii) Using Theorem 28 and Remark 14, it is easy to check that (2.68) is a necessary and sufficient condition for ESMS for the Itô equation

$$dx(t) = a(i)x(t)dt + \sum_{k=1}^{r} g_k(i)x(t)dw_k(t),$$

with $i \in \mathcal{D}$ fixed.

Example 2. Assume that in (2.67) we have $d = 2$, $r = 1$, and

$$Q = \begin{bmatrix} -\alpha & \alpha \\ \alpha & -\alpha \end{bmatrix},$$

with $\alpha > 0$. From Theorem 15, (2.67) defines an ESMS evolution if and only if there exists $K = (K_1, K_2)$, $K_i > 0$, such that

$$2a_i K_i + g_i^2 K_i + \sum_{j=1}^{2} q_{ij} K_j = -\alpha, \ i = 1, 2,$$

where we denoted $a_i = a(i)$, $g_i = g(i)$, and $K_i = K(i)$, $i = 1, 2$. Then, from the above equation, we obtain

$$\left(2a_1 + g_1^2 - \alpha\right)K_1 + \alpha K_2 = -\alpha, \tag{2.69}$$
$$\left(2a_2 + g_2^2 - \alpha\right)K_2 + \alpha K_1 = -\alpha,$$

from which result the necessary conditions for stability:

$$2a_i + g_i^2 - \alpha < 0, \ i = 1, 2.$$

Further, solving (2.69) we get

$$K_1 = \frac{\alpha\left(2a_2 + g_2^2 - 2\alpha\right)}{\alpha\left(2a_1 + g_1^2 + 2a_2 + g_2^2\right) - \left(2a_1 + g_1^2\right)\left(2a_2 + g_2^2\right)},$$

$$K_2 = \frac{\alpha\left(2a_1 + g_1^2 - 2\alpha\right)}{\alpha\left(2a_1 + g_1^2 + 2a_2 + g_2^2\right) - \left(2a_1 + g_1^2\right)\left(2a_2 + g_2^2\right)}.$$

Since $2a_i + g_i^2 - 2\alpha < 0$, it follows that

$$\alpha\left(2a_1 + g_1^2 + 2a_2 + g_2^2\right) - \left(2a_1 + g_1^2\right)\left(2a_2 + g_2^2\right) < 0. \tag{2.70}$$

Then the following cases can occur.

Case 1 If $2a_1 + g_1^2 + 2a_2 + g_2^2 < 0$ the condition (2.70) is accomplished for

$$\alpha > \frac{\left(2a_1 + g_1^2\right)\left(2a_2 + g_2^2\right)}{2a_1 + g_1^2 + 2a_2 + g_2^2}.$$

Case 2 If $2a_1 + g_1^2 + 2a_2 + g_2^2 > 0$, then (2.70) holds for

$$\alpha < \frac{\left(2a_1 + g_1^2\right)\left(2a_2 + g_2^2\right)}{2a_1 + g_1^2 + 2a_2 + g_2^2}. \tag{2.71}$$

Case 2 implies $2a_i + g_i^2 > 0$, $i = 1, 2$. Then (2.71) contradicts the necessary condition $\alpha > 2a_1 + g_1^2$. Therefore, we conclude that Case 2 must be excluded.

Summarizing, the stochastic system (2.67) with $d = 2$ and $r = 1$ considered in this example defines an ESMS evolution if and only if

$$2a_1 + g_1^2 < 0 \text{ and } 2a_2 + g_2^2 < 0$$

(situation considered in Example 1) or if

$$2a_1 + g_1^2 + 2a_2 + g_2^2 < 0 \text{ and}$$

$$\alpha > \max\left\{2a_1 + g_1^2, 2a_2 + g_2^2, \frac{\left(2a_1 + g_1^2\right)\left(2a_2 + g_2^2\right)}{2a_1 + g_1^2 + 2a_2 + g_2^2}\right\}.$$

Example 3. Consider the stochastic system with jump Markov perturbations in which $n = d = 2$:

$$\frac{dx(t)}{dt} = A(\eta(t))x(t), \ t \geq 0, \qquad (2.72)$$

where

$$A_1 := A(1) = \begin{bmatrix} -a\alpha & 0 \\ \alpha & -a\alpha \end{bmatrix},$$

$$A_2 := A(2) = \begin{bmatrix} -a\alpha & \alpha \\ 0 & -a\alpha \end{bmatrix}$$

with $a > 0$ and

$$Q = \begin{bmatrix} -\alpha & \alpha \\ \alpha & -\alpha \end{bmatrix}$$

with $\alpha > 0$. Then, according to Theorem 15 and Remark 13, (2.72) defines an ESMS evolution if and only if there exist

$$X_1 := X(1) = \begin{bmatrix} x_1 & y_1 \\ y_1 & z_1 \end{bmatrix} \text{ and } X_2 := X(2) = \begin{bmatrix} x_2 & y_2 \\ y_2 & z_2 \end{bmatrix}$$

such that $X_1 > 0$, $X_2 > 0$ and

$$A_1^* X_1 + X_1 A_1 + \sum_{j=1}^{2} q_{1j} X_j = -\alpha I_2,$$

$$A_2^* X_2 + X_2 A_2 + \sum_{j=1}^{2} q_{2j} X_j = -\alpha I_2,$$

which are equivalent to

$$\beta x_1 - 2y_1 - x_2 = 1,$$
$$\beta y_1 - z_1 - y_2 = 0,$$
$$\beta z_1 - z_2 = 1,$$
$$\beta x_2 - x_1 = 1,$$
$$\beta y_2 - x_2 - y_1 = 0,$$
$$\beta z_2 - 2y_2 - z_1 = 1,$$

where we denoted $\beta := 2a + 1$. By solving the above system of algebraic equations,

$$z_1 = \frac{\beta + 1}{(\beta^3 - \beta^2 - \beta - 1)(\beta^3 + \beta^2 - \beta + 1)}.$$

Then for $a \to 0$ one obtains that $z_1 \to -\frac{1}{2}$. This shows that although $A(1)$ and $A(2)$ have their eigenvalues in $\mathbf{C}^-$, that is, they are stable in the deterministic sense, the stochastic system (2.72) defines an unstable evolution.

Example 4. We now consider the case $n = d = 2$ and $r = 1$, namely the situation when the stochastic system is subjected to both Markovian jumping and to multiplicative noise:

$$dx(t) = A_0(\eta(t))x(t)dt + A_1(\eta(t))x(t)dw_1(t), \ t \geq 0, \qquad (2.73)$$

where

$$A_0(1) = \begin{bmatrix} -1 & 0 \\ 1 & -1 \end{bmatrix}, \ A_0(2) = \begin{bmatrix} -1 & 1 \\ 0 & -1 \end{bmatrix},$$

$$A_1(1) = \begin{bmatrix} a & 0 \\ 0 & 0 \end{bmatrix}, \ A_1(2) = \begin{bmatrix} 0 & 0 \\ 0 & a \end{bmatrix},$$

and

$$Q = \begin{bmatrix} -1 & 1 \\ 1 & -1 \end{bmatrix}.$$

According to Theorem 20, the necessary and sufficient condition such that $(A_0, A_1; Q)$ defines an ESMS evolution is that the equations

$$A_0^*(i)X(i) + X(i)A_0(i) + A_1^*(i)X(i)A_1(i) + \sum_{j=1}^{2} q_{ij}X(j) = -I_2,$$

$i = 1, 2$, have the solution $X(i) > 0$ with

$$X(i) = \begin{bmatrix} x_i & y_i \\ y_i & z_i \end{bmatrix}, \ i = 1, 2.$$

The above equation leads to

$$\begin{aligned} (3 - a^2)x_1 - 2y_1 - x_2 &= 1, \qquad (2.74) \\ 3y_1 - z_1 - y_2 &= 0, \\ 3z_1 - z_2 &= 1, \\ 3x_2 - x_1 &= 1, \\ 3y_2 - x_2 - y_1 &= 0, \\ (3 - a^2)z_2 - 2y_2 - z_1 &= 1, \end{aligned}$$

from which we deduce that

$$\begin{aligned} (24 - 9a^2)x_2 + (3a^2 - 10)z_1 &= 8 - 2a^2, \qquad (2.75) \\ (3a^2 - 10)x_2 + (24 - 9a^2)z_1 &= 8 - 2a^2. \end{aligned}$$

For $a^2 = \frac{17}{6}$ we obtain that $x_2 + z_1 = -\frac{14}{9}$, which is not admissible since $X(i) > 0$, $i = 1, 2$, imply that $x_2 > 0$ and $z_1 > 0$.

On the other hand, if $a^2 = \frac{7}{3}$, the system (2.75) is incompatible, and if $a^2 \neq \frac{17}{6}$ and $a^2 \neq \frac{7}{3}$, this system has the unique solution

$$x_2 = z_1 = \frac{a^2 - 4}{3a^2 - 7},$$

which gives in (2.74)

$$x_1 = z_2 = -\frac{5}{3a^2 - 7} \quad \text{and} \quad y_1 = y_2 = \frac{a^2 - 4}{2(3a^2 - 7)}.$$

Therefore, $X(1) > 0$ and $X(2) > 0$ if and only if $a^2 < \frac{7}{3}$, from which we conclude that $(A_0, A_1; Q)$ defines an ESMS evolution if and only if $a^2 < \frac{7}{3}$.

2.7 Affine systems

Consider the system

$$dx(t) = [A_0(t, \eta(t))x(t) + f_0(t)]dt + \sum_{k=1}^{r} [A_k(t, \eta(t))x(t) + f_k(t)]dw_k(t),$$

$$(2.76)$$

where $A_k(t, i), 0 \leq k \leq r$, are bounded on $\mathbf{R}_+$ and continuous matrix-valued functions. Denote

$$u(t) = (f_0^*(t), \ f_1^*(t), \dots, f_r^*(t))^*.$$

If $t_0 \geq 0, x_0 \in \mathbf{R}^n$ and $f_k \in L^2_{\eta,w}([t_0, T], \mathbf{R}^n), 0 \leq k \leq r$ for all $T > t_0$ by Theorem 36 of Chapter 1, it follows that there exists a unique solution $x_u(t, t_0, x_0)$ of the system (2.76) with $x_u(t_0, t_0, x_0) = x_0$ and $x_u(\cdot, t_0, x_0) \in L^2_{\eta,w}([t_0, T], \mathbf{R}^n)$, $T > t_0$; that is, all components of the vector x_u are in $L^2_{\eta,w}([t_0, T])$.

Unfortunately the representation formula (1.29) cannot be used to obtain some useful estimates for solutions of system (2.76) as in the deterministic case. Such estimations are obtained in an indirect way using some techniques based on Lyapunov functions.

Theorem 32. *Assume that the system* $(A_0, A_1, \dots, A_r; Q)$ *is stable.*
(i) *There exist* $c \geq 1, \alpha > 0$ *such that*

$$E\big[|x_u(t, t_0, x_0)|^2 |\eta(t_0) = i\big]$$

$$\leq c \left(e^{-\alpha(t-t_0)}|x_0|^2 + \sum_{k=0}^{r} E \left[\int_{t_0}^{t} e^{-\alpha(t-s)} |f_k(s)|^2 \, ds |\eta(t_0) = i \right] \right)$$

for all $t \geq t_0 \geq 0, x_0 \in \mathbf{R}^n, i \in \mathcal{D}$ *and all* $f_k \in L^2_{\eta,w}([t_0, \infty), \mathbf{R}^n), 0 \leq k \leq r$.

(ii) *There exists $\beta > 0$ such that*

$$E\left[\int_{t_0}^{\infty} |x_u(t, t_0, x_0)|^2 |\eta(t_0) = i\right]$$

$$\leq \beta \left(|x_0|^2 + \sum_{k=0}^{r} E\left[\int_{t_0}^{\infty} |f_k(s)|^2 \, ds |\eta(t_0) = i\right]\right)$$

for all $t_0 \geq 0$, $x_0 \in \mathbf{R}^n$, $f_k \in L^2_{\eta,w}([t_0, \infty), \mathbf{R}^n)$, $0 \leq k \leq r$, $i \in \mathcal{D}$.

(iii)

$$\lim_{t \to \infty} E|x_u(t, t_0, x_0)|^2 = 0$$

for all $t_0 \geq 0$, $x_0 \in \mathbf{R}^n$, $f_k \in L^2_{\eta,w}([t_0, \infty), \mathbf{R}^n)$, $0 \leq k \leq r$.

Proof. Since $(A_0, A_1, \ldots, A_r; Q)$ is stable, then by Theorem 12 the Lyapunov-type equation (2.29) has a unique bounded-on-$\mathbf{R}_+$ and uniformly positive solution $\widetilde{K}(t) = (\widetilde{K}(t, 1), \ldots, \widetilde{K}(t, d))$. Therefore, there exist $\alpha_1 > 0$, $\alpha_2 > 0$ such that

$$\alpha_1 J^d \leq \widetilde{K}(t) \leq \alpha_2 J^d, \quad t \geq 0.$$

Let $x_u(t) = x_u(t, t_0, 0), t \geq t_0$. Applying the Itô-type formula (1.16) to the function $v(t, x, i) = x^* \widetilde{K}(t, i)x$ and to the system (2.76), taking into account the equation (2.29) for $\widetilde{K}(t)$, we obtain

$$E\left[v(t, x_u(t), \eta(t))|\eta(t_0) = i\right] = E\left[\int_{t_0}^{t} \left\{ -|x_u(s)|^2 + 2x_u^*(s)\left[\widetilde{K}(s, \eta(s))f_0(s)\right.\right.\right.$$

$$\left.+ \sum_{k=1}^{r} A_k^*(s, \eta(s))\widetilde{K}(s, \eta(s))f_k(s)\right]$$

$$\left.\left.+ \sum_{k=1}^{r} f_k^*(s)\widetilde{K}(s, \eta(s))f_k(s)\right\} ds|\eta(t_0) = i\right].$$

Denote

$$h_i(t) = E[v(t, x_u(t), \eta(t))|\eta(t_0) = i], i \in \mathcal{D},$$

$$m_i(t) = \sqrt{E[|x_u(t)|^2|\eta(t_0) = i]}, \ i \in \mathcal{D},$$

$$g_i(t) = \sqrt{\sum_{k=0}^{r} E[|f_k(t)|^2|\eta(t_0) = i]}, i \in \mathcal{D}.$$

Then we may write

$$h_i'(t)$$

$$= E\left[\left\{ -|x_u(t)|^2 + 2x_u^*(t)\left[\widetilde{K}(t, \eta(t))f_0(t) + \sum_{k=1}^{r} A_k^*(t, \eta(t))\widetilde{K}(t, \eta(t))f_k(t)\right]\right.\right.$$

$$\left.\left.+ \sum_{k=1}^{r} f_k^*(t)\widetilde{K}(t, \eta(t))f_k(t)\right\} |\eta(t_0) = i\right]$$

a.e. $t \geq t_0, i \in \mathcal{D}$.

Since A_k, $\widetilde{K}$ are bounded, there exist $\gamma > 0$, $\delta > 0$ such that

$$h_i'(t) \leq -m_i^2(t) + \gamma\big[m_i(t)g_i(t) + g_i^2(t)\big] \leq -\frac{1}{2}m_i^2(t) + \delta g_i^2(t).$$

Taking into account that $\alpha_1 I_n \leq \widetilde{K}(t, \eta(t)) \leq \alpha_2 I_n$ it follows that

$$\alpha_1 m_i^2(t) \leq h_i(t) \leq \alpha_2 m_i^2(t).$$

Hence $h_i'(t) \leq -\frac{1}{2\alpha_2}h_i(t) + \delta g_i^2(t)$. Since $h_i(t_0) = 0$ we obtain

$$\alpha_1 m_i^2(t) \leq h_i(t) \leq \delta \int_{t_0}^{t} e^{-\alpha(t-s)} g_i^2(s)\,ds, \ t \geq t_0, i \in \mathcal{D} \qquad (2.77)$$

with $\alpha = \frac{1}{2\alpha_2}$. On the other hand,

$$x_u(t, t_0, x_0) = x_u(t, t_0, 0) + \Phi(t, t_0)x_0. \qquad (2.78)$$

Combining (2.77) and (2.78), (i) is proved. Part (ii) follows from (i) and the Fubini Theorem. We now prove (iii). Since

$$\sum_{i=1}^{d} E\left[\int_{t_0}^{\infty} \sum_{k=0}^{r} |f_k(t)|^2 dt\,|\eta(t_0) = i\right] < \infty,$$

it follows that for every $\varepsilon > 0$ there exists $t_\varepsilon > t_0$ such that

$$\sum_{i=1}^{d} \int_{t_\varepsilon}^{\infty} g_i^2(t)dt < \varepsilon.$$

For each $t \geq t_\varepsilon$ we have

$$\int_{t_0}^{t} e^{-\alpha(t-s)} g_i^2(s)\,ds = e^{-\alpha(t-t_\varepsilon)} \int_{t_0}^{t_\varepsilon} e^{-\alpha(t_\varepsilon-s)} g_i^2(s)\,ds + \int_{t_\varepsilon}^{t} e^{-\alpha(t-s)} g_i^2(s)\,ds$$

$$\leq e^{-\alpha(t-t_\varepsilon)} \int_{t_0}^{\infty} g_i^2(s)\,ds + \varepsilon.$$

From this inequality and (2.77) we conclude

$$\lim_{t\to\infty} E\big[|x_u(t, t_0, 0)|^2|\eta(t_0) = i\big] = 0.$$

Finally, using (2.78) we obtain

$$\lim_{t\to\infty} E\big[|x_u(t, t_0, x_0)|^2|\eta(t_0) = i\big] = 0$$

and the proof is complete.   □

Remark 17. If we do not know that the system $(A_0, A_1, \ldots, A_r; Q)$ is stable, then the estimation from Theorem 32(i) is not uniform with respect to $t, t_0 \in \mathbf{R}_+$. In general we may prove that for any compact interval $[t_0, t_1]$ there exists a positive constant c depending upon $t_1 - t_0$ such that

$$E\left[|x_u(t, t_0, x_0)|^2 | \eta(t_0) = i\right] \leq c\left(|x_0|^2 + \sum_{k=0}^{r} E\left[\int_{t_0}^{t_1} |f_k(s)|^2 \, ds | \eta(t_0) = i\right]\right)$$

for all $t \in [t_0, t_1]$, $x_0 \in \mathbf{R}^n$, $i \in \mathcal{D}$ and all $f_k \in L^2_{\eta, w}([t_0, t_1], \mathbf{R}^n)$, $0 \leq k \leq r$.

To this end we notice that since $A_k(t, i), 0 \leq k \leq r, i \in \mathcal{D}$, are bounded on $\mathbf{R}_+$, from (2.76) and Theorem 31 of Chapter 1 it follows easily that there exists an absolute constant $\gamma > 1$ such that for all $t \in [t_0, t_1], i \in \mathcal{D}$ we have

$$E\left[|x_u(t, t_0, x_0)|^2 | \eta(t_0) = i\right]$$
$$\leq \left\{\gamma|x_0|^2 + E\left[\int_{t_0}^{t} |x_u(s, t_0, x_0)|^2 \, ds | \eta(t_0) = i\right]((t_1 - t_0) + 1)\right.$$
$$\left. + \sum_{k=0}^{r} E\left[\int_{t_0}^{t_1} |f_k(s)|^2 \, ds | \eta(t_0) = i\right]((t_1 - t_0) + 1)\right\}.$$

By using the Gronwall Lemma we get

$$\sup_{t_0 \leq t \leq t_1} E\left[|x_u(t, t_0, x_0)|^2 \eta(t_0) = i\right]$$
$$\leq c\left(|x_0|^2 + \sum_{k=0}^{r} E\left[\int_{t_0}^{t_1} |f_k(s)|^2 | \eta(t_0) = i\right]\right),$$

$i \in \mathcal{D}$, where $c > 0$ depends only on $t_1 - t_0$.

Notes and references

In the control literature one can find a large number of papers devoted to the stability of Itô-type differential equation systems. For this reason it is impossible to give an exhaustive bibliography for this subject. We shall limit ourselves to pointing the reader to the monographs [5], [6], [11], [74], [77], [78], [21], which contain many references concerning this subject. Theorem 29 has been proved in [79] for a larger class of systems of linear stochastic differential equations.

The ESMS for stochastic systems of differential equations with Markov perturbations has been introduced and studied for the first time in [73], in which characterizations using Lyapunov-type equations are given.

The results in this chapter concerning time-varying linear differential systems with jump Markov perturbations have been proved in [89]. The mean square exponential stability for time-invariant differential systems with jump Markov perturbations has been investigated in [86], [84], [48], [70], [49], [82], [85].

The ESMS problem for differential equations subjected to both Markov perturbations and multiplicative white noise has been also considered in [83]. In that paper sufficient conditions for stability are given in terms of some M-matrices, and it is proved that ESMS implies almost sure stability. Results concerning the stability and the boundedness of solutions of nonlinear Itô differential systems subjected to Markov perturbations can be also found in [80].

Most of the results included in Sections 2.1–2.5 have been proved in [33].

3

Structural Properties of Linear Stochastic Systems

In this chapter we present the stochastic version of some basic concepts in control theory, namely stabilizability, detectability, observability, and controllability. All these concepts are defined in terms of both Lyapunov operators and stochastic systems. The definitions given in this chapter extend the corresponding definitions from the deterministic time-varying systems. Some examples will show that the stochastic observability does not always imply stochastic detectability, and stochastic controllability does not necessarily imply stochastic stabilizability. As in the deterministic case the concepts of stochastic detectability and observability are used in some criteria of ESMS.

3.1 Stabilizability and detectability of stochastic linear systems

Let us consider the following stochastic input–output system:

$$dx(t) = [A_0(t, \eta(t))x(t) + B_0(t, \eta(t))u(t)] dt$$

$$+ \sum_{k=1}^{r} [A_k(t, \eta(t))x(t) + B_k(t, \eta(t))u(t)] dw_k(t), \qquad (3.1)$$

$$y(t) = C_0(t, \eta(t))x(t),$$

$t \in \mathbf{R}_+$, with the inputs $u \in \mathbf{R}^m$ and the outputs $y \in \mathbf{R}^p$, and denote $\mathbf{A} = (A_0, A_1, \ldots, A_r)$ and $\mathbf{B} = (B_0, B_1, \ldots, B_r)$.

Definition 1. (i) *We say that the system* (3.1) *is* stochastically stabilizable *or equivalently, the triple* $(\mathbf{A}, \mathbf{B}; Q)$ *is* stabilizable *if there exists* $F : \mathbf{R}_+ \to \mathcal{M}_{m,n}^d$ *bounded and continuous function such that the zero solution of the system obtained by taking* $u(t) = F(t, \eta(t))x(t)$, *namely*

$$dx(t) = [A_0(t, \eta(t)) + B_0(t, \eta(t))F(t, \eta(t))]x(t)dt$$

$$+ \sum_{k=1}^{r} [A_k(t, \eta(t)) + B_k(t, \eta(t))F(t, \eta(t))]x(t)dw_k(t),$$

$t \geq 0$, *is ESMS.*

(ii) *We say that the system* (3.1) *is* stochastically detectable, *or equivalently, that the triple* $(C_0, \mathbf{A}; Q)$ *is* detectable *if there exists* $K: \mathbf{R}_+ \to M_{n,p}^d$ *continuous and bounded function such that the zero solution of the system*

$$dx(t) = [A_0(t, \eta(t)) + K(t, \eta(t))C_0(t, \eta(t))]x(t)\,dt + \sum_{k=1}^{r} A_k(t, \eta(t))x(t)dw_k(t)$$

is ESMS.

Remark 1. (i) The above definition of the stochastic detectability would also be stated if the output of the system (3.1) is of the form

$$dy(t) = C_0(t, \eta(t))x(t)dt + \sum_{k=1}^{r} C_k(t, \eta(t))x(t)dw_k(t).$$

(ii) The function $F(t) = (F(t, 1), F(t, 2), \dots, F(t, d))$ and the function $K(t) = (K(t, 1), K(t, 2), \dots, K(t, d))$ from the above definition will be termed *stabilizing feedback gain* and *stabilizing injection*, respectively.

The concepts of stochastic stabilizability and stochastic detectability in the particular cases when the system (3.1) is subjected only to either Markovian jumping (i.e., $A_k = 0$, $B_k = 0$, $1 \le k \le r$) or multiplicative white noise (i.e., $\mathcal{D} = \{1\}$) are obviously defined in the same way. In the case of Markovian jumping systems, we shall say that $(A_0, B_0; Q)$ is stabilizable and $(C_0, A_0; Q)$ is detectable, and in the case of Itô systems we shall say that $(\mathbf{A}, \mathbf{B})$ is stabilizable and $(C_0, \mathbf{A})$ is detectable.

Remark 2. If the system (3.1) is in the stationary case, then the stabilizing feedback gain and the stabilizing injection are supposed to be of the form $F = (F(1), \dots, F(d))$, $H = (H(1), \dots, H(d))$.

In the next chapter we shall show that in the case when the coefficients of the system (3.1) are θ-periodic functions with respect to their first argument, then this system is stochastically stabilizable (stochastically detectable) if and only if there exists a θ-periodic stabilizing feedback gain (a θ-periodic stabilizing injection, respectively). Moreover, if the system (3.1) is in the time-invariant case, then it is stochastically stabilizable (stochastically detectable) if and only if there exists a stabilizing feedback gain $F = (F(1)F(2) \dots F(d))$ (a stabilizing injection $K = (K(1)K(2) \dots K(d))$, respectively).

Let us consider the following numerical example with $n = 2$, $d = 2$, and $r = 1$, where

$$Q = \begin{bmatrix} -1 & 1 \\ 1 & -1 \end{bmatrix}, \; A_0(1) = \begin{bmatrix} -1 & 0 \\ \alpha & \beta \end{bmatrix}, \; A_0(2) = \begin{bmatrix} \gamma & \delta \\ 0 & -1 \end{bmatrix},$$

$$A_1(1) = \begin{bmatrix} a & 0 \\ 0 & 0 \end{bmatrix}, \; A_1(2) = \begin{bmatrix} 0 & 0 \\ 0 & a \end{bmatrix}, \; B(1) = \begin{bmatrix} 0 \\ 1 \end{bmatrix}, \; B(2) = \begin{bmatrix} 1 \\ 0 \end{bmatrix}$$

with $a^2 < 7/3$ and $\alpha, \beta, \gamma, \delta \in \mathbf{R}$. The system $(A_0, A_1, Q; B)$ is stabilizable. Indeed, let $F(1) = \begin{bmatrix} 1 - \alpha & -1 - \beta \end{bmatrix}$, $F(2) = \begin{bmatrix} -1 - \gamma & 1 - \delta \end{bmatrix}$. Then

$$A_0(1) + B(1)F(1) = \begin{bmatrix} -1 & 0 \\ 1 & -1 \end{bmatrix} \text{ and } A_0(2) + B(2)F(2) = \begin{bmatrix} -1 & 1 \\ 0 & -1 \end{bmatrix},$$

from which we deduce, according to *Example* 4 of Section 2.6 that $(A_0 + BF, A_1; Q)$ is stable. Let us remark that the pairs $(A_0(1), B(1))$ and $(A_0(2), B(2))$ are not controllable. One can also remark that if $\beta \geq 1/2$ or $\gamma \geq 1/2$, then the system $(A_0, A_1; Q)$ is not stable since it does not satisfy the necessary conditions of stability, namely the matrices $A_0(i) + \frac{1}{2}q_{ii}I_2$, $i = 1, 2$ being stable.

The next result immediately follows.

Proposition 1. (i) *The system* (3.1) *is stochastically stabilizable if and only if there exists a continuous and bounded function* $F : \mathbf{R}_+ \to \mathcal{M}_{m,n}^d$ *such that the system* $(A_0 + B_0 F, A_1 + B_1 F, \dots, A_r + B_r F; Q)$ *is stable.*

(ii) *The system* (3.1) *is stochastically detectable if and only if there exists a continuous and bounded function* $K : \mathbf{R}_+ \to \mathcal{M}_{n,p}^d$ *such that the system* $(A_0 + KC_0, A_1, \dots, A_r; Q)$ *is stable.* $\square$

From Theorems 19, 24, and 28 of Chapter 2, the following result can be obtained.

Proposition 2. (i) *If the system* (3.1) *is stochastically stabilizable (stochastically detectable, respectively), then the system with Markovian jumping,*

$$\dot{x}(t) = A_0(t, \eta(t))x(t) + B_0(t, \eta(t))u(t),$$
$$y(t) = C_0(t, \eta(t))x(t),$$

is stochastically stabilizable (stochastically detectable, respectively).

(ii) *If the system* (3.1) *is stochastically stabilizable (stochastically detectable, respectively), then, for each* $i \in \mathcal{D}$, *the system described by the Itô differential equations,*

$$dx_i(t) = \left[\widetilde{A}_0(t, i)x_i(t) + B_0(t, i)u(t) \right] dt$$
$$+ \sum_{k=1}^{r} [A_k(t, i)x_i(t) + B_k(t, i)u(t)]dw_k(t),$$
$$y_i(t) = C_0(t, i)x_i(t),$$

is stochastically stabilizable (stochastically detectable, respectively) where $\widetilde{A}_0(t, i) = A_0(t, i) + \frac{1}{2}q_{ii}I_n$. $\square$

Remark 3. It is not difficult to see that the definition of the stochastic stabilizability and stochastic detectability can be stated for triplets $(\mathbf{A}, \mathbf{B}; Q)$ and $(\mathbf{C}, \mathbf{A}; Q)$ in the case when the elements of the matrix Q verify only condition (2.7); $\mathbf{C} = (C_0, C_1, \dots, C_r)$ and A_k, B_k, C_k are continuous matrix-valued functions on a right unbounded interval $\mathcal{I} \subseteq \mathbf{R}$.

More precisely, we have the following definition.

Definition 2. (i) *The triple* $(\mathbf{A}, \mathbf{B}; Q)$ *is stabilizable if there exists a bounded and continuous function* $F : \mathcal{I} \to \mathcal{M}_{m,n}^d$ *such that*

$$\|T_F(t, s)\| \le \beta e^{-\alpha(t-s)}, \forall t \ge s \in \mathcal{I}$$

$(\alpha > 0, \beta > 0$ *being constants);* $T_F(\cdot, \cdot)$ *is the linear evolution operator defined by the linear differential equation over* $\mathcal{S}_n^d$:

$$\frac{d}{dt} S(t) = \mathcal{L}_F(t) S(t),$$

where $\mathcal{L}_F(t) : \mathcal{S}_n^d \to \mathcal{S}_n^d$ *by*

$$(\mathcal{L}_F(t)S)(i) = [A_0(t, i) + B_0(t, i)F(t, i)]S(i) + S(i)[A_0(t, i) + B_0(t, i)F(t, i)]^*$$

$$+ \sum_{k=1}^r [A_k(t, i) + B_k(t, i)F(t, i)]S(i) \tag{3.2}$$

$$\times [A_k(t, i) + B_k(t, i)F(t, i)]^* + \sum_{j=1}^d q_{ji} S(j),$$

$i \in \mathcal{D}, S \in \mathcal{S}_n^d$.

(ii) *The triple* $(\mathbf{C}, \mathbf{A}; Q)$ *is* detectable *if there exists a bounded and continuous function* $K : \mathcal{I} \to \mathcal{M}_{n,p}^d$, *such that* $\|T^K(t, s)\| \le \beta e^{-\alpha(t-s)} \; \forall t \ge s \in \mathcal{I}, \beta > 0,$ $\alpha > 0$ *being constants.* $T^K(t, s)$ *is the linear evolution operator defined by the linear differential equation*

$$\frac{d}{dt} S(t) = \mathcal{L}^K(t) S(t),$$

where $\mathcal{L}^K(t) : \mathcal{S}_n^d \to \mathcal{S}_n^d$ *by*

$$[\mathcal{L}^K(t)S](i) = [A_0(t, i) + K(t, i)C_0(t, i)]S(i) + S(i)[A_0(t, i) + K(t, i)C_0(t, i)]^*$$

$$+ \sum_{k=1}^r [A_k(t, i) + K(t, i)C_k(t, i)]S(i)[A_k(t, i) + K(t, i)C_k(t, i)]^*$$

$$+ \sum_{j=1}^d q_{ji} S(j), \tag{3.3}$$

$i \in \mathcal{D}, S \in \mathcal{S}_n^d$.

The next result easily follows from Theorem 21 of Chapter 2.

Proposition 3. *Assume that the system* (3.1) *is in the time-invariant case. Then the following are equivalent:*

(i) *The system* (3.1) *is stochastically stabilizable.*

(ii) *There exists* $F = (F(1), F(2), \ldots, F(d)) \in \mathcal{M}_{m,n}^d$ *such that the affine Lyapunov equation over* $\mathcal{S}_n^d$,

$$\mathcal{L}_F X + J^d = 0,$$

has a solution $X > 0$.

(iii) *The linear matrix inequalities*

$$\begin{bmatrix} \mathcal{L}(X, \Gamma)(i) & \mathcal{P}(X, \Gamma)(i) \\ \mathcal{P}^*(X, \Gamma)(i) & \mathcal{R}(X)(i) \end{bmatrix} < 0 \tag{3.4}$$

have a solution $(X, \Gamma) \in \mathcal{S}_n^d \times \mathcal{M}_{m,n}^d$, $X > 0$, *where*

$$\mathcal{L}(X, \Gamma)(i) = A_0(i)X(i) + X(i)A_0^*(i) + B_0(i)\Gamma(i) + \Gamma^*(i)B_0^*(i) + \sum_{j=1}^d q_{ji}X(j),$$

$$\mathcal{P}(X, \Gamma)(i) = (A_1(i)X(i) + B_1(i)\Gamma(i)$$
$$A_2(i)X(i) + B_2(i)\Gamma(i) \cdots A_r(i)X(i) + B_r(i)\Gamma(i)),$$

$$\mathcal{R}(X)(i) = \begin{bmatrix} -X(i) & 0 & 0 & \cdots & 0 \\ 0 & -X(i) & 0 & \cdots & 0 \\ 0 & 0 & -X(i) & \cdots & 0 \\ \cdots & \cdots & \cdots & \cdots & \cdots \\ 0 & 0 & 0 & \cdots & -X(i) \end{bmatrix} \in \mathcal{S}_{rn}.$$

Moreover, if $(X, \Gamma) \in \mathcal{S}_n^d \times \mathcal{M}_{m,n}^d$ *is a solution of the linear matrix inequalities* (3.4) *with* $X > 0$, *then* $F = (F(1), F(2), \ldots, F(d))$, *with*

$$F(i) = \Gamma(i)X(i)^{-1}, \tag{3.5}$$

$i \in \mathcal{D}$ *is a stabilizing feedback gain.*   □

In the particular case with $B_k = 0, k = 1, 2, \ldots, r$ we have the following proposition.

Proposition 4. *Assume that the system* (3.1) *is in the time-invariant case and* $B_k(i) = 0$, $i \in \mathcal{D}, k = 1, \ldots, r$; *then the following are equivalent:*

(i) *The system* (3.1) *is stochastically stabilizable.*

(ii) *The system of linear matrix equations*

$$A_0(i)X(i) + X(i)A_0^*(i) + B_0(i)\Gamma(i) + \Gamma^*(i)B_0^*(i)$$
$$+ \sum_{k=1}^r A_k(i)X(i)A_k^*(i) + \sum_{j=1}^d q_{ji}X(j) + I_n = 0, \tag{3.6}$$

$i \in \mathcal{D}$, *has a solution* $(X, \Gamma) \in \mathcal{S}_n^d \times \mathcal{M}_{m,n}^d$, $X > 0$. *Moreover, if* $(X, \Gamma) \in \mathcal{S}_n^d \times \mathcal{M}_{m,n}^d$ *is a solution of the system* (3.6) *with* $X > 0$, *then a stabilizing feedback gain may be obtained as in the previous proposition.*   □

The next result follows easily from Theorem 20 of Chapter 2.

Proposition 5. *Assume that the system* (3.1) *is in the time-invariant case; then the following are equivalent:*
   (i) *The system* (3.1) *is stochastically detectable.*
   (ii) *The system of linear matrix equations*

$$A_0^*(i)Y(i) + Y(i)A_0(i) + \Lambda(i)C_0(i) + C_0(i)^*\Lambda^*(i)$$

$$+ \sum_{k=1}^{r} A_k^*(i)Y(i)A_k(i) + \sum_{j=1}^{d} q_{ij}Y(j) + I_n = 0, \tag{3.7}$$

$i \in \mathcal{D}$, *has a solution* $(Y, \Lambda) \in \mathcal{S}_n^d \times \mathcal{M}_{n,p}^d$, $Y > 0$. *Moreover, if* (Y, Λ) *is a solution of the system* (3.7), *then* $K = (K(1), \ldots, K(d))$, *with*

$$K(i) = Y^{-1}(i)\Lambda(i), \tag{3.8}$$

$i \in \mathcal{D}$, *being a stabilizing injection.*
   (iii) *The system of linear matrix inequalities*

$$A_0^*(i)Y(i) + Y(i)A_0(i) + \Lambda(i)C_0(i) + C_0^*(i)\Lambda^*(i)$$

$$+ \sum_{k=1}^{r} A_k^*(i)Y(i)A_k(i) + \sum_{j=1}^{d} q_{ij}Y(j) < 0, \tag{3.9}$$

$i \in \mathcal{D}$, *has a solution* $(Y, \Lambda) \in \mathcal{S}_n^d \times \mathcal{M}_{n,p}^d$, $Y > 0$. *Moreover, if* (Y, Λ) *is a solution of the system* (3.9) *with* $Y > 0$, *then a stabilizing injection is obtained as in* (3.8). $\square$

Based on Remark 3 we can establish a duality relationship between the stabilizability and detectability in this stochastic framework.

Proposition 6. *Assume that:*
   (i) $A_k : \mathbf{R} \to \mathcal{M}_n^d$, $B_k : \mathbf{R} \to \mathcal{M}_{n,m}^d$ *are continuous and bounded functions,* $k = 0, 1, \ldots, r$.
   (ii) *The elements of the matrix* Q *verify* (2.7).
   Then the triple $(\mathbf{A}, \mathbf{B}; Q)$ *is stabilizable if and only if the triple* $(\mathbf{B}^\sharp, \mathbf{A}^\sharp; Q^\sharp)$ *is detectable, where*

$$\mathbf{A}^\sharp = (A_0^\sharp, A_1^\sharp, \ldots, A_r^\sharp), \quad \mathbf{B}^\sharp = (B_0^\sharp, B_1^\sharp, \ldots, B_r^\sharp),$$

$$A_k^\sharp(t) = (A_k^\sharp(t, 1), A_k^\sharp(t, 2), \ldots, A_k^\sharp(t, d)),$$

$$B_k^\sharp(t) = (B_k^\sharp(t, 1), B_k^\sharp(t, 2), \ldots, B_k^\sharp(t, d)),$$

$$A_k^\sharp(t, i) := A_k^*(-t, i),$$

$$B_k^\sharp(t, i) := B_k^*(-t, i),$$

$$Q^\sharp = Q^*,$$

$t \in \mathbf{R}, i \in \mathcal{D}, k = 0, 1, \ldots, r.$

Proof. If $(\mathbf{A}, \mathbf{B}; Q)$ is stabilizable, then there exists a bounded and continuous function $F : \mathbf{R} \to \mathcal{M}_{m,n}^d$ such that

$$||T_F(t, s)|| \leq \beta e^{-\alpha(t-s)} \tag{3.10}$$

for all $t \geq s, t, s \in \mathbf{R}, \beta > 0, \alpha > 0$ being positive constants, $T_F(\cdot, \cdot)$ being the linear evolution operator defined by linear differential equation over $\mathcal{S}_n^d$,

$$\frac{d}{dt} S(t) = \mathcal{L}_F(t) S(t), \tag{3.11}$$

and $L_F(t)$ being defined as in (3.2).

It is easy to see that $S(t)$ is a solution of the equation (3.11) if and only if $t \to S(-t)$ is a solution of the equation

$$\frac{d}{dt} X(t) + (\mathcal{L}^\sharp(t))^* X(t) = 0, \tag{3.12}$$

where $\mathcal{L}^\sharp(t) : \mathcal{S}_n^d \to \mathcal{S}_n^d$ is defined by

$$\begin{aligned}
(\mathcal{L}^\sharp(t) S)(i) = &\left[A_0^\sharp(t, i) + K^\sharp(t, i) B_0^\sharp(t, i)\right] S(i) \\
&+ S(i)\left[A_0^\sharp(t, i) + K^\sharp(t, i) B_0^\sharp(t, i)\right]^* \\
&+ \sum_{k=1}^r \left[A_k^\sharp(t, i) + K^\sharp(t, i) B_k^\sharp(t, i)\right] S(i) \\
&\times \left[A_k^\sharp(t, i) + K^\sharp(t, i) B_k^\sharp(t, i)\right]^* \\
&+ \sum_{j=1}^d q_{ji}^\sharp S(j), i \in \mathcal{D}, S \in \mathcal{S}_n^d,
\end{aligned}$$

where $A_k^\sharp, B_k^\sharp$ were defined in the statement and $K^\sharp(t, i) = F^*(-t, i), q_{ji}^\sharp = q_{ij}$, $i, j \in \mathcal{D}$. If $T^\sharp(t, s)$ stands for the linear evolution operator over $\mathcal{S}_n^d$ defined by the differential equation

$$\frac{d}{dt} S(t) = \mathcal{L}^\sharp(t) S(t),$$

then we obtain from (3.12) that $S(-t) = (T^\sharp(s, t))^* S(-s)$ for all $t \leq s$, hence $S(t) = (T^\sharp(-s, -t))^* S(s)$ for all $t \geq s$.

On the other hand, $S(t) = T_F(t, s) S(s), t \geq s$. Hence we have $T^\sharp(t, s) = T_F^*(-s, -t)$ $\forall t \geq s$. Finally, invoking (3.10), we deduce that

$$||T^\sharp(t, s)|| \leq \beta e^{-\alpha(t-s)}, \ \forall t \geq s,$$

which shows that $(\mathbf{B}^\sharp, \mathbf{A}^\sharp; Q^\sharp)$ is detectable and the proof is complete. $\square$

Remark 4. (i) In the same way we may prove that $(\mathbf{C}, \mathbf{A}; Q)$ is detectable, if and only if $(\mathbf{A}^\sharp, \mathbf{C}^\sharp; Q^\sharp)$ is stabilizable.

(ii) From Proposition 6 it follows immediately that in the time-invariant case, $(\mathbf{A}, \mathbf{B}; Q)$ is stabilizable if and only if the triple $(\mathbf{B}^*, \mathbf{A}^*; Q^*)$ is detectable.

Now we prove the following theorem, which extends a well-known result from the deterministic framework.

Theorem 7. *Suppose the following.*

(i) *$(C_0, \mathbf{A}; Q)$ is stochastically detectable.*

(ii) *The differential equation*

$$\frac{d}{dt}K(t) + \mathcal{L}^*(t)K(t) + \tilde{C}(t) = 0 \qquad (3.13)$$

has a bounded solution $\tilde{K}: \mathbf{R}_+ \to \mathcal{S}_n^d$, $\tilde{K}(t) = (\tilde{K}(t,1), \ldots, \tilde{K}(t,d))$, $\tilde{K}(t,i) \geq 0$, $t \geq 0, i \in \mathcal{D}$, *where* $\tilde{C}(t) = (\tilde{C}(t,1), \ldots, \tilde{C}(t,d))$, $\tilde{C}(t,i) = C_0^*(t,i)C_0(t,i)$.

Then the solution of the system (1.22) *is mean square exponentially stable (or equivalently, the system* $((A_0, A_1, \ldots, A_r); Q)$ *is stable).*

Proof. Consider $v: \mathbf{R}_+ \times \mathbf{R}^n \times \mathcal{D} \to \mathbf{R}$, $v(t,x,i) = x^*\tilde{K}(t,i)x$. Let $x(t) = x(t,t_0,x_0)$ be a solution of the system (1.22). Applying the identity (1.6) to the function v and to the system (1.22) and taking into account the equation (3.13) we get for all $t \geq t_0$ and $i \in \mathcal{D}$

$$E[v(t,x(t),\eta(t))|\eta(t_0) = i] - x_0^*\tilde{K}(t_0,i)x_0$$
$$= -E\left[\int_{t_0}^t |C_0(s,\eta(s))x(s)|^2 ds|\eta(t_0) = i\right].$$

Hence

$$E\left[\int_{t_0}^\infty |C_0(t,\eta(t))x(t)|^2 dt|\eta(t_0) = i\right] \leq x_0^*K(t_0,i)x_0 \leq \gamma|x_0|^2, \qquad (3.14)$$

$t_0 \geq 0, x_0 \in \mathbf{R}^n, i \in \mathcal{D}$.

We may write

$$dx(t) = \{[A_0(t,\eta(t)) + H(t,\eta(t))C_0(t,\eta(t))]x(t) + f_0(t)\}dt$$
$$+ \sum_{k=1}^r A_k(t,\eta(t))x(t)dw_k(t),$$

where $f_0(t) = -H(t,\eta(t))C_0(t,\eta(t))x(t)$.

Since the system $(A_0+HC_0, A_1 \ldots A_r; Q)$ is stable and since $f_0 \in L^2_{\eta,w}([t_0,\infty) \times \mathbf{R}^n)$ (see (3.14)) we may use Theorem 32(ii) of Chapter 2 to obtain

$$E\left[\int_{t_0}^\infty |\Phi(t,t_0)x_0|^2 dt|\eta(t_0) = i\right] \leq \delta[|x_0|^2 + E\left[\int_{t_0}^\infty |f_0(t)|^2 dt|\eta(t_0) = i\right]$$
$$\leq \beta|x_0|^2$$

for all $t_0 \geq 0, x_0 \in \mathbf{R}^n, i \in \mathcal{D}$.

Using Theorem 19 of Chapter 2 we conclude that $(A_0, A_1, \ldots, A_r; Q)$ is stable and the proof is complete. $\qquad\qquad\qquad\qquad\qquad\qquad\qquad\qquad\qquad\qquad\qquad\square$

Remark 5. If $(\mathbf{C}, \mathbf{A}; Q)$ is detectable, then it follows based on a similar proof that the result remains valid if one replaces $\widetilde{C}(t)$ with $\widetilde{C}(t, i) = \sum_{k=0}^{r} C_k^*(t, i) C_k(t, i)$.

3.2 Stochastic observability

Definition 3. *We say that the system* (3.1) *is* stochastically uniformly observable *(or equivalently, that* $(C_0, \mathbf{A}; Q)$ *is uniformly observable) if there exist* $\tau > 0, \beta > 0$ *such that*

$$\int_{t}^{t+\tau} T^*(s, t)\widetilde{C}(s)ds \geq \beta J^d \qquad (3.15)$$

$\forall t \geq 0$, *where* $\widetilde{C}(s) = \left(\widetilde{C}(s, 1), \; \widetilde{C}(s, 2), \ldots, \widetilde{C}(s, d)\right), \; \widetilde{C}(s, i) = C_0^*(s, i)C_0(s, i),$ $i \in \mathcal{D}, s \geq 0$. *In the time-invariant case we shall say that the system* (3.1) *is* stochastically observable, *or the triple* $(C_0, \mathbf{A}; Q)$ *is* observable.

Remark 6. (i) If in the system (3.1) we have $A_k(t, i) = 0, k = 1, \ldots, r, \mathcal{D} = \{1\}$, then the Lyapunov operator (2.8) is the Lyapunov operator of deterministic framework. In this case (3.15) becomes

$$\int_{t}^{t+\tau} \Phi_0^*(s, t)C_0^*(s)C_0(s)\Phi_0(s, t)ds \geq \beta I_n, \; \forall t \geq 0,$$

where $\Phi_0(\cdot, \cdot)$ is the fundamental matrix solution of the differential equation $\dot{x}(t) = A_0(t)x(t)$.

This shows that the above definition of stochastic uniform observability is a natural extension of the uniform observability used for linear time-varying deterministic systems (see [72]).

(ii) If the system (3.1) is subjected only to Markovian jumping, then the condition (3.15) becomes $\int_{t}^{t+\tau} \widehat{T}^*(s, t)\widetilde{C}(s)ds \geq \beta J^d$. If this is fulfilled we shall say that the triple $(C_0, A_0; Q)$ is *uniformly observable.*

(iii) If the system (3.1) is subjected only to multiplicative white noise and the corresponding inequaliy (3.15) is fulfilled, then we shall say that $(C_0, A_0, A_1, \ldots, A_r)$, or more briefly $(C_0, \mathbf{A})$, is *uniformly observable.*

The following result follows immediately from Theorem 4 of Chapter 2.

Proposition 8. *The system* (3.1) *is stochastically uniformly observable if and only if there exist* $\beta > 0, \tau > 0$ *such that*

$$E\left[\int_{t}^{t+\tau} \Phi^*(s, t)C_0^*(s, \eta(s))C_0(s, \eta(s))\Phi(s, t)ds \middle| \eta(t) = i\right] \geq \beta I_n$$

for all $t \geq 0, i \in \mathcal{D}, \Phi(\cdot, \cdot)$ *being the fundamental matrix solution of the system* (1.22). $\qquad\qquad\qquad\qquad\qquad\qquad\qquad\qquad\qquad\qquad\qquad\qquad\square$

The proof of the next result is based on some preliminary results that develop the ones presented in Section 2.2. First, remark that since

$$\Phi_i(t, t_0) = e^{\frac{1}{2}q_{ii}(t-t_0)}\widehat{\Phi}_i(t, t_0),$$

where $\Phi_i(t, t_0)$ is defined in the proof of Theorem 5 of Chapter 2 and $\widehat{\Phi}_i(t, t_0)$ is the fundamental matrix solution for fixed $i \in \mathcal{D}$ of the linear deterministic system

$$\frac{dx}{dt} = A_0(t, i)x(t),$$

it follows that for each $i \in \mathcal{D}$ the pair $\left(C_0(., i), \widetilde{A}_0(., i)\right)$ is uniformly observable if and only if the pair $(C_0(., i), A_0(., i))$ is uniformly observable, where

$$\widetilde{A}_0(t, i) = A_0(t, i) + \frac{1}{2}q_{ii} I_n.$$

Further, for each $i \in \mathcal{D}$, let

$$\mathcal{L}^i(t) : \mathcal{S}_n \to \mathcal{S}_n$$

be the Lyapunov-type linear operator defined by

$$\mathcal{L}^i(t)M = \widetilde{A}_0(t, i)M + M\widetilde{A}_0^*(t, i) + \sum_{j=1}^{r} A_j(t, i)MA_j^*(t, i), \quad M \in \mathcal{S}_n,$$

and let $T^i(t, t_0)$ be the linear evolution operator on $\mathcal{S}_n$ associated with the operator $\mathcal{L}^i(t)$.

Let $\overline{\mathcal{L}}(t) : \mathcal{S}_n^d \to \mathcal{S}_n^d$ be defined by

$$\left(\overline{\mathcal{L}}(t)H\right)(i) = \mathcal{L}^i(t)H(i), \quad H \in \mathcal{S}_n^d, \ i \in \mathcal{D},$$

and let $\overline{T}(t, t_0)$ be the linear evolution operator on $\mathcal{S}_n^d$ associated with the linear operator $\overline{\mathcal{L}}(t)$. It is easy to prove that

$$(\overline{T}(t, t_0)H)(i) = T^i(t, t_0)H(i), \quad H \in \mathcal{S}_n^d, \ i \in \mathcal{D}.$$

From the definitions $\widehat{T}(t, t_0), T_1(t, t_0)$ (see Section 2.2) easily follow

$$T(t, t_0) \geq \widehat{T}(t, t_0) \geq T_1(t, t_0), \tag{3.16}$$

$$T(t, t_0) \geq \overline{T}(t, t_0).$$

From (3.16), the next proposition immediately follows.

Proposition 9. *We make the following assumptions.*

(i) If for each $i \in \mathcal{D}$, the pair $(C_0(\cdot, i), A_0(\cdot, i))$ is uniformly observable, then the triple $(C_0, A_0; Q)$ is uniformly observable.

(ii) If $(C_0, A_0; Q)$ is uniformly observable, then $(C_0, \mathbf{A}; Q)$ is uniformly observable.

(iii) If for every $i \in \mathcal{D}$, the system $\left(C_0(., i), \widetilde{A}_0(., i), A_1(., i), \ldots, A_r(., i)\right)$ is uniformly observable, then the system $(C_0, \mathbf{A}; Q)$ is uniformly observable, too. □

Proposition 10. *Assume that the system* (3.1) *is in the time-invariant case. Then the following are equivalent:*

(i) *The system* (3.1) *is stochastically observable.*

(ii) *There exists* $\tau > 0$ *such that*

$$\int_0^\tau e^{\mathcal{L}^* s} \widetilde{C} ds > 0.$$

(iii) *There exists* $\tau > 0$ *such that* $X_0(\tau) > 0$, *where* $X_0(t)$ *is the solution of the problem with initial value:*

$$\frac{d}{dt} X_0(t) = \mathcal{L}^* X_0(t) + \widetilde{C}, \quad X_0(0) = 0.$$

Proof. (i) $\Longleftrightarrow$ (ii) follows from (2.16).

Since $X_0(t) = \int_0^t e^{\mathcal{L}^*(t-s)} \widetilde{C} ds = \int_0^t e^{\mathcal{L}^* s} \widetilde{C} ds$, $t \geq 0$, it follows that (iii) $\Longleftrightarrow$ (ii). The proof is complete. $\square$

Proposition 11. *Assume that the system* (3.1) *is in the time-invariant case. Let* $X_0(t)$ *be the solution of the Cauchy problem on* $\mathcal{S}_n^d$,

$$\frac{d}{dt} X_0(t) = \mathcal{L}^* X_0(t) + \widetilde{C}, \ t \geq 0, \quad X_0(0) = 0.$$

If there exists $\tau > 0$, *such that* $X_0(\tau) > 0$, *then* $X_0(t) > 0$ *for all* $t > 0$.

Proof. For each $t > 0$, we write the representation

$$X_0(t) = (X_0(t, 1), \ X_0(t, 2), \ldots, \ X(t, d)) = \int_0^t e^{L^*(t-s)} \widetilde{C} ds.$$

Since $e^{L^*(t-s)} : \mathcal{S}_n^d \to \mathcal{S}_n^d$ is a positive operator, we deduce that $X_0(t) \geq 0$ for all $t \geq 0$. Moreover if $t \geq \tau$ we have $X_0(t) \geq X_0(\tau)$; therefore, if $X_0(\tau) > 0$, we have $X_0(t) > 0$ for all $t \geq \tau$. It remains to show that $X_0(t) > 0, 0 < t < \tau$. To this end we show that $\det X_0(t, i) > 0$, $0 < t < \tau$, $i \in \mathcal{D}$. Indeed, since $\det X_0(t, i) = \det\{\int_0^t e^{\mathcal{L}^*(t-s)} \widetilde{C} ds)(i)\}$, we deduce that $t \to \det X_0(t, i)$ is an analytic function.

The set of its zeros on $[0, \tau]$ has no accumulation point. In this way it will follow that there exists $\tau_1 > 0$ such that $\det X_0(t, i) > 0$ for all $t \in (0, \tau_1]$. Invoking again the monotonicity of the function $t \to X_0(t)$ we conclude that $X_0(t) > 0$ for all $t \geq \tau_1$, and the proof is complete. $\square$

Remark 7. From Propositions 10 and 11 it follows that the stochastic observability for a system (3.1) in the time-invariant case may be checked by using a numerical procedure to compute the solution $X_0(t)$ through a long enough interval of time.

The following two results can be considered as Barbashin–Krasovskii-type theorems [58].

Theorem 12. *Assume that* $(C_0, \mathbf{A}; Q)$ *is uniformly observable and the affine differential equation*

$$\frac{d}{dt}X(t) + \mathcal{L}^*(t)X(t) + \tilde{C}(t) = 0 \tag{3.17}$$

has a bounded and semipositive solution $\tilde{X}(t), t \geq 0$. *Then*

(i) *The system* $(A_0, A_1, \ldots, A_r; Q)$ *is stable.*

(ii) $\tilde{X}(t) \gg 0$.

(iii) *Equation* (3.17) *has only one bounded solution that is uniform positive.*

Proof. From (2.12) it follows that

$$\tilde{X}(t) = T^*(s, t)\tilde{X}(s) + \int_t^s T^*(u, t)\tilde{C}(u)du, \quad s \geq t. \tag{3.18}$$

Since $0 \leq \tilde{X}(s) \leq \beta_0 J^d$ with some $\beta_0 > 0$ and $T(s, t) \geq 0$, one gets $0 \leq \int_t^s T^*(u, t)\tilde{C}(u)du \leq \tilde{X}(t) \leq \beta_0 J^d$ for all $s \geq t \geq 0$. Hence the integral $\hat{X}(t) = \int_t^\infty T^*(s, t)\tilde{C}(s)ds$ is convergent and $0 \leq \hat{X}(t) \leq \beta_0 J^d$, $t \geq 0$.

By (2.12) it follows directly that $\hat{X}$ is a solution of the equation (3.17).

Since $(C_0; A_0, \ldots, A_r, Q)$ is uniformly observable it follows that $\hat{X}$ is uniformly positive. Since $T^*(t + \tau, t)T^*(s, t + \tau) = T^*(s, t)$ we have

$$T^*(t + \tau, t)\hat{X}(t + \tau) = \int_{t+\tau}^\infty T^*(s, t)\tilde{C}(s)ds = \hat{X}(t) - \int_t^{t+\tau} T^*(s, t)\tilde{C}(s)ds.$$

Hence $T^*(t + \tau)\hat{X}(t + \tau) \leq \hat{X}(t) - \beta J^d \leq \left(1 - \frac{\beta}{\beta_0}\right)\hat{X}(t), t \geq 0$. Thus by Theorem 31 of Chapter 2 it follows that the system $(A_0, \ldots, A_r, Q)$ is stable. Hence by Theorem 12(ii) of Chapter 2, $\|T^*(s, t)\| \leq \gamma e^{-\alpha(s-t)}, s \geq t$.

Taking $s \to \infty$ in (3.18) one gets $\tilde{X}(t) = \hat{X}(t), t \geq 0$, and thus the proof is complete. $\qquad\qquad \square$

Corollary 13. *Suppose that* $A_k(t, i) = A_k(i), C_0(t, i) = C(i), t \geq 0, i \in \mathcal{D}, 0 \leq k \leq r$. *Assume that* $(C_0; A_0, \ldots, A_r, Q)$ *is observable and the algebraic equation on* S_n^d,

$$\mathcal{L}^*X + \tilde{C} = 0, \tag{3.19}$$

has a solution $\tilde{X} \geq 0$.

Then:

(i) *The system* $(A_0, A_1, \ldots, A_r, Q)$ *is stable.*

(ii) $\tilde{X} > 0$.

(iii) *The equation* (3.19) *has a unique positive semidefinite solution.* $\qquad \square$

The next result gives sufficient conditions concerning the observability of the system $(C_0; A_0, \ldots, A_r, Q)$.

Theorem 14. *Under the assumption of Proposition* 10 *if the system* $(C_0; A_0, \ldots,$ $A_r, Q)$ *is not observable, then there exist* $x_0 \in \mathbf{R}^n$, $x_0 \neq 0$, *and* $i_0 \in \mathcal{D}$ *such that*

(i) $C_0(i_0)x_0 = 0$.

(ii) $q_{i_0 i} C_0(i) x_0 = 0$ *for all* $i \in \mathcal{D}$.

(iii) $C_0(i_0)(A_0(i_0))^m x_0 = 0$ *for all* $m \geq 1$.

(iv) $q_{i_0 i} q_{ij} C_0(j) x_0 = 0$ *for all* $i \neq i_0$, $j \in \mathcal{D}$.

(v) $C_0(i_0) A_k(i_0) x_0 = 0$, $1 \leq k \leq r$.

Proof. Suppose that $(C_0; A_0, \ldots, A_r, Q)$ is not observable. From Proposition 10 it follows that there exist $x_0 \in \mathbf{R}^n$, $x_0 \neq 0$, and $i_0 \in \mathcal{D}$ such that $x_0^* \int_0^1 (e^{\mathcal{L}^* t} \widetilde{C})$ $(i_0) dt x_0 = 0$. Hence $x_0^* (e^{\mathcal{L}^* t} \widetilde{C})(i_0) x_0 = 0$ for all $t \in [0, 1]$. Since $e^{\mathcal{L}^* t} \geq e^{\widehat{\mathcal{L}}^* t} \geq e^{\mathcal{L}_1^* t}$ (see (3.16) and Remark 3 of Chapter 2) one gets $x_0^* (e^{\widehat{\mathcal{L}}^* t} \widetilde{C})(i_0) x_0 = 0$, $x_0^* (e^{\mathcal{L}_1^* t} \widetilde{C})$ $(i_0) x_0 = 0$, $t \in [0, 1]$. From the last equality we get $C_0(i_0) e^{A_0(i_0) t} x_0 = 0$, $t \in [0, 1)$. Hence differentiating successively we have

$$x_0^* ((\mathcal{L}^*)^m \widetilde{C})(i_0) x_0 = 0, \quad m \geq 0, \tag{3.20}$$

$$C_0(i_0)(A_0(i_0))^m x_0 = 0, \quad m \geq 0, \tag{3.21}$$

$$x_0^* ((\widehat{\mathcal{L}}^*)^m \widetilde{C})(i_0) x_0 = 0, \, x_0^* ((\mathcal{L}_1^*)^m \widetilde{C})(i_0) x_0 = 0 \tag{3.22}$$

for all $m \geq 0$.

Thus (i) and (iii) follow from (3.21)

Now, from (3.20) and (3.22) we have

$$0 = x_0^* (\mathcal{L}^* \widetilde{C})(i_0) x_0 = x_0^* (\mathcal{L}_2^* \widetilde{C})(i_0) x_0 + x_0^* (\widehat{\mathcal{L}}^* \widetilde{C})(i_0) x_0$$

$$= x_0 (\mathcal{L}_2^* \widetilde{C})(i_0) x_0 = x_0^* \sum_{k=1}^r A_k^*(i_0) C_0^*(i_0) C_0(i_0) A_k(i_0) x_0,$$

and thus (v) follows.

Further, by (3.22) we can write

$$0 = x_0^* (\widehat{\mathcal{L}}^* \widetilde{C})(i_0) x_0 = x_0^* (\mathcal{L}_1^* \widetilde{C})(i_0) x_0 + x_0^* (\mathcal{L}_3^* \widetilde{C})(i_0) x_0$$

$$= x_0^* (\mathcal{L}_3^* \widetilde{C})(i_0) x_0 = x_0^* \sum_{j \neq i_0} q_{i_0 j} C_0^*(j) C_0(j) x_0,$$

where $\mathcal{L}, \widehat{\mathcal{L}}, \mathcal{L}_1$ are defined in Section 2.2 and $\mathcal{L}_2 = \mathcal{L} - \widehat{\mathcal{L}}$ and $\mathcal{L}_3 = \widehat{\mathcal{L}} - \mathcal{L}_1$. Then, since $q_{ij} \geq 0$, if $i \neq j$ one gets (ii).

Also from (3.22) it follows that

$$0 = x_0^* ((\widehat{\mathcal{L}}^*)^2 \widetilde{C})(i_0) x_0$$

$$= x_0^* \{ [((\mathcal{L}_1^*)^2 + \mathcal{L}_1^* \mathcal{L}_3^* + \mathcal{L}_3^* \mathcal{L}_1^* + (\mathcal{L}_3^*)^2) \widetilde{C}](i_0) \} x_0$$

$$= x_0^* [(\mathcal{L}_1^* \mathcal{L}_3^* \widetilde{C})(i_0) + (\mathcal{L}_3^* \mathcal{L}_1^* \widetilde{C})(i_0) + ((\mathcal{L}_3^*)^2 \widetilde{C})(i_0)] x_0.$$

But, by using (ii) we can write

$$x_0^*(\mathcal{L}_1^*\mathcal{L}_3^*\widetilde{C})(i_0)x_0 = 2x_0^*\left[A_0^*(i_0) + \frac{1}{2}q_{i_0i_0}I_n\right]\sum_{i\neq i_0} q_{i_0i}C_0^*(i)C_0(i)x_0 = 0,$$

$$x_0^*(\mathcal{L}_3^*\mathcal{L}_1^*\widetilde{C})(i_0)x_0 = 2x_0^*\sum_{i\neq i_0} q_{i_0i}\left(A_0^*(i) + \frac{1}{2}q_{ii}I_n\right)C_0^*(i)C_0(i)x_0 = 0.$$

Hence one gets

$$0 = x_0^*((\mathcal{L}_3^*)^2\widetilde{C})(i_0)x_0 = x_0^*\sum_{i\neq i_0}\sum_{j\neq i} q_{i_0i}q_{ij}C_0^*(j)C_0(j)x_0,$$

and since $q_{i_0i}q_{ij} \geq 0$ for $i \neq i_0$, $j \neq i$, one obtains $q_{i_0i}q_{ij}C(j)x_0 = 0$ for all $i \neq i_0$ and $j \neq i$, and thus by (ii) it follows that (iv) holds and hence the proof is complete. □

Corollary 15. *Under the assumption of Proposition* 10, *if for every* $i \in \mathcal{D}$, *rank* $M(i) = n$, *where*

$$M(i) = \big[C_0^*(i), A_0^*(i)C_0^*(i), \ldots, (A_0^*(i))^{n-1}C_0^*(i),$$
$$q_{i1}C_0^*(1), \ldots, q_{id}C_0^*(d), A_1^*(i)C_0^*(i), \ldots, A_r^*(i)C_0^*(i)\big],$$

then the system $(C_0; A_0, A_1, \ldots, A_r, Q)$ *is observable.* □

In the following examples, the stochastic observability used in this paper is compared with other types of stochastic observability, for example, the one introduced in [70] and [86]. We also show that the stochastic observability used in this paper doesn't imply the stochastic detectability as we would have expected.

Example 1. Let us consider the case of a system with Markovian jumping with $d = 2$, $n = 2$, $p = 1$. Take

$$A_0(1) = A_0(2) = \begin{bmatrix} \alpha & 0 \\ 0 & \alpha \end{bmatrix},$$

$$C_0(1) = [1 \quad 0], C_0(2) = [0 \quad 1], Q = \begin{bmatrix} -q & q \\ q & -q \end{bmatrix}, \alpha \in \mathbf{R}, q > 0.$$

It is obvious that the pairs $(C_0(1), A_0(1))$, $(C_0(2), A_0(2))$ are not observable. Therefore, this system is not stochastically observable, in the sense of [86]. We shall show that this system is stochastically observable in the sense of Definition 3.

To this end we use the implication (iii) $\Longrightarrow$ (i) in Proposition 10. We show that there exists $\tau > 0$ such that $X_1(\tau) > 0$, $X_2(\tau) > 0$, where $X_i(t)$, $i = 1, 2$, is the solution of the Cauchy problem:

$$\frac{d}{dt}X_i(t) = A_0^*(i)X_i(t) + X_i(t)A_0(i) + \sum_{j=1}^{2}q_{ij}X_j(t) + C_0^*(i)C_0(i), \quad (3.23)$$

$$X_i(0) = 0, \quad i = 1, 2.$$

From the representation formula

$$(X_1(t), X_2(t)) = \int_0^t e^{\widehat{\mathcal{L}_0^*}(t-s)}\widetilde{C}ds,$$

it follows that $X_i(t) \geq 0$ for all $t \geq 0$.

Therefore it is sufficient to show that there exists $\tau > 0$ such that $det\, X_i(\tau) > 0$.

Set

$$X_i(t) = \begin{pmatrix} x_i(t) & y_i(t) \\ y_i(t) & z_i(t) \end{pmatrix}, \quad i = 1, 2,$$

and obtain from (3.23) the following system of affine differential equations:

$$x_1'(t) = (2\alpha - q)x_1(t) + qx_2(t) + 1,$$
$$x_2'(t) = qx_1(t) + (2\alpha - q)x_2(t),$$
$$y_1'(t) = (2\alpha - q)y_1(t) + qy_2(t),$$
$$y_2'(t) = \alpha y_1(t) + (2\alpha - q)y_2(t),$$
$$z_1'(t) = (2\alpha - q)z_1(t) + qz_2(t),$$
$$z_2'(t) = qz_1(t) + (2\alpha - q)z_2(t) + 1,$$
$$x_i(0) = y_i(0) = z_i(0) = 0, i = 1, 2.$$

Hence $y_1(t) = y_2(t) = 0, t \geq 0$.

From the uniqueness of the solution of a Cauchy problem it follows that $x_1(t) = z_2(t) = \tilde{x}(t)$ and $x_2(t) = z_1(t) = \tilde{z}(t)$, where $t \to (\tilde{x}(t), \tilde{z}(t))$ is the solution of the problem

$$\frac{d}{dt}\tilde{x}(t) = (2\alpha - q)\tilde{x}(t) + q\tilde{z}(t) + 1,$$

$$\frac{d}{dt}\tilde{z}(t) = q\tilde{x}(t) + (2\alpha - q)\tilde{z}(t),$$

$$\tilde{x}(0) = \tilde{z}(0) = 0.$$

We have $det\, X_i(t) = x_i(t)z_i(t) - y_i^2(t) = x_i(t)z_i(t) = \tilde{x}(t)\tilde{z}(t), t \geq 0$.

But

$$\tilde{x}(t) = \frac{1}{2}\int_0^t \left[e^{\alpha s} + e^{(2\alpha - q)s}\right]ds,$$

$$\tilde{z}(t) = \frac{1}{2}\int_0^t \left[e^{2\alpha s} - e^{2(\alpha - q)s}\right]ds.$$

It is easy to see that for every $\alpha \in \mathbf{R}, q > 0$ we have $\lim_{t \to \infty}\tilde{x}(t)\tilde{z}(t) > 0$.

Remark 8. Let us consider the system of type (3.1) with $n = 2, d = 2, p = 1, r = 1$, and $A_0(1) = A_0(2) = \alpha I_2$, $C_0(1) = [1 \quad 0]$, $C_0(2) = [0 \quad 1]$, $A_1(i)$ 2×2 arbitrary matrix,

$$Q = \begin{bmatrix} -q & q \\ q & -q \end{bmatrix},$$

$\alpha \in \mathbf{R}, q > 0$. Combining the conclusion of Example 1 with Proposition 9 it follows that the system $(C_0, (A_0, A_1); Q)$ is observable.

Example 2. The stochastic observability does not always imply stochastic detectability. Let us consider the system with Markovian jumping with $d = 2, n = 2, p = 1$,

$$A_0(1) = A_0(2) = \frac{q}{2} I_2, \; C_0(1) = [1 \quad 0], \; C_0(2) = [0 \quad 1], \; Q = \begin{bmatrix} -q & q \\ q & -q \end{bmatrix}.$$

$$(3.24)$$

From the previous example we conclude that the system $(C_0, A_0; Q)$ is observable. Invoking (i) $\Leftrightarrow$ (ii) from Proposition 5 we deduce that if the system (3.24) would be stochastically detectable, then there would exist the matrices $X(i) > 0$, and

$$\Lambda(i) = \begin{bmatrix} \lambda_1(i) \\ \lambda_2(i) \end{bmatrix}, \quad i = 1, 2,$$

which verify the following system of linear equations:

$$A_0^*(i)X(i) + X(i)A_0(i) + \Lambda(i)C_0(i)$$

$$+ C_0^*(i)\Lambda^*(i) + \sum_{j=1}^{2} q_{ij} X(j) + I_2 = 0, \quad i = 1, 2,$$

which implies

$$I_2 + \begin{bmatrix} 2\lambda_1(1) & \lambda_2(1) \\ \lambda_2(1) & 0 \end{bmatrix} < 0,$$

which is a contradiction.

Example 3. Let us consider the stochastic system

$$dx(t) = A_0(\eta(t))x(t)dt + A_1(\eta(t))x(t)dw_1(t),$$

$$y(t) = C_0(\eta(t))x(t)$$

$$(3.25)$$

with $n = 2$, $d = 2$, $r = 1$, $p = 1$, $A_0(1) = A_0(2) = \alpha I_2$, $C_0(1) = [1 \quad 0]$, $C_0(2) = [0 \quad 1]$, $A_1(1) = \beta I_2$, $A_1(2)$ is a 2×2 arbitrary matrix,

$$Q = \begin{bmatrix} -q & q \\ q & -q \end{bmatrix},$$

$\alpha \in \mathbf{R}$, $\beta \in \mathbf{R}$, $q > 0$, which satisfy $2\alpha - q + \beta^2 = 0$.

From Remark 8 it follows that the system described by (3.25) is stochastically observable. We show that it is not stochastically detectable. If, on the contrary, the system (3.25) is stochastically detectable, then, again using Proposition 5, we deduce that there exist matrices

$$X(i) > 0, \Lambda(i) = \begin{bmatrix} \lambda_1(i) \\ \lambda_2(i) \end{bmatrix}, \lambda_k(i) \in \mathbf{R},$$

which verify the following system of linear equations:

$$A_0^*(i)X(i) + X(i)A_0(i) + \Lambda(i)C_0(i) + C_0^*(i)\Lambda^*(i)$$
$$+ A_1^*(i)X(i)A_1(i) + \sum_{j=1}^{2} q_{ij}X(j) + I_2 = 0,$$

which leads to the same contradiction as in the previous example.

Remark 9. It can be remarked that the system

$$dx(t) = A_0(\eta(t))x(t)dt + \sum_{k=1}^{r} A_k(\eta(t))x(t)dw_k(t), \qquad (3.26)$$
$$y(t) = C_0(\eta(t))x(t),$$

with $A_0(i), C_0(i)$ as in (3.24) and $A_k(i), k = 1, 2, \ldots, r$, 2×2 arbitrary matrices, is stochastically observable, but it is not stochastically detectable. If, on the contrary, (3.26) would be stochastically detectable, then by Proposition 2 (i) it could follow that the system described by (3.24) would be stochastically detectable, which contradicts the conclusion of Example 2.

From the representation formula in Theorem 4 in Chapter 2, the next result follows.

Proposition 16. *Assume that the system* (3.1) *is in the time-invariant case. Then the triple* $(C_0, \mathbf{A}; Q)$ *is observable if and only if* $\tau > 0, i \in \mathcal{D},$ *and* $x_0 \neq 0$ *do not exist such that*

$$E\big[|y(t, 0, x_0)|^2 | \eta(0) = i\big] = 0$$

$\forall t \in [0, \tau]$, *with* $y(t, 0, x_0) = C_0(\eta(t))x(t, 0, x_0)$, $x(t, 0, x_0)$ *being the solution of* (3.1) *for* $u(t) = 0$ *and having the initial condition* $x(0, 0, x_0) = x_0$. □

In the deterministic framework the analogue of the above statement is one of the usual definitions of observability.

Remark 10. In Definition 3 of observability, no condition on Q is imposed. All the results proved above except Propositions 8 and 16 require only the condition $q_{ij} \geq 0$ for $i \neq j$. The additional condition $\sum_{j=1}^{d} q_{ij} = 0$ is used only in the proof of the two mentioned propositions.

3.3 Stochastic controllability

In this section the controllability of stochastic systems will be introduced. For simplicity we shall consider only the time-invariant case.

Let $A_k(i) \in \mathbf{R}^{n \times n}$, $0 \leq k \leq r, i \in \mathcal{D}$, $B(i) \in \mathbf{R}^{n \times m}$, $Q = [q_{ij}]$, $i, j \in \mathcal{D}$ with $q_{ij} \geq 0$ for $i \neq j$.

Definition 4. *We say that the system* $(A_0, A_1, \ldots, A_r, B; Q)$ *is* controllable *if* $\tau > 0$ *exists such that*

$$\int_0^\tau e^{\mathcal{L}t} \widetilde{B} dt > 0,$$

where $\mathcal{L}$ is defined by (2.15) *and* $\widetilde{B} \in \mathcal{S}_n^d$, $\widetilde{B}(i) = B(i)B^*(i), i \in \mathcal{D}$.

Remark 11. One can easily see that in the deterministic case, namely if $\mathcal{D} = \{1\}$, $q_{11} = 0$, and $A_k(1) = 0$, $1 \leq k \leq r$, the above definition reduces to the definition of controllability of the pair $(A_0(1), B(1))$.

The following result can be directly proved.

Proposition 17. *The system* $(A_0, A_1, \ldots, A_r, B; Q)$ *is controllable if and only if the system* $(B^*, A_0^*, A_1^*, \ldots, A_r^*; Q^*)$ *is observable.*   □

From the above proposition and from Propositions 10 and 11 and Remark 11, the next proposition immediately follows.

Proposition 18. *The following assertions are equivalent:*
   (i) *The system* $(A_0, A_1, \ldots, A_r, B; Q)$ *is controllable.*
   (ii) *There exists $\tau > 0$ such that $K_0(\tau) > 0$ where $K_0(t)$ denotes the solution of the affine equation in the space $\mathcal{S}_n^d$:*

$$\frac{d}{dt} K_0(t) = \mathcal{L} K_0(t) + \widetilde{B}$$

with $K_0(0) = 0$.
   (iii) *For any $t > 0$, $K_0(t) > 0$.*   □

In the following we shall consider the situation when the system is subjected only to white noise perturbations, namely if $\mathcal{D} = \{1\}$, $q_{11} = 0$, $A_k(1) = A_k$, $B(1) = B$. The inequality in Definition 4 becomes

$$\int_0^\tau e^{\bar{\mathcal{L}}t} \widehat{B} dt > 0,$$

where $\bar{\mathcal{L}}$ denotes the linear operator defined on $\mathcal{S}_n$ by (2.23) and $\widehat{B} = BB^*$. If this inequality is fulfilled for some $\tau > 0$ we shall say that the system $(A_0, A_1, \ldots, A_r, B)$ *is controllable*. Therefore, in the case of systems with multiplicative white noise, the proposition above becomes the following proposition.

Proposition 19. *The following assertions are equivalent:*
   (i) *The system* $(A_0, A_1, \ldots, A_r, B)$ *is controllable.*
   (ii) *There exists* $\tau > 0$ *such that* $\widetilde{K}(\tau) > 0$, *where*

$$\frac{d}{dt}\widetilde{K}(t) = A_0\widetilde{K}(t) + \widetilde{K}(t)A_0^* + \sum_{k=1}^{r} A_k\widetilde{K}(t)A_k^* + BB^* \text{ with } \widetilde{K}(0) = 0. \quad (3.27)$$

   (iii) $\widetilde{K}(t) > 0$ *for all* $t > 0$. □

From Remark 7 of Chapter 2 it immediately follows that

$$e^{\overline{\mathcal{L}t}} H = E[\Phi(t, 0)H\Phi^*(t, 0)], \ t \geq 0, \ H \in \mathcal{S}_n,$$

where $\Phi(t, t_0)$, $t \geq t_0$, denotes the fundamental matrix associated with the linear Itô system

$$dx(t) = A_0x(t)dt + \sum_{k=1}^{r} A_kx(t)dw_k(t).$$

Therefore the next result directly follows.

Proposition 20. *The system* $(A_0, A_1, \ldots, A_r, B)$ *is controllable if and only if* $\tau > 0$ *exists such that* $E \int_0^\tau [\Phi(t, 0)BB^*\Phi^*(t, 0)]dt > 0$. □

We shall now give another characterization, in stochastic terms, of the controllability of the system $(A_0, A_1, \ldots, A_r, B)$. Consider the affine Itô system

$$dx(t) = A_0x(t)dt + \sum_{k=1}^{r} A_kx(t)dw_k(t) + Bdv(t), \ t \geq 0, \quad (3.28)$$

where $(w(t), v(t))^*$ is a standard $(r+m)$-dimensional Wiener process. Let $\tilde{x}(t), t \geq 0$, be the solution of (3.28) with $\tilde{x}(0) = 0$. Using the Itô formula (Theorem 33 of Chapter 1), one can easily verify that $\widetilde{K}(t) = E[\tilde{x}(t)\tilde{x}^*(t)]$, $\widetilde{K}$ being defined in Proposition 19. Then the following result is immediately obtained.

Proposition 21. *The system* $(A_0, A_1, \ldots, A_r, B)$ *is controllable if and only if* $E[\tilde{x}(t)\tilde{x}^*(t)] > 0$ *for all* $t > 0$. □

The above characterization has been considered as a definition of controllability of the system $(A_0, A_1, \ldots, A_r, B)$ in [10].

The next result proved in [10] characterizes the controllability of the system $(A_0, A_1, \ldots, A_r, B)$ in terms of invariant subspaces as in the deterministic case $(A_k = 0, 1 \leq k \leq r)$.

Theorem 22. *The system* $(A_0, A_1, \ldots, A_r, B; Q)$ *is controllable if and only if no invariant subspace exists with the dimension less than n of the collection* $A_k, 0 \leq k \leq r$, *containing all columns of* B.

For the proof of the above theorem we need the following lemma.

Lemma 23. *The following two assertions are equivalent:*

(i) *An invariant subspace exists with dimension less than n of the matrices* $A_k, 0 \leq k \leq r$, *containing all columns of B.*

(ii) $\xi \in \mathbf{R}^n$, $\xi \neq 0$, *exists such that* $\xi^* M B = 0$ *for all* $M = A_{i_1}^{s_1} A_{i_2}^{s_2} \ldots A_{i_p}^{s_p}$, *where* $0 \leq i_j \leq r$, *and* $s_j \geq 0$, $1 \leq j \leq p$, $p \geq 1$ *are natural numbers.*

Proof. (i) $\Rightarrow$ (ii) Let S be an invariant subspace of the matrices $A_k, 0 \leq k \leq r$, with dimension less than n containing all columns of B. Denote by $S^{\perp}$ the orthogonal subspace of S. Since $S^{\perp} \neq \{0\}$, consider $\xi \in S^{\perp}$ such that $\xi \neq 0$. Since all the columns of the matrices $M B$ with M as in the statement are included in S, it follows that $\xi^* M B = 0$.

(ii) $\Rightarrow$ (i) Assume that $\xi \neq 0$ exists, satisfying (ii). Let S be the subspace generated by the columns of all matrices $M B$, M being defined as in the statement. Since $\xi \neq 0$ it follows that $S \neq \mathbf{R}^n$. On the other hand, it is easy to check that if $x \in S$, then $A_k x \in S$ for all $0 \leq k \leq r$. Thus the proof is complete. □

Proof of Theorem 22. Necessity. Assume that the system $(A_0, A_1, \ldots, A_r, B)$ is controllable. It follows that $B \neq 0$, and therefore, if $n = 1$ the condition in the statement is automatically accomplished. We now consider the case $n \geq 2$ and that there exists a subspace $S, S \neq \{0\}$, $S \neq \mathbf{R}^n$ invariant of $A_k, 0 \leq k \leq r$, containing all columns of B. Then it follows that a basis in $\mathbf{R}^n$ exists with respect to which the matrices A_k have the structure

$$\tilde{A}_k = \begin{bmatrix} A_{1k} & A_{2k} \\ 0 & A_{3k} \end{bmatrix}, \ 0 \leq k \leq r,$$

and B has the form

$$\tilde{B} = \begin{bmatrix} B_0 \\ 0 \end{bmatrix},$$

where A_{1k} are $s \times s$ matrices with $1 \leq s < n$. Let $\overline{K}(t), t \geq 0$, be the solution of equation (3.27) corresponding to the matrices $\tilde{A}_k$ and $\tilde{B}$ and $\overline{K}(0) = 0$. It is easy to check that if

$$\overline{K}(t) = \begin{bmatrix} \overline{K}_{11}(t) & \overline{K}_{12}(t) \\ \overline{K}_{21}(t) & \overline{K}_{22}(t) \end{bmatrix},$$

then $\overline{K}_{22}(t)$ verifies a linear equation. Since $\overline{K}(0) = 0$ it follows that $\overline{K}_{22}(t) = 0$ for all $t \geq 0$, and therefore $\overline{K}_{22}(t)$ is not positive definite for all $t \geq 0$. Taking into account that $\tilde{K}(t) = T\overline{K}(t)T^*$ with T nonsingular it follows that $\tilde{K}(t)$ is not positive definite, which contradicts the assumption (see Proposition 19).

Sufficiency. We prove that $\tilde{K}(t) > 0$ for all $t > 0$. Indeed, assume that $\tau > 0$ and $\xi \in \mathbf{R}^n$, $\xi \neq 0$, exist such that $\xi^* \tilde{K}(t)\xi = 0$. Then one can easily check that

$$\tilde{K}(t) = \sum_{k=1}^{r} \int_0^t e^{A_0(t-s)} A_k \tilde{K}(s) A_k^* e^{A_0^*(t-s)} ds + \int_0^t e^{A_0 s} B B^* e^{A_0^* s} ds. \qquad (3.29)$$

Since $\widetilde{K}(t) \geq 0$, from (3.29) we successively obtain

$$\widetilde{K}(t) \geq \int_0^t e^{A_0 s} B B^* e^{A_0^* s} ds.$$

$$\widetilde{K}(t) \geq \sum_{i_1=1}^r \int_0^t \left(\int_0^{s_1} e^{A_0(t-s_1)} A_{i_1} e^{A_0 s_0} B B^* A_{i_1}^* e^{A_0^*(t-s_1)} ds_0 \right) ds_1,$$

$$\vdots$$

$$\widetilde{K}(t) \geq \sum_{i_p, i_{p-1}, \dots, i_1} \int_0^t \int_0^{s_p} \int_0^{s_{p-1}} \cdots \int_0^{s_1} e^{A_0(t-s_p)} A_{i_p} e^{A_0(s_p - s_{p-1})} A_{i_{p-1}}$$

$$\cdots e^{A_0(s_2 - s_1)} A_{i_1} e^{A_0 s_0} B B^* A_{i_1}^* e^{A_0^*(s_2 - s_1)}$$

$$\cdots A_{i_p}^* e^{A_0^*(t-s_p)} ds_0 \cdots ds_p.$$

Therefore, $\xi^* e^{A_0 s} B = 0$ for all $0 \leq s \leq \tau$ and

$$\xi^* e^{A_0(\tau - s_p)} A_{i_p} e^{A_0(s_p - s_{p-1})} \cdots e^{A_0(s_2 - s_1)} A_{i_1} e^{A_0 s_0} B = 0$$

for all $\tau > s_p > s_{p-1} > \cdots > s_2 > s_1 > s_0 \geq 0$ and for all $1 \leq i_j \leq r$, $1 \leq j \leq p$. It follows that $\xi^* A_0^k B = 0$, $k \geq 0$, and

$$\xi^* A_0^{k_1} A_{i_p} A_0^{k_2} A_{i_{p-1}} \cdots A_{i_1} A_0^{k_p} B = 0$$

for all $1 \leq i_j \leq r$, $1 \leq j \leq p$ and $k_s \geq 0$, $0 \leq s \leq p$. Therefore, $\xi^* M B = 0$ for all M as in the statement of Lemma 23, and according to this lemma, we obtained a contradiction. Thus the proof of the theorem is complete. $\qquad \square$

From the above theorem a corollary immediately follows.

Corollary 24. *If a pair (A_k, B) is controllable for a certain $k \in \{0, 1, \dots, r\}$, then the system $(A_0, A_1, \dots, A_r, B)$ is controllable.* $\qquad \square$

We shall show below that the converse of the corollary is not usually true. However, in the case $n = 2$, $m = 1$, $r = 1$ such an implication is valid; namely one can prove the following.

Proposition 25. *If $n = 2$, $m = 1$, $r = 1$ and the pairs (A_0, B) and (A_1, B) are not controllable, then the system (A_0, A_1, B) is not controllable.*

Proof. Let

$$A_0 = \begin{bmatrix} a & b \\ c & d \end{bmatrix}, \quad A_1 = \begin{bmatrix} \alpha & \beta \\ \gamma & \delta \end{bmatrix}, \quad \text{and } B = \begin{bmatrix} b_1 \\ b_2 \end{bmatrix}$$

such that (A_0, B) and (A_1, B) are not controllable, that is

$$b_1 b_2 (d - a) = b b_2^2 - b_1^2 c \quad \text{and} \quad b_1 b_2 (\delta - \alpha) = \beta b_2^2 - b_1^2 \gamma. \tag{3.30}$$

According to Proposition 19 the considered system $(A_0, \; B)$ is controllable if and only if $\tilde{K}(t) > 0$ for all $t > 0$, where $\tilde{K}$ verifies (3.27) written for this particular case. Taking

$$\tilde{K} = \begin{bmatrix} x & y \\ y & z \end{bmatrix},$$

(3.27) gives

$$\frac{dx}{dt} = (2a + \alpha^2)x + 2(b + \alpha\beta)y + \beta^2 z + b_1^2,$$

$$\frac{dy}{dt} = (c + \alpha\gamma)x + (a + d + \gamma\beta + \alpha\delta)y + (b + \beta\delta)z + b_1 b_2, \qquad (3.31)$$

$$\frac{dz}{dt} = \gamma^2 x + 2(c + \gamma\delta)y + (2d + \delta^2)z + b_2^2.$$

If $b_1 = 0$ and $b_2 = 0$, it immediately follows that $x(t) = y(t) = z(t) = 0$ for all $t \in \mathbf{R}$.

If $b_1 \neq 0$ and $b_2 = 0$, from (3.30) one obtains that $c = 0$ and $\gamma = 0$, and therefore $z(t) = 0$ for all $t \in \mathbf{R}$.

If $b_1 = 0$ and $b_2 \neq 0$, then (3.30) gives $b = \beta = 0$ and hence $x(t) = 0$ for all $t \in \mathbf{R}$.

Assume that $b_1 \neq 0$ and $b_2 \neq 0$. Using (3.30) one can easily check that $(\bar{x}, \bar{y}, \bar{z})$ verifies (3.31) where

$$\bar{x}(t) = \frac{b_1}{b_2}\bar{y}(t), \quad \bar{z}(t) = \frac{b_2}{b_1}\bar{y}(t),$$

and $\bar{y}(t)$ is the solution of the equation

$$\frac{d\bar{y}}{dt} = \left[(c + \alpha\gamma)\frac{b_1}{b_2} + (a + d + \gamma\beta + \alpha\delta) + (b + \beta\delta)\frac{b_2}{b_1} \right]\bar{y} + b_1 b_2$$

and $\bar{y}(0) = 0$. From the uniqueness of the solution it follows that $x(t) = \bar{x}(t)$, $y(t) = \bar{y}(t)$, $z(t) = \bar{z}(t)$ and therefore $x(t)z(t) - (y(t))^2 = 0$ for all $t \in \mathbf{R}$, and therefore by Proposition 19 (A_0, A_1, B) is not controllable.   □

The next example shows that the converse of Corollary 24 is not generally true; namely it is possible to have a controllable system (A_0, A_1, B) but with the pairs (A_0, B) and (A_1, B) not controllable.

Example. Consider the case $n = 3$, $m = 1$, and $r = 1$ in which

$$A_0 = \begin{bmatrix} 1 & 0 & 0 \\ 0 & -1 & 3 \\ 0 & 0 & 2 \end{bmatrix}, \; A_1 = \begin{bmatrix} 3 & 0 & 0 \\ 2 & 1 & 0 \\ 0 & 0 & -1 \end{bmatrix}, \; B = \begin{bmatrix} 1 \\ 1 \\ 1 \end{bmatrix}.$$

It is easy to check that (A_0, B) and (A_1, B) are not controllable. In this case (3.27) gives for

$$\tilde{K} = \begin{bmatrix} x & y & z \\ y & u & v \\ z & v & q \end{bmatrix},$$

$$\frac{dx}{dt} = 11x + 1,$$

$$\frac{dy}{dt} = 3y + 6x + 3z + 1,$$

$$\frac{dz}{dt} = 1,$$

$$\frac{du}{dt} = -u + 6v + 4x + 4y + 1,$$

$$\frac{dv}{dt} = 3q - 2z + 1,$$

$$\frac{dq}{dt} = 5q + 1,$$

with $x(0) = y(0) = z(0) = u(0) = v(0) = q(0) = 0$. One can directly check that the solution of the above system is given by

$$x(t) = \frac{1}{11}\left(e^{11t} - 1\right),$$

$$y(t) = \frac{3}{44}e^{11t} + \frac{5}{12}e^{3t} - t - \frac{16}{33},$$

$$z(t) = t,$$

$$v(t) = \frac{3}{25}\left(e^{5t} - 1\right) - t^2 + \frac{2}{5}t,$$

$$q(t) = \frac{1}{5}\left(e^{5t} - 1\right),$$

and $u(t)$ has the form

$$u(t) = \frac{17}{132}e^{11t} + \alpha_1 e^{5t} + \alpha_2 e^{3t} + \alpha_3 e^{-t} + \alpha_4 t^2 + \alpha_5 t + \alpha_6.$$

Then it follows that $\lim_{t \to \infty} \det \tilde{K}(t) = \infty$, which implies that $\tilde{K}(t) > 0$ for some $t \geq 0$, and therefore, according to Proposition 19, the system (A_0, A_1, B) is controllable.

Remark 12. We have previously shown that by contrast with the deterministic case, the stochastic controllability of Markovian systems does not imply their stochastic stabilizability. A similar affirmation is valid for the stochastic systems subjected to Itô multiplicative noise.

Indeed the system (A_0, A_1, B) in the above example is controllable, but it is not stabilizable, since in such a situation, according to Proposition 4 applied in this case $(D = \{1\}, q_{11} = 0)$, there exists $(X, \Delta), X > 0$,

$$X = \begin{bmatrix} x & y & z \\ y & u & v \\ z & v & q \end{bmatrix} \text{ and } \Delta = \begin{bmatrix} 2f_1 & f_1 + f_2 & f_1 + f_3 \\ f_1 + f_2 & 2f_2 & f_2 + f_3 \\ f_1 + f_3 & f_2 + f_3 & 2f_3 \end{bmatrix}$$

such that

$$A_0 X + X A_0^* + A_1 X A_1^* + I_3 + \Delta = 0.$$

Therefore

$$\begin{aligned} 11x + 1 + 2f_1 &= 0, \\ 3y + 6x + 3z + f_1 + f_2 &= 0, \\ f_1 + f_3 &= 0, \\ -u + 6v + 4x + 4y + 1 + 2f_2 &= 0, \\ 3q - 2z + f_2 + f_3 &= 0, \\ 5q + 1 + 2f_3 &= 0. \end{aligned}$$

Since $x > 0$ and $q > 0$ it follows that $f_1 < 0$, $f_3 < 0$, which contradicts $f_1 + f_3 = 0$. Hence (A_0, A_1, B) is not stabilizable.

Notes and references

Stochastic controllability for Itô differential equations was introduced in [10]. Theorem 22 can also be found in [10]. The numerical example and Remark 12 appear for the first time in this book.

Other concepts of stochastic controllability have been studied in terms of control which generalize recurrence notions of stochastic processes (see, e.g., [120], [75], [76], [47], [12], [13], [105] for Itô systems and [70] for jump linear Markovian systems). In the present book the concept of stochastic controllability is not used, and therefore a reduced space is devoted to this concept.

The stochastic uniform observability was defined in [88] for Itô systems and in [89] for systems with jump Markovian perturbations. These concepts have been used to solve the linear quadratic problem with infinite horizon for these corresponding systems. The results in this chapter devoted to stochastic stabilizability, detectability, and observability can be found in [33], [31], and [34].

4

The Riccati Equations of Stochastic Control

In many control problems, in both the deterministic and stochastic framework, a crucial role is played by a class of nonlinear matrix differential equations or nonlinear matrix algebraic equations known as *matrix Riccati equations*.

In this chapter we deal with a class of systems of matrix differential equations as well as systems of nonlinear algebraic equations arising in connection with the solution of several control problems, such as linear quadratic optimization, H^2 control, and H^∞ control problems for stochastic systems. These will be called *stochastic generalized Riccati differential equations* (SGRDEs) or *stochastic generalized Riccati algebraic equations* (SGRAEs). It is easy to see that the systems of matrix Riccati differential equations considered in this chapter contain as particular cases many types of matrix Riccati equations that are known in both the deterministic and the stochastic framework. The results derived in this general framework are also applicable to these particular cases. These kinds of SGRDEs are regarded as mathematical objects of interest in themselves, and the proofs avoid any connection with an optimization problem. The proofs are mainly based on positivity properties of linear evolution operators defined by the Lyapunov differential equations. We provide conditions that guarantee the existence and the uniqueness of some global solutions of SGRDEs as maximal solution, minimal solution, and stabilizing solution. We prove that if the coefficients of SGRDEs are periodic functions, then the maximal solution, the minimal solution, and the stabilizing solution are also periodic functions. Moreover, if the coefficients of the SGRDEs do not depend on the parameter t, then the above-mentioned special solutions are constant and they solve the corresponding SGRAE. The necessary and sufficient conditions that guarantee the existence of the maximal solution, the minimal solution, and of the stabilizing solution, respectively, are expressed in terms of solvability of a class of suitable systems of linear matrix inequalities. Finally we shall provide an iterative procedure that allows us to compute these special solutions to the SGRDE and to the SGRAE.

4.1 Preliminaries

In this chapter we study systems of nonlinear matrix differential equations of the following form:

$$
\frac{d}{dt} X(t, i) + A_0^*(t, i) X(t, i) + X(t, i) A_0(t, i)
$$

$$
+ \sum_{k=1}^{r} A_k^*(t, i) X(t, i) A_k(t, i) + \sum_{j=1}^{d} q_{ij} X(t, j)
$$

$$
- \left(X(t, i) B_0(t, i) + \sum_{k=1}^{r} A_k^*(t, i) X(t, i) B_k(t, i) + L(t, i) \right)
$$

$$
\times \left(R(t, i) + \sum_{k=1}^{r} B_k^*(t, i) X(t, i) B_k(t, i) \right)^{-1}
$$

$$
\times \left(B_0^*(t, i) X(t, i) + \sum_{k=1}^{r} B_k^*(t, i) X(t, i) A_k(t, i) + L^*(t, i) \right)
$$

$$
+ M(t, i) = 0.
$$

(4.1)

where $t \to A_k(t, i) : \mathcal{I} \to \mathbf{R}^{n \times n}$, $t \to B_k(t, i) : \mathcal{I} \to \mathbf{R}^{n \times m}$, $0 \le k \le r$, $t \to M(t, i) : \mathcal{I} \to \mathcal{S}_n$, $t \to L(t, i) : \mathcal{I} \to \mathbf{R}^{n \times m}$, $t \to R(t, i) : \mathcal{I} \to \mathcal{S}_m$, $i \in \mathcal{D}$, are bounded, and continuous matrix-valued functions. $\mathcal{I} \subset \mathbf{R}$ is a right unbounded interval. The elements q_{ij} of the matrix Q verify only the weaker assumption $q_{ij} \ge 0$ for $i \ne j$. The assumption $\sum_{j=1}^{d} q_{ij} = 0$ will be used only for the results referring to stochastic observability and detectability. If $A_k(t, i) = 0$, $B_k(t, i) = 0$, $1 \le k \le r$, $(t, i) \in \mathcal{I} \times \mathcal{D}$, the system (4.1) becomes the system of Riccati-type equations intensively investigated in connection with the linear quadratic problem for linear stochastic systems with Markovian jumping. In the particular case $\mathcal{D} = \{1\}$, the system (4.1) reduces to

$$
\frac{d}{dt} X(t) + A_0^*(t) X(t) + X(t) A_0(t) + \sum_{k=1}^{r} A_k^*(t) X(t) A_k(t)
$$

$$
- \left(X(t) B_0(t) + \sum_{k=1}^{r} A_k^*(t) X(t) B_k(t) + L(t) \right)
$$

$$
\times \left(R(t) + \sum_{k=1}^{r} B_k^*(t) X(t) B_k(t) \right)^{-1}
$$

$$
\times \left(B_0^*(t) X(t) + \sum_{k=1}^{r} B_k^*(t) X(t) A_k(t) + L^*(t) \right) + M(t) = 0, \ t \in \mathcal{I},
$$

(4.2)

where we denoted $A_0(t) = A_0(t, 1) + \frac{1}{2} q_{11} I_n$, $A_k(t) = A_k(t, 1)$, $1 \le k \le r$, $B_k(t) = B_k(t, 1)$, $0 \le k \le r$, $M(t) = M(t, 1)$, $L(t) = L(t, 1)$, $R(t) = R(t, 1)$. If $A_k(t) = 0$,

$B_k(t) = 0$, $1 \le k \le r$, $t \in \mathcal{I}$, the equation (4.2) becomes the well-known matrix Riccati differential equation intensively investigated in connection with various types of control problems in the deterministic framework.

In this book the system (4.1) and its particular form (4.2) will be called the SGRDE. The system of differential equations (4.1) will be written in compact form as a nonlinear differential equation on the space $\mathcal{S}_n^d$. To this end we make the following *convention of notation*: if $C \in \mathcal{M}_{p,n}^d$, $B \in \mathcal{M}_{n,m}^d$, $C = (C(1), C(2), \ldots, C(d))$, $B = (B(1), B(2), \ldots, B(d))$, then by $D = CB$ we understand the following element of $\mathcal{M}_{p,m}^d$, $D = (D(1), D(2), \ldots, D(d))$, $D(i) = C(i)B(i)$, $i \in \mathcal{D}$. If $A \in \mathcal{M}_n^d$, $A = (A(1), A(2), \ldots, A(d))$, by A^{-1} we denote the element of $\mathcal{M}_n^d$ defined as follows: $A^{-1} = (A^{-1}(1), A^{-1}(2), \ldots, A^{-1}(d))$ if all matrices $A(i), i \in \mathcal{D}$, are invertible. If $B \in \mathcal{M}_{n,m}^d$, $B = (B(1), B(2), \ldots, B(d))$, then $B^* \in \mathcal{M}_{m,n}^d$, and it is defined by $B^* = (B^*(1), B^*(2), \ldots, B^*(d))$.

With these conventions the system (4.1) can be written as

$$\frac{d}{dt} X(t) + \mathcal{L}^*(t)X(t) - \mathcal{P}^*(t, X(t))\mathcal{R}^{-1}(t, X(t))\mathcal{P}(t, X(t)) + M(t) = 0, \quad (4.3)$$

$\mathcal{L}^*(t)$ being the adjoint operator of $\mathcal{L}(t)$ defined as in (2.8):

$$X \to \mathcal{P}(t, X) : \mathcal{S}_n^d \to \mathcal{M}_{m,n}^d,$$

$$\mathcal{P}(t, X) = (\mathcal{P}_1(t, X), \mathcal{P}_2(t, X), \ldots, \mathcal{P}_d(t, X)),$$

$$\mathcal{P}_i(t, X) = B_0^*(t, i)X(i) + \sum_{k=1}^{r} B_k^*(t, i)X(i)A_k(t, i) + L^*(t, i),$$

$$X \to \mathcal{R}(t, X) : \mathcal{S}_n^d \to \mathcal{S}_m^d \text{ by,}$$

$$\mathcal{R}(t, X) = (\mathcal{R}_1(t, X), \mathcal{R}_2(t, X), \ldots, \mathcal{R}_d(t, X)),$$

$$\mathcal{R}_i(t, X) = R(t, i) + \sum_{k=1}^{r} B_k^*(t, i)X(i)B_k(t, i),$$

$$M(t) = (M(t, 1), M(t, 2), \ldots, M(t, d)) \in \mathcal{S}_n^d.$$

If the coefficients of (4.1) do not depend on t, then the operators $\mathcal{L}, \mathcal{P}, \mathcal{R}$ do not depend on t. In this case we shall use the following algebraic nonlinear equation over $\mathcal{S}_n^d$:

$$\mathcal{L}^* X - \mathcal{P}^*(X)\mathcal{R}^{-1}(X)\mathcal{P}(X) + M = 0. \quad (4.4)$$

Let us remark that equation (4.3) is defined on the set

$$\Gamma = \{(t, X) \in \mathcal{I} \times \mathcal{S}_n^d \mid \det \mathcal{R}_i(t, X) \neq 0, \ \forall i \in \mathcal{D}\}.$$

Definition 1. *A C^1 function $X : \mathcal{I}_1 \to \mathcal{S}_n^d$ ($\mathcal{I}_1 \subseteq \mathcal{I}$ being an interval). $X(t) = (X(t, 1), \ldots, X(t, d))$ is said to be a solution of the equation (4.3) if for every $t \in \mathcal{I}_1$ and $i \in \mathcal{D}$ the matrix $\mathcal{R}_i(t, X(t))$ is invertible and the relations (4.1) hold for all $t \in \mathcal{I}_1$ and $i \in \mathcal{D}$.*

As we can see, SGRDE (4.3) is associated to a quadruple $\Sigma = (\mathbf{A}, \mathbf{B}, \mathcal{V}, \mathcal{Q})$ where, as usual, $\mathbf{A} = (A_0, A_1, \ldots, A_r)$, $\mathbf{B} = (B_0, B_1, \ldots, B_r)$, $\mathcal{V}: \mathcal{I} \rightarrow \mathcal{S}_{n+m}^d$, $\mathcal{V}(t) = (\mathcal{V}(t, 1), \ldots, \mathcal{V}(t, d))$,

$$\mathcal{V}(t, i) = \begin{bmatrix} M(t, i) & L(t, i) \\ L^*(t, i) & R(t, i) \end{bmatrix}. \tag{4.5}$$

If $X: \mathcal{I} \rightarrow \mathcal{S}_n^d$ is a C^1 function we denote

$$\mathcal{N}_i(t, X(t)) = \begin{bmatrix} \dfrac{d}{dt} X(t, i) + \mathcal{L}_i^*(t)(X(t)) + M(t, i) & \mathcal{P}_i^*(t, X(t)) \\ \mathcal{P}_i(t, X(t)) & \mathcal{R}_i(t, X(t)) \end{bmatrix},$$

which will be called the *dissipation matrix*, where

$$\mathcal{L}_i^*(t)(X(t)) = \left(\mathcal{L}^*(t)X(t)\right)(i),$$

$\mathcal{L}^*(t)$ being the adjoint operator associated with the Lyapunov operator as in Section 2.2, and $\mathcal{P}_i$, $\mathcal{R}_i$ are defined above related to equation (4.3). We shall also denote

$$\mathcal{N}(t, X(t)) = (\mathcal{N}_1(t, X(t)), \ldots, \mathcal{N}_d(t, X(t))) \in \mathcal{S}_{n+m}^d.$$

To a quadruple $\Sigma = (\mathbf{A}, \mathbf{B}, \mathcal{V}, \mathcal{Q})$ we associate the following two sets of C^1 functions, which will play an important role in subsequent developments:

$$\mathbf{\Gamma}^\Sigma = \left\{ \widehat{X} \in C_b^1(\mathcal{I}, \mathcal{S}_n^d) | \, \mathcal{N}_i\!\left(t, \widehat{X}(t)\right) \geq 0, \, \mathcal{R}_i\!\left(t, \widehat{X}(t)\right) \gg 0, \, t \in \mathcal{I}, i \in \mathcal{D} \right\} \tag{4.6}$$

and

$$\widetilde{\mathbf{\Gamma}}^\Sigma = \left\{ \widehat{X} \in C_b^1(\mathcal{I}, \mathcal{S}_n^d) | \, \mathcal{N}_i\!\left(t, \widehat{X}(t)\right) \gg 0, \, t \in \mathcal{I}, \, i \in \mathcal{D} \right\}, \tag{4.7}$$

where $C_b^1(\mathcal{I}, \mathcal{S}_n^d) = \left\{ X \in C^1(\mathcal{I}, \mathcal{S}_n) | \, X, \frac{d}{dt} X \text{ are bounded functions} \right\}$. It is obvious that $\mathbf{\Gamma}^\Sigma \supset \widetilde{\mathbf{\Gamma}}^\Sigma$. One can also see that the set $\mathbf{\Gamma}^\Sigma$ contains all bounded solutions $X: \mathcal{I} \rightarrow \mathcal{S}_n^d$ of SGRDE (4.1), which verify the condition

$$\mathcal{R}_i\!\left(t, \widehat{X}(t)\right) \gg 0, \, t \in \mathcal{I}, i \in \mathcal{D}. \tag{4.8}$$

Remark 1. With the exception of some particular cases that will be discussed later, we do not make any assumption concerning the signature of the matrices $\mathcal{V}(t, i)$ in (4.5) and $R(t, i)$.

As we shall see in subsequent developments, the sign of the expression plays an important role in the characterization of SGRDE (4.1):

$$\mathcal{R}_i(t, X(t)) = R(t, i) + \sum_{k=1}^{r} B_k^*(t, i) X(t, i) B_k(t, i).$$

Then in this chapter we consider only the case $\mathcal{R}_i(t, X(t)) > 0$, since this is the case required by the quadratic optimization problem. In Chapter 6 the case $\mathcal{R}_i(t, X(t)) < 0$ will be considered in connection with some Bounded Real Lemma–type results.

At the end of this section we prove an auxiliary result that will used several times in the following developments of this chapter.

Lemma 1. (i) *If* $X(t) = (X(t, 1), \ldots, X(t, d))$ *is a solution of equation* (4.3), *then* $t \mapsto X(t)$ *solves the following equation on* $\mathcal{S}_n^d$:

$$\frac{d}{dt} X(t) + \mathcal{L}_G^*(t) X(t)$$
$$- (F(t) - G(t))^* \mathcal{R}(t, X(t)) (F(t) - G(t)) + M_G(t) = 0 \qquad (4.9)$$

for an arbitrary $G : \mathcal{I} \to \mathcal{M}_{m,n}^d$, *where* $\mathcal{L}_G^*(t)$ *is the adjoint operator of the operator* $\mathcal{L}_G(t) : \mathcal{S}_n^d \to \mathcal{S}_n^d$ *defined as in* (3.2) *and*

$$F(t) = (F(t, 1), \ldots, F(t, d)), \text{ with}$$
$$F(t, i) = -\mathcal{R}_i^{-1}(t, X(t))\mathcal{P}_i(t, X(t)),$$
$$M_G(t) = (M_G(t, 1), \ldots, M_G(t, d)) \text{ with}$$
$$M_G(t, i) = M(t, i) + L(t, i)G(t, i) + G^*(t, i)L^*(t, i) \qquad (4.10)$$
$$+ G^*(t, i)R(t, i)G(t, i),$$

$t \in \mathbf{R}_+$, $i \in \mathcal{D}$.

(ii) *If* $X : \mathcal{I} \to \mathcal{S}_n^d$ *is a solution of* (4.3), *then* $X(t)$ *solves the following Lyapunov-type equation:*

$$\frac{d}{dt} X(t) + \mathcal{L}_F^*(t) X(t) + M_F(t) = 0.$$

(iii) $X(t) : \mathcal{I} \to \mathcal{S}_n^d$ *is a solution of the SGRDE* (4.3) *if and only if* $X(t)$ *is a solution of the following modified SGRDE:*

$$\frac{d}{dt} X(t) + \mathcal{L}_G^*(t) X(t) - \mathcal{P}_G^*(t, X(t)) \mathcal{R}^{-1}(t, X(t)) \mathcal{P}_G(t, X(t)) + M_G(t) = 0$$
$$(4.11)$$

for arbitrary $G : \mathcal{I} \mapsto \mathcal{M}_{m,n}^d$, *where* $X \mapsto \mathcal{P}_G(t, X) : \mathcal{S}_n^d \mapsto \mathcal{M}_{m,n}^d$, *by*

$$\mathcal{P}_G(t, X) = (\mathcal{P}_{G,1}(t, X), \ldots, \mathcal{P}_{G,d}(t, X))$$

with

$$\mathcal{P}_{G,i}(t, X) = B_0^*(t, i) X(i) + \sum_{k=1}^r B_k^*(t, i) X(i)(A_k(t, i) + B_k(t, i)G(t, i))$$
$$+ L^*(t, i) + R(t, i)G(t, i).$$

Proof. (i) It is easy to check that $X(t) = (X(t, 1), \ldots, X(t, d))$ is a solution of the SGRDE (4.3) if and only if $(X(t), F(t))$ solves the system

$$\mathcal{N}_i(t, X(t)) \begin{bmatrix} I_n \\ F(t, i) \end{bmatrix} = 0, \quad i \in \mathcal{D}. \qquad (4.12)$$

Taking into account that the matrix

$$\begin{bmatrix} I_n & F^*(t, i) \\ 0 & I_m \end{bmatrix}$$

is invertible it follows that (4.12) is equivalent to

$$\begin{bmatrix} I_n & F^*(t,i) \\ 0 & I_m \end{bmatrix} \mathcal{N}_i(t, X(t)) \begin{bmatrix} I_n \\ F(t,i) \end{bmatrix} = 0,$$

or equivalently

$$\begin{bmatrix} I_n & G^*(t,i) + (F(t,i) - G(t,i))^* \\ 0 & I_m \end{bmatrix} \mathcal{N}_i(t, X(t))$$

$$\times \begin{bmatrix} I_n \\ G(t,i) + F(t,i) - G(t,i) \end{bmatrix} = 0,$$

from which by direct calculations one obtains (4.9).

(ii) directly follows, taking in (4.9) $G(t,i) = F(t,i)$.

(iii) follows from (i), taking into account that

$$F(t,i) - G(t,i) = -\mathcal{R}_i^{-1}(t, X(t))(\mathcal{P}_i(t, X(t)) + \mathcal{R}_i(t, X(t))G(t,i))$$
$$= -\mathcal{R}_i^{-1}(t, X(t))\mathcal{P}_{G,i}(t, X(t)),$$

and hence the proof is complete.   □

4.2 The maximal solution of SGRDE

In the following developments one will frequently use the next well-known result in connection with *Schur complements* (see, e.g., [9]).

Lemma 2. *Consider the symmetric matrix*

$$M = \begin{bmatrix} M_{11} & M_{12} \\ M_{12}^* & M_{22} \end{bmatrix},$$

where $M_{22} > 0$. Then the following are equivalent:

(i) $M \geq 0$, $(M > 0)$;

(ii) $M_{11} - M_{12}M_{22}^{-1}M_{12}^* \geq 0$ $(M_{11} - M_{12}M_{22}^{-1}M_{12}^* > 0)$.

With the notations from the previous section we introduce the following.

Definition 2. *We say that a solution $\widetilde{X} : \mathcal{I} \to \mathcal{S}_n^d$ of the SGRDE (4.1) is a maximal solution with respect to the set $\mathbf{\Gamma}^\Sigma$, or the maximal solution for short, if $\widetilde{X}(t) \geq \widehat{X}(t)$ for arbitrary $\widehat{X}(\cdot) \in \mathbf{\Gamma}^\Sigma$.*

Theorem 3. *Assume that $(\mathbf{A}, \mathbf{B}; Q)$ is stabilizable. Then the following are equivalent:*

(i) *The set $\mathbf{\Gamma}^\Sigma$ is not empty.*

(ii) *The SGRDE (4.1) has a bounded maximal solution $\widetilde{X} : \mathcal{I} \to \mathcal{S}_n^d$ which verifies (4.8).*

Moreover, if the coefficients of the system (4.1) are θ-periodic functions, then the maximal solution $\widetilde{X}(t)$ is a θ-periodic function too. If the coefficients of the system (4.1) do not depend upon t, then the maximal solution $\widetilde{X}(t)$ is constant and it solves (4.4).

Proof. (ii) ⟹ (i) is obvious since if the SGRDE (4.1) has a maximal solution $\widetilde{X}(t)$ verifying (4.8), then $\widetilde{X}(\cdot) \in \mathbf{\Gamma}^\Sigma$.

(i) $\Rightarrow$ (ii) Since $(\mathbf{A}, \mathbf{B}; Q)$ is stabilizable there exists a feedback gain $\widetilde{F}: \mathcal{I} \rightarrow$ $\mathcal{M}_{m,n}^d$ bounded and continuous function such that the system $(A_0 + B_0\widetilde{F}, A_1 + B_1\widetilde{F}, \ldots, A_r + B_r\widetilde{F}; Q)$ is stable. Let $\widehat{X}(\cdot) \in \Gamma^{\Sigma}$. Then by a Schur complement argument, $\widehat{X}(t)$ is a solution of the following differential inequation on $\mathcal{S}_n^d$:

$$\frac{d}{dt}X(t) + \mathcal{L}^*(t)X(t) - \mathcal{P}^*(t, X(t))\mathcal{R}^{-1}(t, X(t))\mathcal{P}(t, X(t)) + M(t) \geq 0. \quad (4.13)$$

Set

$$\widehat{M}(t) = \mathcal{P}^*(t, \widehat{X}(t))\mathcal{R}^{-1}(t, \widehat{X}(t))\mathcal{P}(t, \widehat{X}(t)) - M(t) - \mathcal{L}^*(t)\widehat{X}(t) - \frac{d}{dt}\widehat{X}(t).$$

Obviously $\widehat{M}(t) \leq 0, t \in \mathcal{I}$ and

$$\frac{d}{dt}\widehat{X}(t, i) + A_0^*(t, i)\widehat{X}(t, i) + \widehat{X}(t, i)A_0(t, i)$$

$$+ \sum_{k=1}^{r} A_k^*(t, i)\widehat{X}(t, i)A_k(t, i) + \sum_{j=1}^{d} q_{ij}\widehat{X}(t, j)$$

$$- \left(\widehat{X}(t, i)B_0(t, i) + \sum_{k=1}^{r} A_k^*(t, i)\widehat{X}(t, i)B_k(t, i) + L(t, i)\right)$$

$$\times \left(R(t, i) + \sum_{k=1}^{r} B_k^*(t, i)\widehat{X}(t, i)B_k(t, i)\right)^{-1}$$

$$\times \left(B_0^*(t, i)\widehat{X}(t, i) + \sum_{k=1}^{r} B_k^*(t, i)\widehat{X}(t, i)A_k(t, i) + L^*(t, i)\right)$$

$$+ M(t, i) + \widehat{M}(t, i) = 0, \quad (4.14)$$

$i \in \mathcal{D}, t \in \mathcal{I}$.

Let $\varepsilon > 0$ be fixed and we define (see Proposition 14 of Chapter 2) $X_0^{\varepsilon}(t) = (X_0^{\varepsilon}(t, 1), \ldots, X_0^{\varepsilon}(t, d))$ as the unique bounded solution of the system of linear equations

$$\frac{d}{dt}X(t, i) + \left(A_0(t, i) + B_0(t, i)\widetilde{F}(t, i)\right)^*X(t, i) + X(t, i)(A_0(t, i)$$

$$+ B_0(t, i)\widetilde{F}(t, i)) + \sum_{k=1}^{r} \left(A_k(t, i) + B_k(t, i)\widetilde{F}(t, i)\right)^*X(t, i)\left(A_k(t, i)\right.$$

$$+ B_k(t, i)\widetilde{F}(t, i)) + \sum_{j=1}^{d} q_{ij}X(t, j) + \widetilde{F}^*(t, i)R(t, i)F(t, i)$$

$$+ \widetilde{F}^*(t, i)L^*(t, i) + L(t, i)\widetilde{F}(t, i) + M(t, i) + \varepsilon I_n = 0, \quad (4.15)$$

$i \in \mathcal{D}, t \in \mathcal{I}$. We show that there exists $\mu > 0$ such that

$$X_0^{\varepsilon}(t, i) - \widehat{X}(t, i) \geq \mu I_n$$

for all $(t, i) \in \mathcal{I} \times \mathcal{D}$.

Indeed, by Lemma 1 the system (4.14) will be written

$$\frac{d}{dt}\widehat{X}(t,i) + \big(A_0(t,i) + B_0(t,i)\widetilde{F}(t,i)\big)^*\widehat{X}(t,i) + \widehat{X}(t,i)\big(A_0(t,i)$$

$$+ B_0(t,i)\widetilde{F}(t,i)\big) + \sum_{k=1}^{r}\big(A_k(t,i) + B_k(t,i)\widetilde{F}(t,i)\big)^*\widehat{X}(t,i)\big(A_k(t,i)$$

$$+ B_k(t,i)\widetilde{F}(t,i)\big) + \sum_{j=1}^{d} q_{ij}\widehat{X}(t,j) + M(t,i) + \widehat{M}(t,i) \tag{4.16}$$

$$+ \widetilde{F}^*(t,i)L^*(t,i) + L(t,i)\widetilde{F}(t,i) + \widetilde{F}^*(t,i)R(t,i)\widetilde{F}(t,i)$$

$$- \big(\widetilde{F}(t,i) - \widehat{F}(t,i)\big)^*\left(R(t,i) + \sum_{k=1}^{r} B_k^*(t,i)\widehat{X}(t,i)B_k(t,i)\right)$$

$$\times \big(\widetilde{F}(t,i) - \widehat{F}(t,i)\big) = 0, \quad i \in \mathcal{D},$$

where $\widehat{F}(t,i) = -\mathcal{R}_i^{-1}(t,\widehat{X}(t))\mathcal{P}(t,\widehat{X}(t))(i)$, with $i \in \mathcal{D}$ and $t \in \mathcal{I}$. Subtracting (4.16) from (4.15), we obtain

$$\frac{d}{dt}\big[X_0^{\varepsilon}(t,i) - \widehat{X}(t,i)\big] + \big[A_0(t,i) + B_0(t,i)\widetilde{F}(t,i)\big]^*\big[X_0^{\varepsilon}(t,i) - \widehat{X}(t,i)\big]$$

$$+ \big[X_0^{\varepsilon}(t,i) - \widehat{X}(t,i)\big]\big[A_0(t,i) + B_0(t,i)\widetilde{F}(t,i)\big]$$

$$+ \sum_{k=1}^{r}\big[A_k(t,i) + B_k(t,i)\widetilde{F}(t,i)\big]^*\big[X_0^{\varepsilon}(t,i) - \widehat{X}(t,i)\big]$$

$$\times \big[A_k(t,i) + B_k(t,i)\widetilde{F}(t,i)\big] + \sum_{j=1}^{d} q_{ij}\big[X_0^{\varepsilon}(t,j) - \widehat{X}(t,j)\big] + \varepsilon I_n - \widehat{M}(t,i)$$

$$+ \big(\widetilde{F}(t,i) - \widehat{F}(t,i)\big)\mathcal{R}_i\big(t,\widehat{X}(t)\big)\big(\widetilde{F}(t,i) - \widehat{F}(t,i)\big) = 0, \ i \in \mathcal{D}, t \in \mathcal{I},$$

which leads to the fact that $t \to X_0^{\varepsilon}(t) - \widehat{X}(t)$ verifies the following linear differential equation on $\mathcal{S}_n^d$:

$$\frac{d}{dt}\big[X_0^{\varepsilon}(t) - \widehat{X}(t)\big] + \mathcal{L}_{\widetilde{F}}^*(t)\big[X_0^{\varepsilon}(t) - \widehat{X}(t)\big] + \varepsilon J^d + \Delta_0(t) = 0, \quad t \in \mathcal{I}, \tag{4.17}$$

where

$$\Delta_0(t) = (\Delta_0(t,1), \ldots, \Delta_0(t,d)),$$

$$\Delta_0(t,i) = \big(\widetilde{F}(t,i) - \widehat{F}(t,i)\big)^*\mathcal{R}_i\big(t,\widehat{X}(t)\big)\big(\widetilde{F}(t,i) - \widehat{F}(t,i)\big) - \widehat{M}(t,i) \geq 0,$$

$i \in \mathcal{D}, t \in \mathcal{I}$. Since $X_0^{\varepsilon}(t) - \widehat{X}(t), t \in \mathcal{I}$, is a bounded function, $\widetilde{F}$ is a stabilizing feedback gain and $\varepsilon J^d + \Delta_0(t) \gg 0$; then by Theorem 14 of this chapter and Theorem 12(i)–(vi) of Chapter 2 it follows that it exists $\mu > 0$ such that

$$X_0^{\varepsilon}(t) - \widehat{X}(t) \geq \mu J^d, \quad \forall t \in \mathcal{I}. \tag{4.18}$$

Combining (4.18) with (4.8) we conclude that $\mathcal{R}_i(t, X_0^{\varepsilon}(t)) \geq \nu I_n > 0 \ \forall t \in \mathcal{I},$ $i \in \mathcal{D}$ for some positive constant ν.

Set $F_0^\varepsilon(t, i) = -\mathcal{R}_i^{-1}(t, X_0^\varepsilon(t))\mathcal{P}_i(t, X_0^\varepsilon(t))$, with $t \in \mathcal{I}$ and $i \in \mathcal{D}$. We prove that $F_0^\varepsilon(t) = (F_0^\varepsilon(t, 1), \ F_0^\varepsilon(t, 2), \ldots, F_0^\varepsilon(t, d))$ is a stabilizing feedback gain.

We rewrite the systems (4.15) and (4.14) as

$$
\frac{d}{dt} X_0^\varepsilon(t, i) + \left[A_0(t, i) + B_0(t, i)F_0^\varepsilon(t, i)\right]^* X_0^\varepsilon(t, i)
$$
$$
+ X_0^\varepsilon(t, i)\left[A_0(t, i) + B_0(t, i)F_0^\varepsilon(t, i)\right]
$$
$$
+ \sum_{k=1}^{r} \left[(A_k(t, i) + B_k(t, i)F_0^\varepsilon(t, i)\right]^* X_0^\varepsilon(t, i)\left[A_k(t, i) + B_k(t, i)F_0^\varepsilon(t, i)\right]
$$
$$
+ \sum_{j=1}^{d} q_{ij} X_0^\varepsilon(t, j) + M(t, i) + \varepsilon I_n + (F_0^\varepsilon(t, i))^* R(t, i)F_0^\varepsilon(t, i)
$$
$$
+ L(t, i)F_0^\varepsilon(t, i) + (F_0^\varepsilon(t, i))^* L^*(t, i) + \left[F_0^\varepsilon(t, i) - \tilde{F}(t, i)\right]^*
$$
$$
\times \mathcal{R}_i(t, X_0^\varepsilon(t))\left[F_0^\varepsilon(t, i) - \tilde{F}(t, i)\right] = 0,
$$

$t \in \mathcal{I}, i \in \mathcal{D},$

$$
\frac{d}{dt} \widehat{X}(t, i) + \left[A_0(t, i) + B_0(t, i)F_0^\varepsilon(t, i)\right]^* \widehat{X}(t, i)
$$
$$
+ \widehat{X}(t, i)\left[A_0(t, i) + B_0(t, i)F_0^\varepsilon(t, i)\right]
$$
$$
+ \sum_{k=1}^{r} \left[A_k(t, i) + B_k(t, i)F_0^\varepsilon(t, i)\right]^* \widehat{X}(t, i)\left[A_k(t, i) + B_k(t, i)F_0^\varepsilon(t, i)\right]
$$
$$
+ \sum_{j=1}^{d} q_{ij} \widehat{X}(t, j) + M(t, i) + \widehat{M}(t, i) + (F_0^\varepsilon(t, i))^* R(t, i)F_0^\varepsilon(t, i)
$$
$$
+ L(t, i)F_0^\varepsilon(t, i) + (F_0^\varepsilon(t, i))^* L^*(t, i)
$$
$$
- \left[F_0^\varepsilon(t, i) - \widehat{F}(t, i)\right]^* \mathcal{R}_i(t, X_0^\varepsilon(t))\left[F_0^\varepsilon(t, i) - \widehat{F}(t, i)\right] = 0.
$$

We get

$$
\frac{d}{dt} \left[X_0^\varepsilon(t, i) - \widehat{X}(t, i)\right] + \left[A_0(t, i) + B_0(t, i)F_0^\varepsilon(t, i)\right]^* \left[X_0^\varepsilon(t, i) - \widehat{X}(t, i)\right]
$$
$$
+ \left[X_0^\varepsilon(t, i) - \widehat{X}(t, i)\right]\left[A_0(t, i) + B_0(t, i)F_0^\varepsilon(t, i)\right]
$$
$$
+ \sum_{k=1}^{r} \left[A_k(t, i) + B_k(t, i)F_0^\varepsilon(t, i)\right]^* \left[X_0^\varepsilon(t, i) - \widehat{X}(t, i)\right]
$$
$$
\times \left[A_k(t, i) + B_k(t, i)F_0^\varepsilon(t, i)\right] + \sum_{j=1}^{d} q_{ij}\left[X_0^\varepsilon(t, j) - \widehat{X}(t, j)\right]
$$
$$
+ \varepsilon I_n - \widehat{M}(t, i) + \left[F_0^\varepsilon(t, i) - \tilde{F}(t, i)\right]^* \mathcal{R}_i(t, X_0^\varepsilon(t))\left[F_0^\varepsilon(t, i) - \tilde{F}(t, i)\right]
$$
$$
+ \left[F_0^\varepsilon(t, i) - \widehat{F}(t, i)\right]^* \mathcal{R}_i(t, X_0^\varepsilon(t))\left[F_0^\varepsilon(t, i) - \widehat{F}(t, i)\right] = 0, i \in \mathcal{D}, t \in \mathcal{I}.
$$

$$
(4.19)
$$

From (4.18) and (4.19) we deduce that $t \to X_0^{\varepsilon}(t) - \widehat{X}(t)$ is the bounded and uniform positive solution of the differential inequality on $\mathcal{S}_n^d$:

$$\frac{d}{dt}X(t) + \mathcal{L}_{F_0^{\varepsilon}}^*(t)X(t) + \varepsilon J^d \leq 0.$$

Applying Theorem 12(viii)–(i) of Chapter 2 we deduce that the system $(A_0 + B_0F_0^{\varepsilon}, A_1 + B_1F_0^{\varepsilon}, \ldots, A_r + B_rF_0^{\varepsilon}; Q)$ is stable.

Using $(X_0^{\varepsilon}(t), F_0^{\varepsilon}(t))$ as an initial step we shall construct iteratively bounded functions $X_p^{\varepsilon}(t) = (X_p^{\varepsilon}(t, 1), \ldots, X_p^{\varepsilon}(t, d))$, $F_p^{\varepsilon}(t) = (F_p^{\varepsilon}(t, 1), \ldots, F_p^{\varepsilon}(t, d))$, $p = 0, 1, 2 \ldots$ with the following properties:

(a)

$$X_p^{\varepsilon}(t) \gg \widehat{X}(t), \quad t \in \mathcal{I};$$

(b) the system $(A_0 + B_0F_p^{\varepsilon}, A_1 + B_1F_p^{\varepsilon}, \ldots, A_r + B_rF_p^{\varepsilon}; Q)$ is stable for all $p = 0, 1, 2 \ldots$;

(c)

$$X_{p-1}^{\varepsilon}(t) \geq X_p^{\varepsilon}(t), \quad t \in \mathcal{I}.$$

If the system $(A_0 + B_0F_{p-1}^{\varepsilon}, A_1 + B_1F_{p-1}^{\varepsilon}, \ldots, A_r + B_rF_{p-1}^{\varepsilon}; Q)$ is stable we construct (see Proposition 14 of Chapter 2) $X_p^{\varepsilon}(t) = (X_p^{\varepsilon}(t, 1), \ldots, X_p^{\varepsilon}(t, d))$ as the unique bounded-on-$\mathcal{I}$ solution of the following system of linear differential equations:

$$\frac{d}{dt}X_p^{\varepsilon}(t, i) + \left[A_0(t, i) + B_0(t, i)F_{p-1}^{\varepsilon}(t, i)\right]^* X_p^{\varepsilon}(t, i)$$

$$+ X_p^{\varepsilon}(t, i)\left[A_0(t, i) + B_0(t, i)F_{p-1}^{\varepsilon}(t, i)\right] + \sum_{k=1}^{r}\left[A_k(t, i) + B_k(t, i)F_{p-1}^{\varepsilon}(t, i)\right]^*$$

$$\times X_p^{\varepsilon}(t, i)\left[A_k(t, i) + B_k(t, i)F_{p-1}^{\varepsilon}(t, i)\right] + \sum_{j=1}^{d}q_{ij}X_p^{\varepsilon}(t, j) + M(t, i) + \varepsilon I_n$$

$$+ \left(F_{p-1}^{\varepsilon}(t, i)\right)^* R(t, i)F_{p-1}^{\varepsilon}(t, i) + L(t, i)F_{p-1}^{\varepsilon}(t, i) + \left(F_{p-1}^{\varepsilon}(t, i)\right)^* L^*(t, i) = 0,$$
$$(4.20)$$

$t \in \mathcal{I}, i \in \mathcal{D}$. We show that $X_p^{\varepsilon}(t) - \widehat{X}(t) \geq \mu_p J_n \; \forall t \in \mathcal{I}$ for positive constant μ_p. By Lemma 1, the system (4.14) may be rewritten as

$$\frac{d}{dt}\widehat{X}(t, i) + \left[A_0(t, i) + B_0(t, i)F_{p-1}^{\varepsilon}(t, i)\right]^*\widehat{X}(t, i) + \widehat{X}(t, i)\big(A_0(t, i)$$

$$+ B_0(t, i)F_{p-1}^{\varepsilon}(t, i)\big) + \sum_{k=1}^{r}\left[A_k(t, i) + B_k(t, i)F_{p-1}^{\varepsilon}(t, i)\right]^*\widehat{X}(t, i)$$

$$\times \left[A_k(t, i) + B_k(t, i)F_{p-1}^{\varepsilon}(t, i)\right] + \sum_{j=1}^{d}q_{ij}\widehat{X}(t, j) + M(t, i) + \widehat{M}(t, i)$$

$$+ \left(F_{p-1}^{\varepsilon}(t, i)\right)^* R(t, i)F_{p-1}^{\varepsilon}(t, i) + L(t, i)F_{p-1}^{\varepsilon}(t, i) + \left(F_{p-1}^{\varepsilon}(t, i)\right)^* L^*(t, i)$$

$$- \left(F_{p-1}^{\varepsilon}(t, i) - \widehat{F}(t, i)\right)R_i\left(t, \widehat{X}(t)\right)\left(F_{p-1}^{\varepsilon}(t, i) - \widehat{F}(t, i)\right) = 0. \qquad (4.21)$$

Subtracting (4.21) from (4.20) we get

$$\frac{d}{dt}\big[X_p^\varepsilon(t,i) - \widehat{X}(t,i)\big] + \big[A_0(t,i) + B_0(t,i)F_{p-1}^\varepsilon(t,i)\big]^*$$
$$\times \big[X_p^\varepsilon(t,i) - \widehat{X}(t,i)\big] + \big(X_p^\varepsilon(t,i) - \widehat{X}(t,i)\big)\big[A_0(t,i) + B_0(t,i)F_{p-1}^\varepsilon(t,i)\big]$$
$$+ \sum_{k=1}^r \big(A_k(t,i) + B_k(t,i)F_{p-1}^\varepsilon(t,i)\big)^*\big(X_p^\varepsilon(t,i) - \widehat{X}(t,i)\big)$$
$$\times \big(A_k(t,i) + B_k(t,i)F_{p-1}^\varepsilon(t,i)\big) + \sum_{j=1}^d q_{ij}\big[X_p^\varepsilon(t,j) - \widehat{X}(t,j)\big] + \varepsilon I_n - \widehat{M}(t,i)$$
$$+ \big[F_{p-1}^\varepsilon(t,i) - \widehat{F}(t,i)\big]^*\mathcal{R}_i\big(t,\widehat{X}(t)\big)\big(F_{p-1}^\varepsilon(t,i) - \widehat{F}(t,i)\big) = 0,$$

$i \in \mathcal{D}, t \in \mathcal{I}$.

Hence $t \to X_p^\varepsilon(t) - \widehat{X}(t)$ is a bounded-on-$\mathcal{I}$ solution of the linear equation on $\mathcal{S}_n^d$:

$$\frac{d}{dt}X(t) + \mathcal{L}_{F_{p-1}^\varepsilon}^*(t)X(t) + \varepsilon J^d + \Delta_{p-1}(t) = 0, \qquad (4.22)$$

where

$$\Delta_{p-1}(t) = (\Delta_{p-1}(t,1), \ldots, \Delta_{p-1}(t,d)),$$
$$\Delta_{p-1}(t,i) = -\widehat{M}(t,i) + \big(F_{p-1}^\varepsilon(t,i) - \widehat{F}(t,i)\big)^*\mathcal{R}_i\big(t,\widehat{X}(t)\big)\big(F_{p-1}^\varepsilon(t,i) - \widehat{F}(t,i)\big);$$

$\Delta_{p-1}(t,i) \geq 0\ \forall i \in \mathcal{D},\ t \in \mathcal{I}$.

Since $X_p^\varepsilon(t) - \widehat{X}(t),\ t \in \mathcal{I}$, is a bounded function, F_{p-1}^ε is a stabilizing feedback gain and $\varepsilon J^d + \Delta_{p-1}(t) \gg 0,\ t \in \mathcal{I}$, based on Theorems 14 and 12 of Chapter 2, we conclude that there exists $\gamma > 0$ (possible depending upon p) such that

$$X_p^\varepsilon(t) - \widehat{X}(t) \geq \gamma J^d \qquad (4.23)$$

$\forall t \in \mathcal{I}$.

Therefore we showed that $X_p(t)$ satisfies condition (a).

From (4.23) and (4.8) it follows that

$$\mathcal{R}_i\big(t, X_p^\varepsilon(t)\big) \geq \hat{\gamma} J^d$$

$\forall t \in \mathcal{I}$, for some $\hat{\gamma} > 0$.

Define $F_p^\varepsilon(t) = \big(F_p^\varepsilon(t,1), \ldots, F_p^\varepsilon(t,d)\big)$ by

$$F_p^\varepsilon(t,i) = -\mathcal{R}_i^{-1}\big(t, X_p^\varepsilon(t)\big)\mathcal{P}_i\big(t, X_p^\varepsilon(t)\big).$$

Then $F_p^\varepsilon(t)$ is a stabilizing feedback gain. Indeed, we have to check that the system $\big(A_0 + B_0 F_p^\varepsilon, \ldots, A_r + B_r F_p^\varepsilon; Q\big)$ is stable. To this end we rewrite the systems (4.20) and (4.21):

$$\frac{d}{dt}X_p^\varepsilon(t,i) + \left[A_0(t,i) + B_0(t,i)F_p^\varepsilon(t,i)\right]^* X_p^\varepsilon(t,i)$$
$$+ X_p^\varepsilon(t,i)\left[A_0(t,i) + B_0(t,i)F_p^\varepsilon(t,i)\right]$$
$$+ \sum_{k=1}^{r}\left[A_k(t,i) + B_k(t,i)F_p^\varepsilon(t,i)\right]^* X_p^\varepsilon(t,i)\left[A_k(t,i) + B_k(t,i)F_p^\varepsilon(t,i)\right]$$
$$+ \sum_{j=1}^{d}q_{ij}X_p^\varepsilon(t,j) + \varepsilon I_n + M(t,i) + \left(F_p^\varepsilon(t,i)\right)^* R(t,i)F_p^\varepsilon(t,i)$$
$$+ L(t,i)F_p^\varepsilon(t,i) + \left(F_p^\varepsilon(t,i)\right)^* L^*(t,i)$$
$$+ \left(F_p^\varepsilon(t,i) - F_{p-1}^\varepsilon(t,i)\right)^* \mathcal{R}_i\left(t, X_p^\varepsilon(t)\right)\left(F_p^\varepsilon(t,i) - F_{p-1}^\varepsilon(t,i)\right) = 0,$$

$i \in \mathcal{D}, t \in \mathcal{I}.$

$$\frac{d}{dt}\widehat{X}(t,i) + \left[A_0(t,i) + B_0(t,i)F_p^\varepsilon(t,i)\right]^* \widehat{X}(t,i)$$
$$+ \widehat{X}(t,i)\left[A_0(t,i) + B_0(t,i)F_p^\varepsilon(t,i)\right]$$
$$+ \sum_{k=1}^{r}\left[A_k(t,i) + B_k(t,i)F_p^\varepsilon(t,i)\right]^* \widehat{X}(t,i)\left[A_k(t,i) + B_k(t,i)F_p^\varepsilon(t,i)\right]$$
$$+ \sum_{j=1}^{d}q_{ij}\widehat{X}(t,j) + M(t,i) + \widehat{M}(t,i) + \left(F_p^\varepsilon(t,i)\right)^* R(t,i)F_p^\varepsilon(t,i)$$
$$+ L(t,i)F_p^\varepsilon(t,i) + \left(F_p^\varepsilon(t,i)\right)^* L^*(t,i)$$
$$- \left(F_p^\varepsilon(t,i) - \widehat{F}(t,i)\right)^* \mathcal{R}_i\left(t, \widehat{X}(t)\right)\left(F_p^\varepsilon(t,i) - \widehat{F}(t,i)\right) = 0;$$

hence

$$\frac{d}{dt}\left[X_p^\varepsilon(t,i) - \widehat{X}(t,i)\right] + \left[A_0(t,i) + B_0(t,i)F_p^\varepsilon(t,i)\right]^* \left(X_p^\varepsilon(t,i) - \widehat{X}(t,i)\right)$$
$$+ \left(X_p^\varepsilon(t,i) - \widehat{X}(t,i)\right)\left(A_0(t,i) + B_0(t,i)F_p^\varepsilon(t,i)\right)$$
$$+ \sum_{k=1}^{r}\left(A_k(t,i) + B_k(t,i)F_p^\varepsilon(t,i)\right)^* \left(X_p^\varepsilon(t,i) - \widehat{X}(t,i)\right)$$
$$\times \left(A_k(t,i) + B_k(t,i)F_p^\varepsilon(t,i)\right) + \sum_{j=1}^{d}q_{ij}\left(X_p^\varepsilon(t,j) - \widehat{X}(t,j)\right) + \varepsilon I_n$$
$$- \widehat{M}(t,i) + \left(F_p^\varepsilon(t,i) - F_{p-1}^\varepsilon(t,i)\right)^* \mathcal{R}_i\left(t, X_p^\varepsilon(t)\right)\left(F_p^\varepsilon(t,i) - F_{p-1}^\varepsilon(t,i)\right)$$
$$+ \left(F_p^\varepsilon(t,i) - \widehat{F}(t,i)\right)\mathcal{R}_i\left(t, \widehat{X}(t)\right)\left(F_p^\varepsilon(t,i) - \widehat{F}(t,i)\right) = 0, i \in \mathcal{D}, t \in \mathcal{I}.$$

Hence $t \to X_p^\varepsilon(t) - \widehat{X}(t)$ is a solution of the linear differential inequality:

$$\frac{d}{dt}X(t) + \mathcal{L}_{F_p^\varepsilon}^*(t)X(t) + \varepsilon J^d \leq 0.$$

Taking into account (4.23) and Theorem 12 of Chapter 2 we obtain that the system $\left(A_0 + B_0 F_p^\varepsilon, A_1 + B_1 F_p^\varepsilon, \ldots, A_r + B_r F_p^\varepsilon; Q\right)$ is stable. Thus we have shown that (b) is fulfilled.

Writing the system of linear differential equations corresponding to $X_{p-1}^\varepsilon(t, i)$ in the form

$$\frac{d}{dt}X_{p-1}^\varepsilon(t, i) + \left[A_0(t, i) + B_0(t, i)F_{p-1}^\varepsilon(t, i)\right]^* X_{p-1}^\varepsilon(t, i)$$
$$+ X_{p-1}^\varepsilon(t, i)\left[A_0(t, i) + B_0(t, i)F_{p-1}^\varepsilon(t, i)\right]$$
$$+ \sum_{k=1}^r \left[A_k(t, i) + B_k(t, i)F_{p-1}^\varepsilon(t, i)\right]^* X_{p-1}^\varepsilon(t, i)\left[A_k(t, i) + B_k(t, i)F_{p-1}^\varepsilon(t, i)\right]$$
$$+ \sum_{j=1}^d q_{ij}X_{p-1}^\varepsilon(t, j) + \varepsilon I_n + M(t, i) + \left(F_{p-1}^\varepsilon(t, i)\right)^* R(t, i)F_{p-1}^\varepsilon(t, i)$$
$$+ L(t, i)F_{p-1}^\varepsilon(t, i) + \left(F_{p-1}^\varepsilon(t, i)\right)^* L^*(t, i) + \left(F_{p-1}^\varepsilon(t, i) - F_{p-2}^\varepsilon(t, i)\right)^*$$
$$\times \mathcal{R}_i\left(t, X_{p-1}^\varepsilon(t)\right)\left(F_{p-1}^\varepsilon(t, i) - F_{p-2}^\varepsilon(t, i)\right) = 0, \ i \in \mathcal{D},$$

we deduce

$$\frac{d}{dt}\left[X_{p-1}^\varepsilon(t, i) - X_p^\varepsilon(t, i)\right] + \left[A_0(t, i) + B_0(t, i)F_{p-1}^\varepsilon(t, i)\right]^*$$
$$\times \left[X_{p-1}^\varepsilon(t, i) - X_p^\varepsilon(t, i)\right] + \left[X_{p-1}^\varepsilon(t, i) - X_p^\varepsilon(t, i)\right]$$
$$\times \left[A_0(t, i) + B_0(t, i)F_{p-1}^\varepsilon(t, i)\right] + \sum_{k=1}^r \left[A_k(t, i) + B_k(t, i)F_{p-1}^\varepsilon(t, i)\right]^*$$
$$\times \left[X_{p-1}^\varepsilon(t, i) - X_p^\varepsilon(t, i)\right]\left[A_k(t, i) + B_k(t, i)F_{p-1}^\varepsilon(t, i)\right]$$
$$+ \sum_{j=1}^d q_{ij}\left[X_{p-1}^\varepsilon(t, j) - X_p^\varepsilon(t, j)\right] + \left(F_{p-1}^\varepsilon(t, i) - F_{p-2}^\varepsilon(t, i)\right)^* \mathcal{R}_i\left(t, X_{p-1}^\varepsilon(t)\right)$$
$$\times \left(F_{p-1}^\varepsilon(t, i) - F_{p-2}^\varepsilon(t, i)\right) = 0, \ i \in \mathcal{D}, \ t \in \mathcal{I}.$$

Since the system $\left(A_0 + B_0 F_{p-1}^\varepsilon, A_1 + B_1 F_{p-1}^\varepsilon, \ldots, A_r + B_r F_{p-1}^\varepsilon; Q\right)$ is stable, it follows by Proposition 14 of Chapter 2 that $X_{p-1}^\varepsilon(t, i) - X_p^\varepsilon(t, i) \geq 0 \ \forall i \in \mathcal{D}, t \in \mathcal{I}$, and (c) is fulfilled.

From (a) and (c) it follows that the sequence $\{X_p^\varepsilon(t, i)\}$ is convergent.
Set $X^\varepsilon(t, i) = \lim_{p \to \infty} X_p^\varepsilon(t, i)$.

By standard arguments we now obtain that $t \to X^\varepsilon(t) = (X^\varepsilon(t, 1), \ldots, X^\varepsilon(t, d))$ is a bounded solution of the system of differential equations:

$$\frac{d}{dt} X(t, i) + A_0(t, i) X(t, i) + X(t, i) A_0(t, i) + \sum_{k=1}^{r} A_k^*(t, i) X(t, i) A_k(t, i)$$

$$+ \sum_{j=1}^{d} q_{ij} X(t, j) - \left[X(t, j) B_0(t, i) + \sum_{k=1}^{r} A_k^*(t, i) X(t, i) B_k(t, i) + L(t, i) \right]$$

$$\times \left[R(t, i) + \sum_{k=1}^{r} B_k^*(t, i) X(t, i) B_k(t, i) \right]^{-1}$$

$$\times \left[B_0^*(t, i) X(t, i) + \sum_{k=1}^{r} B_k^*(t, i) X(t, i) A_k(t, i) + L^*(t, i) \right]$$

$$+ M(t, i) + \varepsilon I_n = 0, \tag{4.24}$$

$i \in \mathcal{D}$. Moreover we have

$$X^\varepsilon(t, i) \geq \widehat{X}(t, i), i \in \mathcal{D}, t \in \mathcal{I}, \varepsilon > 0. \tag{4.25}$$

Since the construction of $X_p^\varepsilon(t, i)$ does not depend upon the choice of $\widehat{X}$ we conclude that (4.25) still holds if $\widehat{X}(t)$ is replaced by any bounded solution in $\mathbf{\Gamma}^\Sigma$.

From (4.25) we obtain that $\mathcal{R}_i(t, X^\varepsilon(t)) \gg 0$, and therefore the feedback gain $F^\varepsilon(t) = (F^\varepsilon(t, 1), \ldots, F^\varepsilon(t, d))$ is well defined by

$$F^\varepsilon(t, i) = -\mathcal{R}_i^{-1}(t, X^\varepsilon(t)) \mathcal{P}_i(t, X^\varepsilon(t))(i).$$

We prove that $\varepsilon \to X^\varepsilon(t)$ is an increasing function. Take $\varepsilon_1 < \varepsilon_2$. By Lemma 1 we obtain that the system (4.24) for $\varepsilon = \varepsilon_1$ may be written

$$\frac{d}{dt} X^{\varepsilon_1}(t, i) + \left[A_0(t, i) + B_0(t, i) F_{p-1}^{\varepsilon_2}(t, i) \right]^* X^{\varepsilon_1}(t, i)$$

$$+ X^{\varepsilon_1}(t, i) \left[A_0(t, i) + B_0(t, i) F_{p-1}^{\varepsilon_2}(t, i) \right]$$

$$+ \sum_{k=1}^{r} \left[A_k(t, i) + B_k(t, i) F_{p-1}^{\varepsilon_2}(t, i) \right]^* X^{\varepsilon_1}(t, i) \left[A_k(t, i) + B_k(t, i) F_{p-1}^{\varepsilon_2}(t, i) \right]$$

$$+ \sum_{j=1}^{d} q_{ij} X^{\varepsilon_1}(t, j) + \left(F_{p-1}^{\varepsilon_2}(t, i) \right)^* R(t, i) F_{p-1}^{\varepsilon_2}(t, i) + L(t, i) F_{p-1}^{\varepsilon_2}(t, i)$$

$$+ F_{p-1}^{\varepsilon_2}(t, i)^* L^*(t, i) - \left(F_{p-1}^{\varepsilon_2}(t, i) - F^{\varepsilon_1}(t, i) \right) \mathcal{R}_i(t, X^{\varepsilon_1}(t))$$

$$\times \left(F_{p-1}^{\varepsilon_2}(t, i) - F^{\varepsilon_1}(t, i) \right) + M(t, i) + \varepsilon_1 I_n = 0, i \in \mathcal{D}. \tag{4.26}$$

From (4.26) and (4.20), for $\varepsilon = \varepsilon_2$ we obtain

$$\frac{d}{dt}\big[X_p^{\varepsilon_2}(t,i) - X^{\varepsilon_1}(t,i)\big] + \big[A_0(t,i) + B_0(t,i)F_{p-1}^{\varepsilon_2}(t,i)\big]^*\big[X_p^{\varepsilon_2}(t,i) - X^{\varepsilon_1}(t,i)\big]$$

$$+ \big[X_p^{\varepsilon_2}(t,i) - X^{\varepsilon_1}(t,i)\big]\big[A_0(t,i) + B_0(t,i)F_{p-1}^{\varepsilon_2}(t,i)\big]$$

$$+ \sum_{k=1}^{r}\big[A_k(t,i) + B_k(t,i)F_{p-1}^{\varepsilon_2}(t,i)\big]^*\big[X_p^{\varepsilon_2}(t,i) - X^{\varepsilon_1}(t,i)\big]$$

$$\times \big[A_k(t,i) + B_k(t,i)F_{p-1}^{\varepsilon_2}(t,i)\big] + \sum_{j=1}^{d}q_{ij}\big[X_p^{\varepsilon_2}(t,j) - X^{\varepsilon_1}(t,j)\big]$$

$$+ \big(F_{p-1}^{\varepsilon_2}(t,i) - F^{\varepsilon_1}(t,i)\big)^*\mathcal{R}_i\big(t, X^{\varepsilon_1}(t)\big)\big(F_{p-1}^{\varepsilon_2}(t,i) - F^{\varepsilon_1}(t,i)\big)$$

$$+ (\varepsilon_2 - \varepsilon_1)I_n = 0, \quad i \in \mathcal{D}, \quad p = 1, 2, \ldots,$$

which leads to $X_p^{\varepsilon_2}(t,i) - X^{\varepsilon_1}(t,i) \geq 0, i \in \mathcal{D}, t \in \mathcal{I}, p \in \mathbf{N}$. Taking the limit for $p \to \infty$ we get

$$X^{\varepsilon_2}(t,i) \geq X^{\varepsilon_1}(t,i), \forall t \in \mathcal{I}, i \in \mathcal{D}. \tag{4.27}$$

Let $\varepsilon_k, k \in \mathbf{N}$ be a sequence of positive real numbers, $\varepsilon_k > \varepsilon_{k+1}$ and $\lim_{k\to\infty}\varepsilon_k = 0$.

From (4.25) and (4.27) we have $X^{\varepsilon_k}(t,i) \geq X^{\varepsilon_{k+1}}(t,i) \geq \widehat{X}(t,i) \ \forall t \in \mathcal{I}$, $i \in \mathcal{D}, k \in \mathbf{N}$.

Therefore the function $\widetilde{X}(t,i)$ is well defined by $\widetilde{X}(t,i) = \lim_{k\to\infty}X^{\varepsilon_k}(t,i)$, $t \in \mathcal{I}, i \in \mathcal{D}$.

By a standard argument we can show that $\widetilde{X}(t) = \big(\widetilde{X}(t,1) \quad \widetilde{X}(t,2)\ldots\widetilde{X}(t,d)\big)$ is a bounded solution of the equation (4.3) and the proof of the implication (i) $\Rightarrow$ (ii) is complete.

According to Proposition 14 of Chapter 2 it follows that for each $p = 0, 1, 2, \ldots, t \ \to \ X_p^{\varepsilon}(t)$ considered in the proof of the implication (i) $\Rightarrow$ (ii) are θ-periodic functions. Hence $X^{\varepsilon}(t) = \lim_{p\to\infty}X_p^{\varepsilon}(t)$ is a θ-periodic function, and finally $\widetilde{X}(t) = \lim_{\varepsilon\to 0}X^{\varepsilon}(t)$ is a θ-periodic function and the proof of Theorem 3 is complete. $\qquad\square$

Corollary 4. *Assume the following.*
   (i) $(\mathbf{A}, \mathbf{B}; Q)$ *is stabilizable;*
   (ii) $R(t,i) \geq \rho^2 I_n, \ (t,i) \in \mathcal{I} \times \mathcal{D}$.
   (iii) $M(t,i) - L(t,i)R^{-1}(t,i)L^*(t,i) \geq 0, (t,i) \in \mathcal{I} \times \mathcal{D}$.
Under these conditions the equation (4.3) has a bounded solution $\widetilde{X}(t) \geq 0$. *Moreover,* $\widetilde{X}(t) \geq \widehat{X}(t)$ *for any bounded and semipositive solution* $\widehat{X}(t)$ *of the equation (4.3).*

Proof. Under the considered assumptions, $\widehat{X}(t) = 0$ solves the differential inequality $\mathcal{N}_i(t, X(t)) \geq 0, (t,i) \in \mathcal{I} \times \mathcal{D}$ and condition (4.8), and thus the assumptions of Theorem 3 are fulfilled. $\qquad\square$

With the same technique as in Theorem 3 we may prove the following dual result:

Theorem 5. *Assume that*
   (i) $(\mathbf{A}, \mathbf{B}; Q)$ *is stabilizable;*
   (ii) *the differential inequality*

$$\mathcal{N}(t, X(t)) \leq 0, \tag{4.28}$$

$$\mathcal{N}(t, X(t)) = \begin{bmatrix} \dfrac{d}{dt} X(t) + \mathcal{L}^*(t) X(t) & \mathcal{P}^*(t, X(t)) \\ \mathcal{P}(t, X(t)) & \mathcal{R}(t, X(t)) \end{bmatrix}$$

has a bounded solution $\widehat{X}(t)$, *which verifies*

$$\mathcal{R}(t, \widehat{X}(t)) \ll 0. \tag{4.29}$$

Under these conditions the differential equation (4.3) *has a bounded solution* $\widetilde{X}(t)$ *which verifies* $\widetilde{X}(t) \leq \check{X}(t)$ *for any bounded solutions* $\check{X}(t)$ *of the inequality* (4.28), *which verifies* (4.29). $\qquad\qquad\Box$

4.3 Stabilizing solution of the SGRDE

In this section we investigate some aspects concerning the stabilizing solution of the SGRDE (4.1). First we show that the SGRDE (4.1) has at most one bounded and stabilizing solution. The uniqueness of the stabilizing solution is proved without any assumption concerning the sign of $\mathcal{R}_i(t, X(t))$. Further, we provide a necessary and sufficient condition which guarantees the existence of the bounded and stabilizing solution of (4.1) satisfying the additional condition (4.8).

Definition 3. *A solution* $\widetilde{X} : \mathcal{I} \to \mathcal{S}_n^d$ *of the equation* (4.1) *is called a* stabilizing *solution if it has the following properties:*
   (i)

$$\inf_{t \in \mathcal{I}} \left| det \left[R(t, i) + \sum_{k=1}^{r} B_k^*(t, i) \widetilde{X}(t, i) B_k(t, i) \right] \right| > 0, \quad i \in \mathcal{D}.$$

(ii) *The system*

$$\left(A_0 + B_0 \widetilde{F}, A_1 + B_1 \widetilde{F}, \ldots, A_r + B_r \widetilde{F}; Q \right)$$

is stable in the sense of Definition 4 of Chapter 2, where

$$\widetilde{F}(t) = \left(\widetilde{F}(t, 1), \widetilde{F}(t, 2), \ldots, \widetilde{F}(t, d) \right), \tag{4.30}$$

$$\widetilde{F}(t, i) = - \left[R(t, i) + \sum_{k=1}^{r} B_k^*(t, i) \widetilde{X}(t, i) B_k(t, i) \right]^{-1}$$

$$\times \left[B_0(t, i) \widetilde{X}(t, i) + \sum_{k=1}^{r} B_k^*(t, i) \widetilde{X}(t, i) A_k(t, i) + L^*(t, i) \right].$$

Remark 2. (i) The condition (i) in Definition 3 is assumed in order to be sure that the stabilizing feedback gain in (4.30) is bounded.

(ii) The solution $\widetilde{X}(t)$ of the system (4.1) is a stabilizing solution if the control $u(t) = \widetilde{F}(t, \eta(t))x(t)$ stabilizes the system

$$dx(t) = \left[A_0(t, \eta(t))x(t) + B_0(t, \eta(t))u(t)\right] dt$$

$$+ \sum_{k=1}^{r} \left[A_k(t, \eta(t))x(t) + B_k(t, \eta(t))u(t)\right] dw_k(t).$$

Theorem 6. (i) *The system of generalized matrix Riccati differential equations (4.1) has at most one stabilizing and bounded-on-$\mathcal{I}$ solution.*

(ii) *If the coefficients of the system (4.1) are θ-periodic functions, then the stabilizing and bounded solution $\widetilde{X}(t)$ (if it exists) is a θ-periodic function too.*

(iii) *If the coefficients of the system (4.1) do not depend upon t, then its stabilizing and bounded solution $\widetilde{X}(t)$ is constant and solves the following system of nonlinear algebraic equations:*

$$A_0^*(i)X(i) + X(i)A_0(i) + \sum_{k=1}^{r} A_k^*(i)X(i)A_k(i)$$

$$+ \sum_{j=1}^{d} q_{ij}X(j) - \left(X(i)B_0(i) + \sum_{k=1}^{r} A_k^*(i)X(i)B_k(i) + L(i)\right)$$

$$\times \left(R(i) + \sum_{k=1}^{r} B_k^*(i)X(i)B_k(i)\right)^{-1}$$

$$\times \left(B_0^*(i)X(i) + \sum_{k=1}^{r} B_k^*(i)X(i)A_k(i) + L^*(i)\right) + M(i) = 0, \; i \in \mathcal{D}. \quad (4.31)$$

Proof. (i) Let us suppose that the differential equation (4.3) has two bounded and stabilizing solutions, $X_l : \mathcal{I} \to \mathcal{S}_n^d, l = 1, 2$; hence the systems $(A_0 + B_0 F_l, A_1 + B_1 F_l, \ldots, A_r + B_r F_l; Q), l = 1, 2$, are stable, the stabilizing feedback gain being defined as in (4.30). By direct computation we obtain that

$$\frac{d}{dt}X_l(t, i) + \left[A_0(t, i) + B_0(t, i)F_1(t, i)\right]^* X_l(t, i) + X_l(t, i)$$

$$\times \left[A_0(t, i) + B_0(t, i)F_2(t, i)\right] + \sum_{k=1}^{r} \left[A_k(t, i) + B_k(t, i)F_1(t, i)\right]^*$$

$$\times X_l(t, i)\left[A_k(t, i) + B_k(t, i)F_2(t, i)\right] + \sum_{j=1}^{d} q_{ij}X_l(t, j)$$

$$+ F_1^*(t, i)R(t, i)F_2(t, i) + M(t, i) + L(t, i)F_2(t, i) + F_1^*(t, i)L^*(t, i) = 0,$$

$l = 1, 2, \; i \in \mathcal{D}, \; t \in \mathcal{I}.$

Set $\widehat{X}(t, i) = X_1(t, i) - X_2(t, i)$, $i \in \mathcal{D}$, $t \in \mathcal{I}$, and obtain that

$$\widehat{X}(t) = (\widehat{X}(t, 1), \ldots, \widehat{X}(t, d))$$

is a bounded solution of the system

$$\frac{d}{dt}\widehat{X}(t, i) + \left[A_0(t, i) + B_0(t, i)F_1(t, i)\right]^* \widehat{X}(t, i) + \widehat{X}(t, i)$$

$$\times \left[A_0(t, i) + B_0(t, i)F_2(t, i)\right] + \sum_{k=1}^{r}\left[A_k(t, i) + B_k(t, i)F_1(t, i)\right]^*$$

$$\times \widehat{X}(t, i)\left[A_k(t, i) + B_k(t, i)F_2(t, i)\right] + \sum_{j=1}^{d} q_{ij}\widehat{X}(t, j) = 0, \qquad (4.32)$$

$i \in \mathcal{D}$, $t \in \mathcal{I}$. It is easy to see that (4.32) is equivalent to the following linear equation on $\mathcal{S}_{2n}^d$:

$$\frac{d}{dt}\widehat{X}_e(t) + \mathcal{L}_e^*(t)\widehat{X}_e(t) = 0, \qquad (4.33)$$

where $\mathcal{L}_e(t) : \mathcal{S}_{2n}^d \to \mathcal{S}_{2n}^d$, $i \in \mathcal{D}, t \in \mathcal{I}$,

$$A_{k,e}(t, i) = \begin{bmatrix} A_k(t, i) + B_k(t, i)F_1(t, i) & 0 \\ 0 & A_k(t, i) + B_k(t, i)F_2(t, i) \end{bmatrix},$$

$k = 0, 1, \ldots, r.$

$$\widehat{X}_e(t, i) = \begin{bmatrix} 0 & \widehat{X}(t, i) \\ \widehat{X}(t, i) & 0 \end{bmatrix}.$$

From Theorem 12 of Chapter 2 we deduce that there exist the $\mathbf{C}^1$ functions K_j : $\mathcal{I} \to \mathcal{S}_n^d$, $K_j(t) \gg 0$ which are bounded on $\mathcal{I}$ and verify the linear differential equations

$$\frac{d}{dt}K_j(t) + \mathcal{L}_j^*(t)K_j(t) + J^d = 0, \quad j = 1, 2,$$

where $\mathcal{L}_j$ are the Lyapunov operators associated with $(A_0 + B_0 F_j, \ldots, A_r + B_r F_j; Q)$, $j = 1, 2$. Set

$$K_e(t) = \begin{pmatrix} K_1(t) & 0 \\ 0 & K_2(t) \end{pmatrix}.$$

It is easy to see that $K_e(t)$ is a solution of the linear differential equation on $\mathcal{S}_{2n}^d$:

$$\frac{d}{dt}K_e(t) + \mathcal{L}_e^*(t)K_e(t) + J^{2d} = 0. \qquad (4.34)$$

From Theorem 12(v)–(i) of Chapter 2 we conclude that the augmented system $(A_{0,e}, \ldots, A_{r,e}; Q)$ is stable. Applying Proposition 14 of Chapter 2 we deduce that equation (4.33) has a unique bounded solution. Therefore $\widehat{X}_e(t) = 0$ and hence $X_1(t, i) = X_2(t, i)$ for all $(t, i) \in \mathcal{I} \times \mathcal{D}$, and the proof of part (i) is complete.

(ii) Let $\widetilde{X}(t) = (\widetilde{X}(t, 1), \ldots, \widetilde{X}(t, d))$ be the bounded and stabilizing solution of the equation (4.3). Let $\widehat{X}(t) = (\widehat{X}(t, 1), \ldots, \widehat{X}(t, d))$ be defined by $\widehat{X}(t, i) = \widetilde{X}(t + \theta, i)$. It is easy to see that $t \to \widehat{X}(t)$ is a bounded solution of the equation (4.3).

Let $\widetilde{F}(t) = (\widetilde{F}(t, 1), \ldots, \widetilde{F}(t, d))$, $\widehat{F}(t) = (\widehat{F}(t, 1) \ldots \widehat{F}(t, d))$ defined by

$$\widetilde{F}(t, i) = -\mathcal{R}_i^{-1}(t, \widetilde{X}(t))\mathcal{P}_i(t, \widetilde{X}(t))(i),$$

$$\widehat{F}(t, i) = -\mathcal{R}_i^{-1}(t, \widehat{X}(t))\mathcal{P}_i(t, \widehat{X}(t))(i), \quad i \in \mathcal{D}, \ t \in \mathcal{I}.$$

Denote $\widetilde{T}(t, t_0)$ and $\widehat{T}(t, t_0)$, respectively, by the linear evolution operators over $\mathcal{S}_n^d$ defined by the linear differential equations

$$\frac{d}{dt} S(t) = \mathcal{L}_{\widetilde{F}}(t) S(t),$$

$$\frac{d}{dt} S(t) = \mathcal{L}_{\widehat{F}}(t) S(t),$$

respectively, where the operators $\mathcal{L}_{\widetilde{F}}(t)$ and $\mathcal{L}_{\widehat{F}}(t)$ are defined as in (3.2).

By uniqueness arguments we get

$$\widehat{T}(t, t_0) = \widetilde{T}(t + \theta, t_0 + \theta) \tag{4.35}$$

for all $t \geq t_0, t, t_0 \in \mathcal{I}$. Since $\widetilde{X}(t)$ is a stabilizing solution of the equation (4.3) we have $\|\widetilde{T}(t, t_0)\| \leq \beta e^{-\alpha(t-t_0)}$ for all $t \geq t_0, t, t_0 \in \mathcal{I}$ with some $\beta \geq 1, \alpha > 0$.

From (4.35) we deduce that

$$\|\widehat{T}(t, t_0)\| \leq \beta e^{-\alpha(t-t_0)}, \quad t \geq t_0,$$

which shows that $t \to \widehat{X}(t)$ is also a stabilizing solution of the equation (4.3).

Using part (i) we get that $\widehat{X}(t) = \widetilde{X}(t)$ for all $t \in \mathcal{I}$, hence $\widetilde{X}(t + \theta) = \widetilde{X}(t)$.

(iii) From part (ii) it follows that in the time-invariant case the stabilizing and bounded solution is periodic with any period $\theta > 0$ and therefore it is constant. □

A result concerning the existence of a stabilizing solution of SGRDE (4.1) is given by the next theorem.

Theorem 7. *The following are equivalent:*

(i) *The triple $(\mathbf{A}, \mathbf{B}; Q)$ is stabilizable and there exists a C^1 function $\widehat{X} : \mathcal{I} \to \mathcal{S}_n^d$ bounded, with bounded derivative such that differential inequality*

$$\mathcal{N}\left(t, \widehat{X}(t)\right) \gg 0. \tag{4.36}$$

(ii) *The differential equation on $\mathcal{S}_n^d$ (4.3) has a bounded-on-$\mathcal{I}$ and stabilizing solution $\widetilde{X}(t)$ which verifies $\mathcal{R}(t, \widetilde{X}(t)) \gg 0, t \in \mathcal{I}$.*

Proof. (i) $\Rightarrow$ (ii) Let $\widehat{X}$ be a bounded-on-$\mathcal{I}$ solution of (4.36). Hence $\widehat{X} \in \widetilde{\boldsymbol{\Gamma}}^{\Sigma} \subset \boldsymbol{\Gamma}^{\Sigma}$. Based on Theorem 3 we deduce that the equation (4.3) has a bounded solution

$\widetilde{X}: \mathcal{I} \to \mathcal{S}_n^d$ which verifies $\widetilde{X}(t) \geq \widehat{X}(t)$. We show that $\widetilde{X}(t)$ is a stabilizing solution of the equation (4.3).

Set

$$\widehat{M}(t) = \mathcal{P}^*\big(t, \widehat{X}(t)\big)\mathcal{R}^{-1}\big(t, \widehat{X}(t)\big)\mathcal{P}\big(t, \widehat{X}(t)\big) - M(t) - \mathcal{L}^*(t)\widehat{X}(t) - \frac{d}{dt}\widehat{X}(t)$$

and $\widetilde{F}(t) = -\mathcal{R}^{-1}(t, \widetilde{X}(t))\mathcal{P}(t, \widetilde{X}(t))$. It is obvious that $\widehat{M}(t) \ll 0$.

By direct calculation we get

$$\frac{d}{dt}\widetilde{X}(t, i) + \big[A_0(t, i) + B_0(t, i)\widetilde{F}(t, i)\big]^* \widetilde{X}(t, i)$$

$$+ \widetilde{X}(t, i)\big[A_0(t, i) + B_0(t, i)\widetilde{F}(t, i)\big] + \sum_{k=1}^{r}\big[A_k(t, i) + B_k(t, i)\widetilde{F}(t, i)\big]^*$$

$$\times \widetilde{X}(t, i)\big[A_k(t, i) + B_k(t, i)\widetilde{F}(t, i)\big] + \sum_{j=1}^{d} q_{ij}\widetilde{X}(t, j) + \widetilde{F}^*(t, i)R(t, i)\widetilde{F}(t, i)$$

$$+ L(t, i)\widetilde{F}(t, i) + \widetilde{F}(t, i)L^*(t, i) + M(t, i) = 0.$$

Since $\widehat{X}$ verifies (4.16) one gets

$$\frac{d}{dt}\big[\widetilde{X}(t, i) - \widehat{X}(t, i)\big] + \big[A_0(t, i) + B_0(t, i)\widetilde{F}(t, i)\big]^* \big(\widetilde{X}(t, i) - \widehat{X}(t, i)\big)$$

$$+ \big(\widetilde{X}(t, i) - \widehat{X}(t, i)\big)\big[A_0(t, i) + B_0(t, i)\widetilde{F}(t, i)\big]$$

$$+ \sum_{k=1}^{r}\big[A_k(t, i) + B_k(t, i)\widetilde{F}(t, i)\big]^* \big(\widetilde{X}(t, i) - \widehat{X}(t, i)\big)$$

$$\times \big(A_k(t, i) + B_k(t, i)\widetilde{F}(t, i)\big) + \sum_{j=1}^{d} q_{ij}\big(\widetilde{X}(t, j) - \widehat{X}(t, j)\big)$$

$$+ \big(\widetilde{F}(t, i) - \widehat{F}(t, i)\big)^* \mathcal{R}_i\big(t, \widehat{X}(t)\big)\big(\widetilde{F}(t, i) - \widehat{F}(t, i)\big) - \widehat{M}(t, i) = 0. \quad (4.37)$$

Since $\big(\widetilde{F}(t, i) - \widehat{F}(t, i)\big)^*\mathcal{R}_i\big(t, \widehat{X}(t)\big)\big(\widetilde{F}(t, i) - \widehat{F}(t, i)\big) - \widehat{M}(t, i) \gg 0$ from Proposition 13 of Chapter 2, we deduce that the system $\big(A_0 + B_0\widetilde{F}, A_1 + B_1\widetilde{F}, \ldots, A_r + B_r\widetilde{F}; Q\big)$ is stable, hence $\widetilde{X}(t)$ is a stabilizing solution of equation (4.3).

(ii) $\to$ (i) If the equation (4.3) has a stabilizing solution $\widetilde{X}(t)$, then the triple $(\mathbf{A}, \mathbf{B}; Q)$ is stabilizable. Let $\widetilde{X}: \mathcal{I} \to \mathcal{S}_n^d$ be the bounded stabilizing solution of (4.3), which verifies $\mathcal{R}(t, \widetilde{X}(t)) \gg 0$, $t \in \mathcal{I}$. Let $F(t)$ be the stabilizing feedback gain defined by $F(t) = -\mathcal{R}^{-1}(t, \widetilde{X}(t))\mathcal{P}(t, \widetilde{X}(t))$. Define $\mathcal{P}_F(t, X): \mathcal{S}_n^d \to \mathcal{M}_{m,n}^d$

by

$$\mathcal{P}_F(t, X)(i) = B_0^*(t, i) X(i) + \sum_{k=1}^{r} B_k^*(t, i) X(i)(A_k(t, i) + B_k(t, i) F(t, i))$$

$$+ L^*(t, i) + R(t, i) F(t, i), \ i \in \mathcal{D},$$

and

$$M_F(t) = (M_F(t, 1), \dots, M_F(t, d)),$$
$$M_F(t, i) = M(t, i) + L(t, i) F(t, i) + F^*(t, i) L^*(t, i) + F^*(t, i) R(t, i) F(t, i),$$

$i \in \mathcal{D}, \ t \in \mathcal{I}$, where $F(t, i) = F(t)(i)$.

Let $T_F(t, t_0)$ be the linear evolution operator defined by the equation

$$\frac{d}{dt} S(t) = \mathcal{L}_F(t) S(t).$$

Since F is a stabilizing feedback gain we have $\|T_F(t, t_0)\| \leq \beta e^{-\alpha(t-t_0)}$ for all $t \geq t_0$, $t_0 \in \mathcal{I}$, with some $\alpha > 0, \beta \geq 1$. Let $C(\mathcal{I}, \mathcal{S}_n^d)$ be the Banach space of all bounded and continuous functions defined on $\mathcal{I}$ with values in $\mathcal{S}_n^d$. Since $\mathcal{R}(t, \widetilde{X}(t)) \gg 0, t \in \mathcal{I}$, there exists an open set $\mathcal{U} \subset C(\mathcal{I}, \mathcal{S}_n^d)$ such that $\widetilde{X} \in \mathcal{U}$ and $\mathcal{R}(t, X(t)) \gg 0, t \in \mathcal{I}$, for all $X \in \mathcal{U}$.

Consider the operator $\Psi : \mathcal{U} \times \mathbf{R} \rightarrow C(\mathcal{I}, \mathcal{S}_n^d)$ defined by

$$\Psi(X, \delta)(t) = \int_t^\infty T_F^*(s, t)[M_F(s) + \delta J^d$$

$$- \mathcal{P}_F^*(s, X(s)) \mathcal{R}^{-1}(s, X(s)) \mathcal{P}_F(s, X(s))] \, ds - X(t),$$

$t \in \mathcal{I}$.

We shall apply the implicit function theorem to the equation

$$\Psi(X, \delta) = 0 \tag{4.38}$$

to show that there exists a function $X_\delta \in \mathcal{U}$ such that

$$X_\delta(t) = \int_t^\infty T_F^*(s, t) \left[M_F(s) + \delta J^d - \mathcal{P}_F^*(s, X_\delta(s)) \mathcal{R}^{-1}(s, X_\delta(s)) \mathcal{P}_F(s, X_\delta(s)) \right] ds$$

for $|\delta|$ small enough.

It is easy to verify that $(\widetilde{X}, 0)$ is a solution of (4.38).

We show that $\partial_1 \Psi(\widetilde{X}, 0): C(\mathcal{I}, \mathcal{S}_n^d) \rightarrow C(\mathcal{I}, \mathcal{S}_n^d)$ is an isomorphism, $\partial_1 \Psi$ being the derivative of Ψ with respect to the first argument.

Since $\partial_1 \Psi(\widetilde{X}, 0) Y = \lim_{\varepsilon \to 0} \frac{1}{\varepsilon} (\Psi(\widetilde{X} + \varepsilon Y, 0) - \Psi(\widetilde{X}, 0))$ and $\mathcal{P}_F(t, \widetilde{X}(t)) = 0$, one can easily verify that $\partial_1 \Psi(\widetilde{X}, 0) Y = -Y$ and therefore $\partial_1 \Psi(\widetilde{X}, 0) = -\widetilde{J}_n$, $\widetilde{J}_n$ being the identity operator on $\mathcal{S}_n^d$. Also $\partial_1 \Psi(X, \delta)$ is continuous. Applying the implicit function theorem [103] we deduce that there exist $\tilde{\delta} > 0$ and a continuous function $\delta \rightarrow X_\delta : (-\tilde{\delta}, \tilde{\delta}) \rightarrow \mathcal{U}$ which solves $\Psi(X_\delta, \delta) = 0$. It is easy to see that for $\delta \in (-\tilde{\delta}, 0)$, $X_\delta(t)$ will be a solution of the inequality (4.36) with required properties, and the proof is complete. □

Corollary 8. *If the equation* (4.3) *has a stabilizing and bounded-on-$\mathcal{I}$ solution $\widetilde{X}$ which verifies* (4.8), *then $\widetilde{X}(t)$ is the maximal solution with respect to Γ^Σ of* (4.3).

Proof. Suppose that (4.3) has a stabilizing and bounded-on-$\mathcal{I}$ solution $\widetilde{X}$. Then by Theorem 7 it follows that the assumptions of Theorem 3 are fulfilled. Therefore there exists a bounded solution $\widehat{X}$ of (4.3) with the maximality property in Theorem 3. From the proof of Theorem 7 it follows that $\widehat{X}$ is stabilizing. Hence by Theorem 6 we have $\widetilde{X} = \widehat{X}$, and thus the proof is complete.   □

The counterpart of the above theorem for the periodic case is as follows.

Theorem 9. *Assume that the coefficients of* (4.3) *are θ-periodic functions. Then the following are equivalent:*

(i) *$(\mathbf{A}, \mathbf{B}; Q)$ is stabilizable and the differential inequality* (4.36) *has a θ-periodic solution.*

(ii) *The equation* (4.3) *has a stabilizing θ-periodic solution $\widetilde{X}(t)$ which verifies* (4.8).

Proof. (i) → (ii) Applying Theorem 7(i)–(ii) we deduce that the equation (4.3) has a stabilizing and bounded-on-$\mathcal{I}$ solution $\widetilde{X}(t)$ which verifies (4.8). Using Theorem 6(ii) we conclude that $\widetilde{X}(t)$ is a θ-periodic function too.

(ii) → (i) If the equation (4.3) has a stabilizing solution, it follows that the triple $(\mathbf{A}, \mathbf{B}; Q)$ is stabilizable. From the proof of Theorem 7(ii)–(i) it follows that there exists $\delta < 0$ such that

$$\frac{d}{dt}X(t) + \mathcal{L}^*(t)X(t) - \mathcal{P}^*(t, X(t))\mathcal{R}^{-1}(t, X(t))\mathcal{P}(t, X(t)) + M(t) + \delta J^d = 0$$

has a bounded-on-$\mathcal{I}$ solution verifying (4.8). Further, again applying Theorem 7(i)–(ii) for the equation

$$\frac{d}{dt}X(t) + \mathcal{L}^*(t)X(t) - \mathcal{P}^*(t, X(t))\mathcal{R}^{-1}(t, X(t))\mathcal{P}(t, X(t)) + M(t) + \frac{\delta}{2}J^d = 0,$$

we deduce that the above equation has a bounded and stabilizing solution $\widehat{X}(t)$ verifying (4.8). Then by Theorem 6 this solution is periodic. It is not difficult to see that $\widehat{X}(t)$ verifies (4.36) and the proof is complete.   □

With the same proof as in the previous theorem we get the time-invariant counterpart of Theorem 7.

Theorem 10. *Assume that the coefficients of* (4.1) *do not depend upon t. Then the following are equivalent:*

(i) *The triple $(\mathbf{A}, \mathbf{B}; Q)$ is stabilizable and there exists $\widehat{X} \in \mathcal{S}_n^d$ such that $\mathcal{N}(\widehat{X}) > 0$.*

(ii) *The system of generalized Riccati algebraic equations* (4.31) *has a stabilizing solution $\widetilde{X}$ which verifies $\mathcal{R}_i(\widetilde{X}) > 0$ for all $i \in \mathcal{D}$.*   □

Let us consider the following system of nonlinear matrix differential equations:

$$\frac{d}{dt}X(t,i) + A_0^*(t,i)X(t,i) + X(t,i)A_0(t,i) + \sum_{k=1}^{r} A_k^*(t,i)X(t,i)A_k(t,i)$$

$$+ \sum_{j=1}^{d} q_{ij}X(t,j) - \left[X(t,i)B_0(t,i) + \sum_{k=1}^{r} A_k^*(t,i)X(t,i)B_k(t,i) \right]$$

$$\times \left[I_m + \sum_{k=1}^{r} B_k^*(t,i)X(t,i)B_k(t,i) \right]^{-1}$$

$$\times \left[B_0^*(t,i)X(t,i) + \sum_{k=1}^{r} B_k^*(t,i)X(t,i)A_k(t,i) \right] + I_n = 0. \qquad (4.39)$$

$A_k(t,i), B_k(t,i), k = 0,1,\ldots,r$ are continuous and bounded functions. The system (4.39) is a particular case of the system (4.1) taking $M(t,i) = I_n$, $L(t,i) = 0$, $R(t,i) = I_m$. Obviously, in this case $\widehat{X}(t,i) \equiv 0$ verifies (4.36), and therefore by Theorems 7 and 6 it follows that the next result holds.

Corollary 11. *Assume that* $(\mathbf{A}, \mathbf{B}; Q)$ *is stabilizable. Then the system* (4.39) *has a bounded and stabilizing solution* $\widetilde{X}(t) = (\widetilde{X}(t,1),\ldots,\widetilde{X}(t,d))$, $\widetilde{X}(t,i) \geq 0$. *Moreover, if* $A_k(\cdot,i), B_k(\cdot,i)$ *are* θ-*periodic functions, then* $\widetilde{X}(\cdot)$ *is a* θ-*periodic function too, and if* $A_k(t,i) = A_k(i)$, $B_k(t,i) = B_k(i)$, $(t,i) \in \mathcal{I} \times \mathcal{D}, k \in \{0,1,\ldots,r\}$, *then* $\widetilde{X}(t,i) = \widetilde{X}(i)$, $(t,i) \in \mathcal{I} \times \mathcal{D}$. □

Remark 3. (i) From the above corollary we conclude that if $A_k(\cdot,i), B_k(\cdot,i), k = 0,1,\ldots,r$, are continuous θ-periodic functions and the triple $(\mathbf{A}, \mathbf{B}; Q)$ is stabilizable, then there exists a stabilizing feedback gain $\widetilde{F}(t) = (\widetilde{F}(t,1),\ldots,\widetilde{F}(t,d))$ which is a θ-periodic function. Also, if $A_k(t,i) = A_k(i)$, $B_k(t,i) = B_k(i), k = 0,1,\ldots,r$, $(t,i) \in \mathcal{I} \times \mathcal{D}$, and $(\mathbf{A}, \mathbf{B}, ; Q)$ is stabilizable, then there is a stabilizing feedback gain, $F = (F(1),\ldots,F(d))$. Therefore we may conclude, without loss of generality, that in the case of periodic coefficients the triple $(\mathbf{A}, \mathbf{B}; Q)$ is stabilizable if and only if there exists a stabilizing feedback gain $F(t)$ which is a θ-periodic function; in the time-invariant case the triple $(\mathbf{A}, \mathbf{B}; Q)$ is stabilizable if and only if there exists a stabilizing feedback gain $F = (F(1),\ldots,F(d))$ not depending upon t.

(ii) Combining the result in Corollary 11 and Remark 4 of Chapter 3, we may conclude that if $A_k(\cdot,i), C_k(\cdot,i), k = 0,1,\ldots,r$, are θ-periodic functions defined on $\mathbf{R} \times \mathcal{D}$, then the triple $(\mathbf{C}, \mathbf{A}; Q)$ is detectable if and only if there exists a stabilizing injection $K(t)$ which is a continuous θ-periodic function, and in the time-invariant case the triple $(\mathbf{C}, \mathbf{A}; Q)$ is detectable, if and only if there exists a stabilizing injection $K = (K(1),\ K(2),\ldots,K(d)) \in \mathcal{M}_{n.p}^d$.

We point out that Corollary 11 and Remark 3 hold when the elements of the matrix Q verify only the condition $q_{ij} \geq 0$ for $i \neq j$.

4.4 The case $0 \in \Gamma^\Sigma$

In the following we focus our attention on the case when the coefficients of the system (4.1) (and equivalently of the equation (4.4)) satisfy the additional conditions:

$$R(t, i) \geq \rho I_n > 0, \tag{4.40}$$

$$M(t, i) - L(t, i)R^{-1}(t, i)L^*(t, i) \geq 0$$

for all $(t, i) \in \mathcal{I} \times \mathcal{D}, \rho > 0$ not depending upon (t, i). From (4.6) we see that conditions (4.40) are equivalent with the fact that $\widetilde{X}(t) \equiv 0$ belongs to Γ^Σ.

Lemma 12. *Assume that (4.40) holds. Then*

(i) *Let $X : \mathcal{I}_1 \subseteq \mathcal{I} \to \mathcal{S}_n^d$ be a solution of the equation (4.3). If there exists $\tau \in \mathcal{I}_1$ such that $X(\tau, i) \geq 0, i \in \mathcal{D}$, then $X(t, i) \geq 0$ for all $t \in \mathcal{I}_1 \cap (-\infty, \tau]$.*

(ii) *Let $\widehat{X} : \mathcal{I}_1 \subset \mathcal{I} \to \mathcal{S}_n^d, \check{X} : \mathcal{I}_1 \subset \mathcal{I} \to \mathcal{S}_n^d$ be two solutions of the equation (4.3).*

If there exists $\tau \in \mathcal{I}_1$ such that $\check{X}(\tau) \geq \widehat{X}(\tau) \geq 0$, then $\check{X}(t) \geq \widehat{X}(t)$ for all $t \in \mathcal{I}_1 \cap (-\infty, \tau]$.

Proof. (i) Let $F(t) = (F(t, 1), F(t, 2), \ldots, F(t, d))$,

$$F(t, i) = -\mathcal{R}_i^{-1}(t, X(t))\mathcal{P}_i(t, X(t)), \quad t \in \mathcal{I}_1, i \in \mathcal{D}.$$

From Lemma 1 one obtains that the equation (4.3) verified by $X(t)$ may be written as follows:

$$\frac{d}{dt}X(t) + \mathcal{L}_F^*(t)X(t) + \widetilde{M}(t) = 0, \tag{4.41}$$

$t \in \mathcal{I}_1$, where $\widetilde{M}(t) = (\widetilde{M}(t, 1) \ldots \widetilde{M}(t, d))$,

$$\widetilde{M}(t, i) = M(t, i) - L(t, i)R^{-1}(t, i)L^*(t, i) + [R(t, i)F(t, i) + L^*(t, i)]^*$$
$$\times R^{-1}(t, i)[R(t, i)F(t, i) + L^*(t, i)], \quad (t, i) \in \mathcal{I}_1 \times \mathcal{D}.$$

From (4.40) it follows that $\widetilde{M}(t) \geq 0, t \in \mathcal{I}_1$. If $T_F(t, t_0)$ is a linear evolution operator over $\mathcal{S}_n^d$ defined by the linear differential equation

$$\frac{d}{dt}S(t) = \mathcal{L}_F(t)S(t),$$

then we obtain from (4.41) and (2.13) that

$$X(t) = T_F^*(\tau, t)X(\tau) + \int_t^\tau T_F^*(s, t)\widetilde{M}(s)ds$$

$\forall t \in \mathcal{I}_1 \cap (-\infty, \tau]$. Since $T_F^*(s, t) : \mathcal{S}_n^d \to \mathcal{S}_n^d$ is a positive operator we conclude that $X(t) \geq 0, t \leq \tau$.

(ii) Set $\check{F}(t) = \left(\check{F}(t, 1), \check{F}(t, 2), \ldots, \check{F}(t, d)\right)$ and

$$\widehat{F}(t) = \left(\widehat{F}(t, 1) \ \widehat{F}(t, 2) \ldots \widehat{F}(t, d)\right),$$

where

$$\check{F}(t, i) = -\mathcal{R}_i^{-1}\left(t, \check{X}(t)\right)\mathcal{P}_i\left(t, \check{X}(t)\right)$$

and

$$\widehat{F}(t, i) = -\mathcal{R}_i^{-1}\left(t, \widehat{X}(t)\right)\mathcal{P}_i\left(t, \widehat{X}(t)\right)(i).$$

Let $Y(t)$ be defined by $Y(t) = \check{X}(t) - \widehat{X}(t)$, $t \in \mathcal{I}_1$. By using Lemma 1, one concludes that $Y(t)$ is a solution of the affine differential equation on $\mathcal{S}_n^d$:

$$\frac{d}{dt}Y(t) + \mathcal{L}_{\check{F}}^*(t)Y(t) + \check{M}(t) = 0, \quad t \in \mathcal{I}_1,$$

where $\check{M}(t) = \left(\check{M}(t, 1) \ldots \check{M}(t, d)\right)$,

$$\check{M}(t, i) = \left[\check{F}(t, i) - \widehat{F}(t, i)\right]^* \mathcal{R}_i\left(t, \widehat{X}(t)\right)\left[\check{F}(t, i) - \widehat{F}(t, i)\right], \quad (t, i) \in \mathcal{I}_1 \times \mathcal{D}.$$

Based on part (i) of this lemma, we deduce that $\widehat{X}(t) \geq 0$, and hence $\mathcal{R}_i\left(t, \widehat{X}(t)\right) \geq 0$, $t \in \mathcal{I}_1 \cap (-\infty, \tau], i \in \mathcal{D}$, and therefore $\check{M}(t) \geq 0$. Let $\check{T}(t, t_0)$ be the linear evolution operator on $\mathcal{S}_n^d$ defined by the linear differential equation

$$\frac{d}{dt}S(t) = \mathcal{L}_{\check{F}}(t)S(t).$$

We obtain the representation formula

$$Y(t) = \check{T}^*(\tau, t)Y(\tau) + \int_t^\tau \check{T}(s, t)\check{M}(s)\,ds.$$

The conclusion follows taking into account that $\check{T}(s, t)$ is a positive operator on $\mathcal{S}_n^d$.   □

For each $\tau \in \mathcal{I}$ we denote be $X_\tau(\cdot)$ the solution of the equation (4.3) that verifies the condition $X_\tau(\tau, i) = 0, i \in \mathcal{D}$.

Proposition 13. *Assume that* $(\mathbf{A}, \mathbf{B}; Q)$ *is stabilizable and* (4.40) *is fulfilled. Then:*
(i) for each $\tau \in \mathcal{I}$, *the solution* $X_\tau(\cdot)$ *is defined on* $\mathcal{I} \cap (-\infty, \tau]$. *Moreover there exists* $c > 0$, *such that* $0 \leq X_\tau(t) \leq cJ^d \ \forall t \leq \tau, t \in \mathcal{I}$;
(ii) $X_{\tau_1}(t) \leq X_{\tau_2}(t) \ \forall t \leq \tau_1 < \tau_2, t \in \mathcal{I}$.

Proof. (i) Let $\mathcal{I}_\tau \subset (-\infty, \tau] \cap \mathcal{I}$ be the maximal interval on which $X_\tau(\cdot)$ is defined.
From part (i) of Lemma 12 we have that $X_\tau(t) \geq 0, t \in \mathcal{I}_\tau$. Since $(\mathbf{A}, \mathbf{B}; Q)$ is stabilizable, there exist $F^0 : \mathcal{I} \to \mathcal{M}_{m,n}^d$ a continuous and bounded function, such

that the system $(A_0 + B_0 F^0, A_1 + B_1 F^0, \ldots, A_r + B_r F^0; Q)$ is stable. Let $X^0(t)$ be the unique bounded-on-$\mathcal{I}$ solution of the affine Lyapunov-type differential equation:

$$\frac{d}{dt} X^0(t) + \mathcal{L}^*_{F^0}(t) X^0(t) + M^0(t) = O,$$

where $M^0(t) = (M^0(t, 1), M^0(t, 2) \ldots M^0(t, d))$,

$$M^0(t, i) = M(t, i) + L(t, i) F^0(t, i) + (F^0(t, i))^* L^*(t, i) + (F^0(t, i))^* R(t, i) F^0(t, i).$$

Since (4.40) is fulfilled, we obtain that $M^0(t) \geq 0, t \in \mathcal{I}$. Hence by Proposition 14 of Chapter 2 there exists $c > 0$ such that $0 \leq X^0(t) \leq cJ^d$ for all $t \in \mathcal{I}$. By direct computation we obtain that $X^0(t) - X_\tau(t)$ verifies the affine differential equation of Lyapunov type:

$$\frac{d}{dt}(X^0(t) - X_\tau(t)) + \mathcal{L}^*_{F^0}(t)(X^0(t) - X_\tau(t)) + \tilde{M}^0(t) = 0, \tag{4.42}$$

$t \in \mathcal{I}_\tau$ where $\tilde{M}^0(t) = (\tilde{M}^0(t, 1), \tilde{M}^0(t, 2), \ldots, \tilde{M}^0(t, d))$

$$\tilde{M}^0(t, i) = (F^0(t, i) - F_\tau(t, i))^* \mathcal{R}_i(t, X_\tau(t))(F^0(t, i) - F_\tau(t, i)),$$

$(t, i) \in \mathcal{I} \times \mathcal{D}$. Since $X_\tau(t) \geq 0$ we get $\tilde{M}^0(t) \geq 0, t \in \mathcal{I}_\tau$.

From (4.42) we deduce that

$$X^0(t) - X_\tau(t) \geq 0 \tag{4.43}$$

for all $t \in \mathcal{I}_\tau$ which leads to $0 \leq X_\tau(t) \leq X^0(t) \leq cJ_n \; \forall t \in \mathcal{I}_\tau$.

Thus $t \to X_\tau(t)$ is bounded and we conclude that $\mathcal{I}_\tau = (-\infty, \tau] \cap \mathcal{I}$.

(ii) follows immediately from Lemma 12 and the proof is complete.   □

Now we are able to prove the following theorem.

Theorem 14. *Assume that* $(\mathbf{A}, \mathbf{B}; Q)$ *is stabilizable and the condition* (4.40) *is fulfilled. Under these assumptions the equation* (4.3) *has two bounded solutions* $\tilde{X} : \mathcal{I} \to \mathcal{S}^d_n, \tilde{\tilde{X}} : \mathcal{I} \to \mathcal{S}^d_n$ *with the property* $\tilde{X}(t) \geq \hat{X}(t) \geq \tilde{\tilde{X}}(t) \geq 0$ *for all* $t \in \mathcal{I}, \hat{X}(t)$ *being any bounded and semipositive solution of the equation* (4.3).

Proof. The existence of the maximal solution $\tilde{X}(t)$ is guaranteed by Corollary 4. It remains to prove the existence of the minimal solution $\tilde{\tilde{X}}(t)$. To this end we shall use the results of Proposition 13. We define $\tilde{\tilde{X}}(t) = \lim_{\tau \to \infty} X_\tau(t), t \in \mathcal{I}$. Invoking the result of Proposition 13 we obtain that this limit exists.

Since $X_\tau(t)$ is a bounded solution of (4.3), by the standard argument based on Lebesgue's Theorem we conclude that $\tilde{\tilde{X}}(t)$ is a solution of the equation (4.3).

To check the minimality of $\tilde{\tilde{X}}(t)$ in the class of semipositive solutions of the equation (4.3) we shall use Lemma 12. If $\hat{X}(\cdot)$ is a semipositive and bounded solution of the equation (4.3), then for each $\tau \in \mathcal{I}$ we have $\hat{X}(\tau) \geq 0 = X_\tau(\tau)$. Therefore $X_\tau(t) \leq \hat{X}(t)$ for all $t \leq \tau, t \in \mathcal{I}$.

Taking the limit for $\tau \to \infty$, we deduce that $\tilde{\tilde{X}}(t) \leq \hat{X}(t), t \in \mathcal{I}$, and the proof ends.   □

To solve the linear quadratic problems, a crucial role is played by the minimal solution stabilizing solution, respectively, of the following system of matrix nonlinear differential equations:

$$
\frac{d}{dt}X(t,i) + A_0^*(t,i)X(t,i) + X(t,i)A_0(t,i) + \sum_{k=1}^{r} A_k^*(t,i)X(t,i)A_k(t,i)
$$

$$
+ \sum_{j=1}^{d} q_{ij}X(t,j) - \left[X(t,i)B_0(t,i) + \sum_{k=1}^{r} A_k^*(t,i)X(t,i)B_k(t,i) \right] \quad (4.44)
$$

$$
\times \left[R(t,i) + \sum_{k=1}^{r} B_k^*(t,i)X(t,i)B_k(t,i) \right]^{-1}
$$

$$
\times \left[B_0^*(t,i)X(t,i) + \sum_{k=1}^{r} B_k^*(t,i)X(t,i)A_k(t,i) \right] + C_0^*(t,i)C_0(t,i)
$$

$$
= 0,
$$

$t \geq 0, i \in \mathcal{D}$, where $R(t,i) = D_0^*(t,i)D_0(t,i)$, which is a particular form of (4.1) obtained for $M(t,i) = C_0^*(t,i)C_0(t,i)$, $L(t,i) = 0$, $R(t,i) = D_0^*(t,i)D_0(t,i)$, $\mathcal{I} = \mathbf{R}_+$.

We have the following lemma.

Lemma 15. *Assume the following.*
   (i) *There exists $\rho > 0$ such that $D_0^*(t,i)D_0(t,i) \geq \rho I_m$ for all $t \geq 0, i \in \mathcal{D}$.*
   (ii) *The triple $(C_0, \mathbf{A}; Q)$ is detectable.*
   (iii) *The elements of matrix Q verify $q_{ij} \geq 0, i \neq j, \sum_{j=1}^{d} q_{ij} = 0, i \in \mathcal{D}$.*
   Under these assumptions any semipositive and bounded solution of the system (4.44) is stabilizing.

Proof. Let $\widetilde{X}(t) = \left(\widetilde{X}(t,1), \widetilde{X}(t,2), \dots, \widetilde{X}(t,d) \right)$ be a bounded and semipositive solution of the system (4.44). By direct calculation we obtain:

$$
\frac{d}{dt}\widetilde{X}(t,i) + \left[A_0(t,i) + B_0(t,i)\widetilde{F}(t,i) \right]^* \widetilde{X}(t,i)
$$

$$
+ \widetilde{X}(t,i)\left[A_0(t,i) + B_0(t,i)\widetilde{F}(t,i) \right] + \sum_{k=1}^{r} \left(A_k(t,i) + B_k(t,i)\widetilde{F}(t,i) \right)^*
$$

$$
\times \widetilde{X}(t,i)\left(A_k(t,i) + B_k(t,i)\widetilde{F}(t,i) \right) + \sum_{j=1}^{d} q_{ij}\widetilde{X}(t,j) + C_0^*(t,i)C_0(t,i)
$$

$$
+ \widetilde{F}^*(t,i))R(t,i)\widetilde{F}(t,i) = 0, \quad (4.45)
$$

$$
\widetilde{F}(t,i) = -\mathcal{R}_i^{-1}\left(t, \widetilde{X}(t)\right)\mathcal{P}_i\left(t, \widetilde{X}(t)\right)(i), \quad (t,i) \in \mathcal{I} \times \mathcal{D},
$$

or, in compact form, as a Lyapunov-type equation on $\mathcal{S}_n^d$:

$$
\frac{d}{dt}\widetilde{X}(t,i) + \mathcal{L}_{\widetilde{F}}^*(t)\widetilde{X}(t) + \widetilde{C}(t) = 0, \quad (4.46)
$$

where $\mathcal{L}_{\widetilde{F}}^*$ is defined as in Lemma 1, with $\widetilde{F}$ instead of G, and

$$\widetilde{C}(t) = (\widetilde{C}(t, 1), \ldots, \widetilde{C}(t, d))$$

with

$$\widetilde{C}(t, i) = C_0^*(t, i)C_0(t, i) + \widetilde{F}^*(t, i)R(t, i)\widetilde{F}(t, i).$$

With the same reasoning as in Theorem 7 of Chapter 3 applied to (4.46), we deduce that there exists $\gamma > 0$ such that

$$E\left[\int_{t_0}^{\infty} |C_0(t, \eta(t))x(t)|^2 \, dt \mid \eta(t_0) = i\right] \leq \gamma |x_0|^2 \qquad (4.47)$$

and

$$E\left[\int_{t_0}^{\infty} |\widetilde{F}(t, \eta(t))x(t)|^2 \, dt \mid \eta(t_0) = i\right] \leq \gamma |x_0|^2 \qquad (4.48)$$

for all $t_0 \geq 0$, $i \in \mathcal{D}$, and $x_0 \in \mathbf{R}^n$, where $x(t)$ is the solution of the problem

$$dx(t) = \widetilde{A}_0(t, \eta(t))x(t)dt + \sum_{k=1}^{r} \widetilde{A}_k(t, \eta(t))x(t)dw_k(t)$$

$$x(t_0) = x_0,$$

where

$$\widetilde{A}_k(t, i) = A_k(t, i) + B_k(t, i)\widetilde{F}(t, i), \quad k = 0, 1 \ldots, r.$$

According with assumption (ii) it exists $H(t, i)$ such that the system

$$(A_0 + HC_0, A_1, \ldots, A_r; Q)$$

is stable. We may write

$$dx(t) = \{[A_0(t, \eta(t)) + H(t, \eta(t))C_0(t, \eta(t))]x(t) + f_0(t)\} dt$$

$$+ \sum_{k=1}^{r} [A_k(t, \eta(t))x(t) + f_k(t)] dw_k(t),$$

where

$$f_0(t) = \left[-H(t, \eta(t))C_0(t, \eta(t)) + B_0(t, \eta(t))\widetilde{F}(t, \eta(t))\right]x(t)$$

and

$$f_k(t) = B_k(t, \eta(t))\widetilde{F}(t, \eta(t))x(t), \quad k = 1, \ldots, r.$$

Since the system $(A_0 + HC_0, A_1, \ldots, A_r; Q)$ is stable, based on (4.47), (4.48), and Theorem 32(ii) of Chapter 2, we deduce that there exists $\beta > 0$ (independent of t_0 and x_0) such that

$$E\left[\int_{t_0}^{\infty} |x(t)|^2 \, dt \mid \eta(t_0) = i\right] \leq \beta |x_0|^2$$

for all $t_0 \geq 0$, $i \in \mathcal{D}$, and $x_0 \in \mathbf{R}^n$. Therefore, from Theorem 19 of Chapter 2 we conclude that $(\widetilde{A}_0, \ldots, \widetilde{A}_r; Q)$ is stable and the proof is complete.   □

Proposition 16. *Suppose that the assumptions* (i) *and* (iii) *in Lemma 15 hold and that* $(C_0, A_0, A_1, \ldots, A_r; Q)$ *is uniformly observable. Then if K is a positive semidefinite and bounded-on-R_+ solution of system* (4.44) *we have that*
   (i) *K is uniform positive;*
   (ii) *K is a stabilizing solution.*

Proof. Let K be a positive semidefinite and bounded-on-R_+ solution of system (4.44). Set

$$F_K(t, i) = -\mathcal{R}_i^{-1}(t, K(t)) \mathcal{P}_i(t, K(t)),$$
$$\widetilde{A}_k(t, i) = A_k(t, i) + B_k(t, i) F_K(t, i), \quad 0 \le k \le r,$$

and $\widetilde{X}(t, t_0)$ be the fundamental matrix solution associated with the linear system

$$dx(t) = \widetilde{A}_0(t, \eta(t))x(t)dt + \sum_{k=1}^{r} \widetilde{A}_k(t, \eta(t))x(t)dw_k(t).$$

We have to prove that $(\widetilde{A}_0, \widetilde{A}_1, \ldots, \widetilde{A}_r; Q)$ is stable.

Let $\tau > 0$ and $\beta > 0$, verifying the inequality in Proposition 8 of Chapter 3. Define

$$G(t, i) = E\left[\int_t^{t+\tau} \widetilde{X}^*(s, t)\left[C_0^*(s, \eta(s))C_0(s, \eta(s)) + F_K^*(s, \eta(s))R(s, \eta(s)) \right.\right.$$

$$\left.\left. F_K(s, \eta(s)) \right]\widetilde{X}(s, t)ds|\eta(t) = i \right], t \ge 0, \ i \in \mathcal{D}).$$

We shall prove $\inf\{x^*G(t, i)x; |x| = 1, t \ge 0, i \in \mathcal{D}\} > 0$. Suppose on the contrary that for every $\varepsilon > 0$ there exist $x_\varepsilon \in R^n, |x_\varepsilon| = 1, t_\varepsilon \ge 0$, and $i_\varepsilon \in \mathcal{D}$ such that $x_\varepsilon^* G(t_\varepsilon, i_\varepsilon)x_\varepsilon < \varepsilon$.

Let $x_\varepsilon(t) = \widetilde{X}(t, t_\varepsilon)x_\varepsilon$ and $u_\varepsilon(t) = F_K(t, \eta(t))x_\varepsilon(t)$.
We can write

$$\varepsilon > x_\varepsilon^* G(t_\varepsilon, i_\varepsilon) \ge E\left[\int_{t_\varepsilon}^{t_\varepsilon+\tau} u_\varepsilon^*(t)R(t, \eta(t))u_\varepsilon(t)dt|\eta(t_\varepsilon) = i_\varepsilon \right]$$

$$\ge \delta E\left[\int_{t_\varepsilon}^{t_\varepsilon+\tau} |u_\varepsilon(t)|^2 dt|\eta(t_\varepsilon) = i_\varepsilon \right]$$

with some $\delta > 0$. But $x_\varepsilon(t) = \Phi(t, t_\varepsilon)x_\varepsilon + \hat{x}_\varepsilon(t), \ t \ge t_\varepsilon$, where $\widehat{x}_\varepsilon(t_\varepsilon) = 0$ and

$$d\widehat{x}_\varepsilon(t) = (A_0(t, \eta(t))\widehat{x}_\varepsilon(t) + B_0(t, \eta(t))u_\varepsilon(t))dt$$

$$+ \sum_{k=1}^{r} [A_k(t, \eta(t))\widehat{x}_\varepsilon(t) + B_k(t, \eta(t))u_\varepsilon(t)]dw_k(t).$$

Hence, by Remark 17 of Chapter 2 there exists $\gamma_0 > 0$ such that

$$E\left[|\widehat{x}_\varepsilon(t)|^2|\eta(t_\varepsilon) = i_\varepsilon\right] \le \gamma_0 E\left[\int_{t_\varepsilon}^{t_\varepsilon+\tau} |u_\varepsilon(t)|^2 dt|\eta(t_\varepsilon) = i_\varepsilon \right] \le \delta_1\varepsilon.$$

Further, we can write

$$\varepsilon > x_\varepsilon^* G(t_\varepsilon, i_\varepsilon) x_\varepsilon \geq E\left[\int_{t_\varepsilon}^{t_\varepsilon+\tau} |C_0(t, \eta(t)) x_\varepsilon(t)|^2 dt \mid \eta(t_\varepsilon) = i_\varepsilon\right]$$

$$= E\left[\int_{t_\varepsilon}^{t_\varepsilon+\tau} |C_0(t, \eta(t))\Phi(t, t_\varepsilon) x_\varepsilon + C_0(t, \eta(t))\widehat{x}_\varepsilon(t)|^2 dt \mid \eta(t_\varepsilon) = i_\varepsilon\right]$$

$$\geq \frac{1}{2} E\left[\int_{t_\varepsilon}^{t_\varepsilon+\tau} |C_0(t, \eta(t))\Phi(t, t_\varepsilon) x_\varepsilon|^2 dt \mid \eta(t_\varepsilon) = i_\varepsilon\right]$$

$$- E\left[\int_{t_\varepsilon}^{t_\varepsilon+\tau} |C_0(t, \eta(t))\widehat{x}_\varepsilon(t)|^2 dt \mid \eta(t_\varepsilon) = i_\varepsilon\right]$$

$$\geq \frac{1}{2}\beta - \delta_2\varepsilon, \quad \varepsilon > 0,$$

and thus we get a contradiction, since $\beta > 0$. Hence, there exists $\beta_1 > 0$ such that $G(t, i) \geq \beta_1 I_n$, $t \geq 0, i \in \mathcal{D}$. Applying the identity (1.6) to the function $v(t, x, i) = x^* K(t, i)x$ and to the system

$$dx(t) = \widetilde{A}_0(t, \eta(t)) x(t)dt + \sum_{k=1}^{r} \widetilde{A}_k(t, \eta(t)) x(t)dw_k(t)$$

and taking into account the equation (4.44) for $K(t, i)$, we get

$$x_0^* E\left[\widetilde{X}^*(t+\tau, t)K(t+\tau, \eta(t+\tau))\widetilde{X}(t+\tau, t) \mid \eta(t) = i\right] x_0 - x_0^* K(t, i)x_0$$
$$= -x_0^* G(t, i)x_0, \quad t \geq 0, \ x_0 \in R^n, i \in \mathcal{D}.$$

Therefore

$$\beta_1|x_0|^2 \leq x_0^* K(t, i)x_0 \leq \beta_2|x_0|^2, \quad t \geq 0, i \in \mathcal{D}, x_0 \in R^n.$$

Thus K is a uniform positive function and

$$E\left[\widetilde{X}^*(t+\tau, t)K(t+\tau, \eta(t+\tau))\widetilde{X}(t+\tau, t) \mid \eta(t) = i\right] \leq \left(1 - \frac{\beta_1}{\beta_2}\right) K(t, i).$$

By virtue of Theorems 31 and 4 of Chapter 2, it follows that $\left(\widetilde{A}_0, \widetilde{A}_1, \ldots, \widetilde{A}_r; Q\right)$ is stable and thus the proof is complete. $\square$

Theorem 17. *Assume the following.*

   (i) *Assumptions (i) and (iii) of Lemma 15 hold.*

   (ii) *The triple $(\mathbf{A}, \mathbf{B}; Q)$ is stabilizable.*

   (iii) *The system $(C_0; A_0, A_1, \ldots, A_r, Q)$ is either detectable or uniformly observable.*

   Then the Riccati-type system (4.44) has a unique positive semidefinite and bounded-on-$\mathbf{R}_+$ solution. Moreover, this solution is stabilizing.

Proof. The proof follows immediately from Theorem 14, Proposition 16, Lemma 15, and Theorem 6. $\square$

In the particular case when $\mathcal{D} = \{1\}$, the system (4.44) becomes

$$
\frac{d}{dt}X(t) + A_0^*(t)X(t) + X(t)A_0(t) + \sum_{k=1}^{r} A_k^*(t)X(t)A_k(t)
$$

$$
- \left[X(t)B_0(t) + \sum_{k=1}^{r} A_k^*(t)X(t)B_k(t) \right] \tag{4.49}
$$

$$
\times \left[R(t) + \sum_{k=1}^{r} B_k^*(t)X(t)B_k(t) \right]^{-1}
$$

$$
\times \left[B_0^*(t)X(t) + \sum_{k=1}^{r} B_k^*(t)X(t)A_k(t) \right] + C_0^*(t)C_0(t) = 0.
$$

A direct consequence of Theorem 17 is the following corollary.

Corollary 18. *Assume the following.*
  (i) *There exists $\rho > 0$ such that $D_0^*(t)D_0(t) \geq \rho I_m$ for all $t \geq 0$.*
  (ii) *The pair $(\mathbf{A}, \mathbf{B})$ is stabilizable.*
  (iii) *The pair $(C_0, \mathbf{A})$ is either detectable or uniformly observable.*
  Then the Riccati-type equation (4.49) *has a unique positive semidefinite and bounded-on-$\mathbf{R}_+$ solution. Moreover, this solution is stabilizing.* □

Remark 4. Based on Theorem 14 one obtains that under the assumption that $(\mathbf{A}, \mathbf{B}; Q)$ is stabilizable, the SGRDE (4.44) has two remarkable semipositive definite solutions. We refer to the maximal solution $\widetilde{X}(t)$ and to the minimal solution $\widetilde{\widetilde{X}}$, respectively. If additionally $(C_0, \mathbf{A}; Q)$ is either detectable or uniformly observable, then these two solutions coincide, namely $\widetilde{X}(t) \equiv \widetilde{\widetilde{X}}(t)$.

However, in the absence of detectability and uniform observability, $\widetilde{X}(t)$ does not always coincide with $\widetilde{\widetilde{X}}$. This can be seen in the following numerical example.
Numerical example. Consider $n = 2$, $d = 1$, $r = 1$, $p = 1$, $m = 1$. In this case (4.44) reduces to

$$
\frac{d}{dt}X(t) + A_0^*(t)X(t) + X(t)A_0(t) + A_1^*(t)X(t)A_1(t) \tag{4.50}
$$

$$
- \left[X(t)B_0(t) + A_1^*(t)X(t)B_1(t) \right] \left[R(t) + B_1^*(t)X(t)B_1(t) \right]^{-1}
$$

$$
\times \left[B_0^*(t)X(t) + B_1^*(t)X(t)A_1(t) \right] + C_0^*(t)C_0(t) = 0.
$$

Choose

$$
A_0(t) = \begin{bmatrix} 1 & 0 \\ 0 & 3 \end{bmatrix}, \quad A_1(t) = I_2, \quad B_0(t) = \begin{bmatrix} 2 \\ 1 \end{bmatrix}, \quad B_1(t) = \begin{bmatrix} 0 \\ 0 \end{bmatrix},
$$

$$
C_0(t) = \begin{bmatrix} 1 & 0 \end{bmatrix}, \quad R(t) = 1.
$$

One can see (see Propositions 5 and 25 of Chapter 3) that in the stochastic case the pair $(C_0; A_0, A_1)$ is neither detectable nor observable. The maximal solution of the equation (4.50) is

$$\tilde{X}(t) = \begin{bmatrix} 8 & -21 \\ -21 & 63 \end{bmatrix} > 0$$

and the minimal solution is

$$\tilde{\tilde{X}}(t) = \begin{bmatrix} 1 & 0 \\ 0 & 0 \end{bmatrix} \geq 0.$$

Indeed, by Theorem 15(iv) of Chapter 2, $\tilde{X}$ is the stabilizing solution of (4.50), and based on Corollary 8 it coincides with the maximal solution.

On the other hand, if $X_\tau(\cdot)$ is the solution of (4.50) with the given final condition $X_\tau(\tau) = 0$, one obtains that

$$X_\tau(t) = \begin{bmatrix} X(t) & 0 \\ 0 & 0 \end{bmatrix},$$

where

$$X(t) = \frac{1 - e^{-5(\tau - t)}}{1 + 4e^{-5(\tau - t)}} \quad \text{for all } t \leq \tau.$$

Therefore,

$$\lim_{\tau \to \infty} X_\tau(t) = \begin{bmatrix} 1 & 0 \\ 0 & 0 \end{bmatrix} = \tilde{\tilde{X}},$$

and thus one obtains that $\tilde{\tilde{X}}$ is the minimal semipositive definite solution of (4.50). Obviously in this case $\tilde{X} \neq \tilde{\tilde{X}}$.

4.5 The filtering Riccati equation

In this section we focus our attention on the so-called *stochastic generalized filtering Riccati equation* (SGFRE) for stochastic systems. We shall restrict our investigation only to the time-invariant case.

Consider the SGFRE:

$$A_0(i)Y(i) + Y(i)A_0^*(i) + \sum_{k=1}^{r} A_k(i)Y(i)A_k^*(i) + \sum_{j=1}^{d} q_{ji} Y(j)$$

$$- \left(Y(i)C_0^*(i) + \sum_{k=1}^{r} A_k(i)Y(i)C_k^*(i) + \tilde{L}^*(i) \right)$$

$$\times \left(\tilde{R}(i) + \sum_{k=1}^{r} C_k(i)Y(i)C_k^*(i) \right)^{-1}$$

$$\times \left(Y(i)C_0^*(i) + \sum_{k=1}^{r} A_k(i)Y(i)C_k^*(i) + \tilde{L}^*(i) \right) + \tilde{M}(i) = 0 \quad (4.51)$$

with the unknown variables $(Y(1), \ldots, Y(d)) \in \mathcal{S}_n^d$ and $A_k(i) \in \mathbf{R}^{n \times n}$, $C_k(i) \in \mathbf{R}^{p \times n}$, $k = 0, \ldots, r$, $\tilde{L}(i) \in \mathbf{R}^{n \times p}$, $\tilde{R}(i) \in \mathcal{S}_p$, $\tilde{M}(i) \in \mathcal{S}_n$. If $\mathcal{D} = \{1\}$, $A_k(i) = 0$,

$C_k(i) = 0$, $k = 1, 2, \ldots, r$, then (4.51) reduces to the well-known Bucy–Kalman [117] filtering algebraic Riccati equation.

The system (4.51) can be rewritten in compact form as a nonlinear equation in $\mathcal{S}_n^d$ as follows:

$$\mathcal{L}Y - \widetilde{\mathcal{P}}(Y)\widetilde{\mathcal{R}}^{-1}(Y)\widetilde{\mathcal{P}}^*(Y) + \widetilde{M} = 0, \qquad (4.52)$$

where $\mathcal{L}$ is the Lyapunov operator defined by the system $(A_0, A_1, \ldots, A_r; Q)$, $\widetilde{\mathcal{P}} : \mathcal{S}_n^d \to \mathcal{M}_{np}^d$ by

$$\widetilde{\mathcal{P}}(Y) = \left(\widetilde{\mathcal{P}}_1(Y), \ldots, \widetilde{\mathcal{P}}_d(Y)\right),$$

$$\widetilde{\mathcal{P}}_i(Y) = Y(i)C_0^*(i) + \sum_{k=1}^{r} A_k(i)Y(i)C_k^*(i) + \widetilde{L}(i), \ i \in \mathcal{D},$$

$$\widetilde{\mathcal{R}} : \mathcal{S}_n^d \to \mathcal{S}_p \text{ by}$$

$$\widetilde{\mathcal{R}}(Y) = \left(\widetilde{\mathcal{R}}_1(Y), \ldots, \widetilde{\mathcal{R}}_d(Y)\right),$$

$$\widetilde{\mathcal{R}}_i(Y) = \widetilde{\mathcal{R}}(i) + \sum_{k=1}^{r} C_k(i)Y(i)C_k^*(i), \ i \in \mathcal{D},$$

$$\widetilde{M} = \left(\widetilde{M}(1), \ldots, \widetilde{M}(d)\right).$$

Equation (4.52) is defined on a subset of $\mathcal{S}_n^d$ consisting of $Y = (Y(1), \ldots, Y(d))$ such that $\det \widetilde{\mathcal{R}}_i(Y) \neq 0$.

The dissipation matrix corresponding to the filtering Riccati equation under investigation is defined as follows:

$$\widetilde{\mathcal{N}}(Y) = \left(\widetilde{\mathcal{N}}_1(Y), \ldots, \widetilde{\mathcal{N}}_d(Y)\right), \text{ where}$$

$$\widetilde{\mathcal{N}}_i(Y) = \begin{bmatrix} (\mathcal{L}Y)(i) + \widetilde{M}(i) & \widetilde{\mathcal{P}}_i(Y) \\ \widetilde{\mathcal{P}}_i^*(Y) & \widetilde{\mathcal{R}}_i(Y) \end{bmatrix}$$

for all $Y \in \mathcal{S}_n^d$, $i \in \mathcal{D}$.

Definition 4. *A solution* $\widetilde{Y} = \left(\widetilde{Y}(1), \ldots, \widetilde{Y}(d)\right)$ *of (4.52) is a stabilizing solution if the system* $\left(A_0 + C_0\widetilde{K}, A_1 + C_1\widetilde{K}, \ldots, A_r + C_r\widetilde{K}; Q\right)$ *is stable in the sense of Definition 4 of Chapter 2, where* $\widetilde{K} = \left(\widetilde{K}(1), \ldots, \widetilde{K}(d)\right)$,

$$\widetilde{K}(i) = -\widetilde{\mathcal{P}}_i(\widetilde{Y})\widetilde{\mathcal{R}}_i^{-1}(\widetilde{Y}), \ i \in \mathcal{D}. \qquad (4.53)$$

Recalling that $\mathbf{A} = (A_0, \ldots, A_r)$ and $\mathbf{C} = (C_0, \ldots, C_r)$ we prove the following result.

Theorem 19. *The following are equivalent.*

(i) $(\mathbf{C}, \mathbf{A}; Q)$ *is detectable and there exists* $\widehat{Y} = (\widehat{Y}(1), \ldots, \widehat{Y}(d)) \in \mathcal{S}_n^d$ *satisfying* $\widetilde{\mathcal{N}}(\widehat{Y}) > 0$.

(ii) *The equation (4.52) has a stabilizing solution* $\widetilde{Y}$ *which verifies* $\widetilde{\mathcal{R}}_i(\widetilde{Y}) > 0$.

Proof. It is easy to see that equation (4.52) is an equation of type (4.3) associated with the triple $(\mathbf{A}^\sharp, \mathbf{C}^\sharp; Q^\sharp)$, where $\mathbf{A}^\sharp = (A_0^\sharp, \ldots, A_r^\sharp)$, $\mathbf{C}^\sharp = (C_0^\sharp, \ldots, C_r^\sharp)$, and $Q^\sharp = Q^*$, $A_k^\sharp = (A_k^*(1), \ldots, A_k^*(d))$, $C_k^\sharp = (C_k^*(1), \ldots, C_k^*(d))$, $k = 0, \ldots, r$.

From Remark 4 of Chapter 3 it follows that $(\mathbf{A}^\sharp, \mathbf{C}^\sharp; Q^\sharp)$ is stabilizable if and only if $(\mathbf{C}, \mathbf{A}; Q)$ is detectable. The result in the statement follows then from Theorem 7. □

4.6 Iterative procedures

In the first part of this section we present an iterative procedure to compute the maximal solution $\widetilde{X}(t)$ of the equation (4.3), or equivalently the maximal solution of the system (4.1). This procedure may also provide a proof of the implication (i) → (ii) in Theorem 3.

We have the following lemma.

Lemma 20. *Assume that the system (4.1) is stochastically stabilizable. Let $\widetilde{F}_0(t) = (\widetilde{F}_0(t, 1), \ \widetilde{F}_0(t, 2), \ldots, \widetilde{F}_0(t, d))$ be a stabilizing feedback gain and let $X_0(t) = (X_0(t, 1), \ \ldots, X_0(t, d))$ be a bounded with bounded derivative solution of the linear differential inequality on $\mathcal{S}_n^d$:*

$$\frac{d}{dt} X_0(t) + \mathcal{L}_{\widetilde{F}_0}^*(t) X_0(t) + M_0(t) \leq 0, \tag{4.54}$$

where $M_0(t) = (M_0(t, 1), \ M_0(t, 2), \ldots, M_0(t, d))$, $M_0(t, i) = M(t, i) + \varepsilon I_n + L(t, i)\widetilde{F}_0(t, i) + \widetilde{F}_0^*(t, i)L^*(t, i) + \widetilde{F}_0(t, i)R(t, i)\widetilde{F}_0(t, i)$, $\varepsilon > 0$ fixed.

Under the considered assumptions, we have

$$X_0(t) - \widehat{X}(t) \gg 0 \tag{4.55}$$

for arbitrary $\widehat{X}(t) \in \Gamma^\Sigma$ of (4.36), which verifies the condition (4.8).

Proof. If $\widehat{X}(t) \in \Gamma^\Sigma$ is a bounded solution of (4.36) that verifies (4.8), we define $\widehat{M}(t) = (\widehat{M}(t, 1), \ \widehat{M}(t, 2), \ldots, \widehat{M}(t, d))$ by

$$\widehat{M}(t) = \frac{d}{dt} \widehat{X}(t) + \mathcal{L}^*(t)\widehat{X}(t) - \mathcal{P}^*(t, \widehat{X}(t))\mathcal{R}^{-1}(t, \widehat{X}(t))\mathcal{P}(t, \widehat{X}(t)) + M(t), t \in \mathbf{R}_+. \tag{4.56}$$

Clearly $\widehat{M}(t) \geq 0$. By Lemma 1 we verify that

$$\frac{d}{dt} \widehat{X}(t) + \mathcal{L}_{\widetilde{F}_0}^*(t)\widehat{X}(t) + M(t) + L(t)\widetilde{F}_0(t) + \widetilde{F}_0^*(t)L(t) + \widetilde{F}_0^*(t)R(t)\widetilde{F}_0(t)$$
$$- \widehat{M}(t) - (\widehat{F}(t) - \widetilde{F}_0(t))^*\mathcal{R}(t, \widehat{X}(t))(\widehat{F}(t) - \widetilde{F}_0(t)) = 0, \tag{4.57}$$

where $\widehat{F}(t) = (\widehat{F}(t, 1), \ \widehat{F}(t, 2), \ldots, \widehat{F}(t, d))$ with

$$\widehat{F}(t, i) = -\mathcal{R}_i^{-1}(t, \widehat{X}(t))\mathcal{P}_i(t, \widehat{X}(t)), \quad t \in \mathcal{I}, \ i \in \mathcal{D}. \tag{4.58}$$

From (4.57) and (4.54) we get

$$\frac{d}{dt}\left(X_0(t) - \widehat{X}(t)\right) + \mathcal{L}^*_{\widetilde{F}_0}(t)\left(X_0(t) - \widehat{X}(t)\right) + \left(\widetilde{F}_0(t) - \widehat{F}(t)\right)^* \mathcal{R}\left(t, \widehat{X}(t)\right)$$
$$\times \left(\widetilde{F}_0(t) - \widehat{F}(t)\right) + \varepsilon J^d + \widehat{M}(t) \leq 0, \quad t \geq 0.$$

This allows us, by Proposition 14 of Chapter 2, to conclude that $X_0(t) - \widetilde{X}(t) \geq Y(t)$, where $t \to Y(t) = (Y(t, 1), Y(t, 2), \ldots, Y(t, d))$ is the unique bounded solution of the Lyapunov-type equation

$$\frac{d}{dt}Y(t) + \mathcal{L}^*_{\widetilde{F}_0}(t)Y(t) + \varepsilon J^d = 0. \tag{4.59}$$

Let $T_0(t, s)$ be the linear evolution operator on $\mathcal{S}^d_n$, defined by the linear differential equation:

$$\frac{d}{dt}S(t) = \mathcal{L}_{\widetilde{F}_0}(t)S(t).$$

Since $\widetilde{F}_0(t)$ is a stabilizing feedback gain, then there exist positive constants β_0, α_0 such that $\|T_0(t, s)\| \leq \beta_0 e^{-\alpha_0(t-s)}$ $\forall t \geq s, \ t, s \in \mathcal{I}$. Therefore the unique bounded solution of (4.59) is uniform positive, and the proof is complete. □

Remark 5. Based on Remark 3 it follows that if the coefficients of system (4.1) are θ-periodic functions, then a stabilizing feedback gain that is a θ-periodic function may be chosen. Therefore in the periodic case the inequality (4.54) has a periodic solution with the same period as the coefficients. Moreover, if the coefficients of the system (4.1) do not depend upon t, we may choose constant solutions of (4.54), $X_0 = (X_0(1), X_0(2), \ldots, X_0(d))$. Detailing (4.54) in the time-invariant case, it follows that X_0 may be obtained as a solution of the following LMI system:

$$\left[A_0(i) + B_0(i)\widetilde{F}_0(i)\right]^* X_0(i) + X_0(i)\left[A_0(i) + B_0(i)\widetilde{F}_0(i)\right]$$
$$+ \sum_{k=1}^{r}\left[A_k(i) + B_k(i)\widetilde{F}_0(i)\right]^* X_0(i)\left[A_k(i) + B_k(i)\widetilde{F}_0(i)\right]$$
$$+ \sum_{j=1}^{d} q_{ij}X_0(j) + M(i) + \varepsilon I_n + L(i)\widetilde{F}_0(i) + \widetilde{F}^*_0(i)L^*(i)$$
$$+ \widetilde{F}^*_0(i)R(i)\widetilde{F}_0(i) \leq 0, i \in \mathcal{D}. \tag{4.60}$$

Based on (4.55) we deduce that there exists $\mu_0 > 0$ such that $\mathcal{R}_i(t, X_0(t)) \geq \mu_0 I_n, t \in \mathcal{I}, i \in \mathcal{D}$. Hence the feedback gain $F_0(t) = (F_0(t, 1), \ldots, F_0(t, d))$ is well defined by

$$F_0(t, i) = -\mathcal{R}_i^{-1}(t, X_0(t))\mathcal{P}_i(t, X_0(t)), i \in \mathcal{D}, t \in \mathcal{I}. \tag{4.61}$$

We will show that $F_0(t)$ is a stabilizing feedback gain for the triple $(\mathbf{A}, \mathbf{B}; Q)$.

To this end we consider $\widehat{X}(t) \in \Gamma^\Sigma$. By direct computation and using (4.56) and (4.61) we get

$$\frac{d}{dt}\widehat{X}(t) + \mathcal{L}_{F_0}^*(t)\widehat{X}(t) + M(t) + L(t)F_0(t) + F_0^*(t)L^*(t) + F_0^*(t)R(t)F_0(t)$$
$$- \left(\widehat{F}(t) - F_0(t)\right)^* \mathcal{R}\left(t, \widehat{X}(t)\right)\left(\widehat{F}(t) - F_0(t)\right) - \widehat{M}(t) = 0. \qquad (4.62)$$

Further, (4.54) may be rewritten as

$$\frac{d}{dt}X_0(t) + \mathcal{L}_{F_0}^*(t)X_0(t) + M(t) + L(t)F_0(t) + F_0^*(t)L^*(t) + F_0^*(t)R(t)F_0(t)$$
$$+ \left(F_0(t) - \widetilde{F}_0(t)\right)^* \mathcal{R}\left(t, X_0(t)\right)\left(F_0(t) - \widetilde{F}_0(t)\right) + \varepsilon J^d \leq 0. \qquad (4.63)$$

From (4.62), (4.63), and (4.55) we deduce that $t \to X_0(t) - \widehat{X}(t)$ is a bounded and uniform positive solution of the linear differential inequation on $\mathcal{S}_n^d$:

$$\frac{d}{dt}X(t) + \mathcal{L}_{F_0}^*(t)X(t) + \frac{\varepsilon}{2}J^d \ll 0.$$

Using Theorem 12(vii) $\to$ (i) of Chapter 2 we deduce that the system $(A_0 + B_0 F_0, A_1 + B_1 F_0, \ldots, A_r + B_r F_0; Q)$ is stable, which shows that $F_0(t)$ is a stabilizing feedback gain. As a consequence we deduce that for each $i \in \mathcal{D}$, the zero state equilibrium of the linear differential equation on $\mathbf{R}^n$,

$$\frac{d}{dt}X(t) = \left(A_0(t, i) + \frac{1}{2}q_{ii} I_n + B_0(t, i)F_0(t, i)\right)X(t),$$

is exponentially stable.

Particularly in the time-invariant case it follows that the eigenvalues of the matrices $A_0(i) + \frac{1}{2}q_{ii} I_n + B_0(i)F_0(i)$ are located in the half-plane $Re\lambda < 0$.

Taking $X_0(t), F_0(t)$ as a first step, we iteratively construct the sequences $\{X_l(t, i)\}_{l \geq 0}, \{F_l(t, i)\}_{l \geq 0}, i \in \mathcal{D}$, as follows: $t \to X_{l+1}(t, i)$ is the unique bounded solution of the Lyapunov equation

$$\frac{d}{dt}X_{l+1}(t, i) + \left[\widetilde{A}_0(t, i) + B_0(t, i)F_l(t, i)\right]^* X_{l+1}(t, i)$$
$$+ X_{l+1}(t, i)\left[\widetilde{A}_0(t, i) + B_0(t, i)F_l(t, i)\right] + M_{l+1}(t, i) = 0, \qquad (4.64)$$

where $M_{l+1}(t) = \left(M_{l+1}(t, 1), \ldots, M_{l+1}(t, d)\right)$ with

$$M_{l+1}(t, i) = M(t, i) + \frac{\varepsilon}{l+2}I_n + L(t, i)F_l(t, i) + F_l^*(t, i)L^*(t, i)$$
$$+ F_l^*(t, i)R(t, i)F_l(t, i) + \sum_{k=1}^{r}[A_k(t, i) + B_k(t, i)F_l(t, i)]^*$$
$$\times X_l(t, i)[A_k(t, i) + B_k(t, i)F_l(t, i)] + \sum_{j \neq i}q_{ij}X_l(t, j),$$

$$\tilde{A}_0(t, i) = A_0(t, i) + \frac{1}{2}q_{ii}I_n$$

$$F_{l+1}(t, i) = -\left(R(t, i) + \sum_{k=1}^{r} B_k^*(t, i)X_l(t, i)B_k(t, i)\right)^{-1} \tag{4.65}$$

$$\times\left(B_0^*(t, i)X_{l+1}(t, i) + \sum_{k=1}^{r} B_k^*(t, i)X_l(t, i)A_k(t, i) + L^*(t, i)\right),$$

$$l \geq 0, i \in \mathcal{D}.$$

Further, we show that

(a) $X_l(t, i) - \widehat{X}(t, i) \geq \mu_l I_n > 0$ for all integers $l \geq 0, i \in \mathcal{D}, t \in \mathcal{I}, \widehat{X}(t) = (\widehat{X}(t, 1) \ldots \widehat{X}(t, d))$ being an arbitrary bounded function in $\mathbf{\Gamma}^\Sigma$ and μ_l is a positive constant that does not depend upon $\widehat{X}(t)$.

(b) The zero state equilibrium of the linear differential equation on $\mathbf{R}^n$,

$$\frac{d}{dt}x(t) = \left[\tilde{A}_0(t, i) + B_0(t, i)F_l(t, i)\right]x(t),$$

is exponentially stable for each $i \in \mathcal{D}, l \geq 0$.

(c) $X_l(t, i) \geq X_{l+1}(t, i) \forall l \geq 0, (t, i) \in \mathcal{I} \times \mathcal{D}$. We remark that the properties (a) and (b) have been proved for $l = 0$. We shall verify by induction that (a), (b), (c) are fulfilled for every $l \geq 0$.

Let us assume that (a), (b), (c) are fulfilled for the first $l - 1$ terms of the sequences defined by (4.64) and (4.65). By direct computation we obtain that if $\widehat{X}(t) \in \mathbf{\Gamma}^\Sigma$, then

$$\frac{d}{dt}\widehat{X}(t, i) + \left[\tilde{A}_0(t, i) + B_0(t, i)F_{l-1}(t, i)\right]^*\widehat{X}(t, i)$$

$$+ \widehat{X}(t, i)\left[\tilde{A}_0(t, i) + B_0(t, i)F_{l-1}(t, i)\right]$$

$$+ \sum_{k=1}^{r}\left[A_k(t, i) + B_k(t, i)F_{l-1}(t, i)\right]^*\widehat{X}(t, i)\left[A_k(t, i) + B_k(t, i)F_{l-1}(t, i)\right]$$

$$+ \sum_{j=1, j\neq i}^{d} q_{ij}\widehat{X}(t, j) + M(t, i) + L(t, i)F_{l-1}(t, i) + F_{l-1}^*(t, i)L^*(t, i)$$

$$+ F_{l-1}^*(t, i)R(t, i)F_{l-1}(t, i) - \widehat{M}(t, i)$$

$$- \left[\widehat{F}(t, i) - F_{l-1}(t, i)\right]^*\mathcal{R}_i\left(t, \widehat{X}(t)\right)\left[\widehat{F}(t, i) - F_{l-1}(t, i)\right] = 0,$$

$\widehat{M}(t, i), \widehat{F}(t, i)$ being defined in (4.56) and (4.58), respectively.

Using (4.64) with l replaced by $l - 1$ we get

$$\frac{d}{dt}\left[X_l(t, i) - \widehat{X}(t, i)\right] + \left[\tilde{A}_0(t, i) + B_0(t, i)F_{l-1}(t, i)\right]^*\left[X_l(t, i) - \widehat{X}(t, i)\right]$$

$$+ \left[X_l(t, i) - \widehat{X}(t, i)\right]\left[\tilde{A}_0(t, i) + B_0(t, i)F_{l-1}(t, i)\right]$$

$$+ \frac{\varepsilon}{l+1}I_n + \Delta_l(t, i) = 0, \tag{4.66}$$

where

$$\Delta_l(t, i) = \sum_{k=1}^{r}[A_k(t, i) + B_k(t, i)F_{l-1}(t, i)]^*[X_{l-1}(t, i) - \widehat{X}(t, i)]$$

$$\times [A_k(t, i) + B_k(t, i)F_{l-1}(t, i)] + \sum_{j=1, j\neq i}^{d} q_{ij}[X_{l-1}(t, j) - \widehat{X}(t, j)]$$

$$+ \widehat{M}(t, i) + [\widehat{F}(t, i) - F_{l-1}(t, i)]^* \mathcal{R}_i(t, \widehat{X}(t))[\widehat{F}(t, i) - F_{l-1}(t, i)].$$

Since $X_{l-1}(t, i) - \widehat{X}(t, i) \geq \mu_{l-1}I_n$ we get $\Delta_l(t, i) \geq 0$. Taking into account that $\widetilde{A}_0(t, i) + B_0(t, i)F_{l-1}(t, i)$ generates an exponentially stable evolution, we may conclude that the equation (4.66) has a unique, bounded solution which is uniform positive definite. Hence there exists $\mu_l > 0$, such that $X_l(t, i) - \widehat{X}(t, i) \geq \mu_l I_n$ and thus (a) is fulfilled. Further we have that $\mathcal{R}_i(t, X_l(t)) \geq \nu_l I_m > 0$.

Using (4.65) we write

$$\frac{d}{dt}X_l(t, i) + [\widetilde{A}_0(t, i) + B_0(t, i)F_l(t, i)]^* X_l(t, i)$$

$$+ X_l(t, i)[\widetilde{A}_0(t, i) + B_0(t, i)F_l(t, i)] + \sum_{k=1}^{r}[A_k(t, i) + B_k(t, i)F_l(t, i)]^*$$

$$\times X_{l-1}(t, i)[A_k(t, i) + B_k(t, i)F_l(t, i)] + \sum_{j=1, j\neq i}^{d} q_{ij}X_{l-1}(t, j) + M(t, i)$$

$$+ \frac{\varepsilon}{l+1}I_n + L(t, i)F_l(t, i) + F_l^*(t, i)L^*(t, i) + F_l^*(t, i)R(t, i)F_l(t, i)$$

$$+ [F_l(t, i) - F_{l-1}(t, i)]^* \mathcal{R}_i(t, X_{l-1}(t))[F_l(t, i) - F_{l-1}(t, i)] = 0. \qquad (4.67)$$

It is easy to see that $t \to \widehat{X}(t, i)$ verifies

$$\frac{d}{dt}\widehat{X}(t, i) + [\widetilde{A}_0(t, i) + B_0(t, i)F_l(t, i)]^* \widehat{X}(t, i)$$

$$+ \widehat{X}(t, i)[\widetilde{A}_0(t, i) + B_0(t, i)F_l(t, i)] + \sum_{k=1}^{r}(A_k(t, i) + B_k(t, i)F_l(t, i))^*$$

$$\times \widehat{X}(t, i)(A_k(t, i) + B_k(t, i)F_l(t, i)) + \sum_{j=1, j\neq i}^{d} q_{ij}\widehat{X}(t, j) + M(t, i)$$

$$+ F_l^*(t, i)L^*(t, i) + L(t, i)F_l(t, i) + F_l^*(t, i)R(t, i)F_l(t, i) - \widehat{M}(t, i)$$

$$- [\widehat{F}(t, i) - F_l(t, i)]^* \mathcal{R}_i(t, \widehat{X}(t))[\widehat{F}(t, i) - F_l(t, i)] = 0.$$

Thus we obtain that for each $i \in \mathcal{D}, t \to X_l(t, i) - \widehat{X}(t, i)$ is a bounded and uniformly positive definite solution of the linear differential inequality

$$\frac{d}{dt}Y(t, i) + [\widetilde{A}_0(t, i) + B_0(t, i)F_l(t, i)]^* Y(t, i)$$

$$+ Y(t, i)[\widetilde{A}_0(t, i) + B_0(t, i)F_l(t, i)] + \frac{\varepsilon}{2(l+1)}I_n < 0,$$

which allow us to conclude that the zero state equilibrium of the linear differential equation

$$\frac{d}{dt}x(t) = \left(\tilde{A}_0(t, i) + B_0(t, i)F_l(t, i)\right)x(t) \tag{4.68}$$

is exponentially stable and (b) is fulfilled.

Subtracting (4.64) from (4.67) we get that $t \to X_l(t, i) - X_{l+1}(t, i)$ is a bounded solution of the equation

$$\frac{d}{dt}(X_l(t, i) - X_{l+1}(t, i)) + \left(\tilde{A}_0(t, i) + B_0(t, i)F_l(t, i)\right)^*(X_l(t, i) - X_{l+1}(t, i))$$
$$+ (X_l(t, i) - X_{l+1}(t, i))\left(\tilde{A}_0(t, i) + B_0(t, i)F_l(t, i)\right) + \hat{\Delta}_l(t, i) = 0, \tag{4.69}$$

where

$$\hat{\Delta}_l(t, i) = \frac{\varepsilon}{(l+1)(l+2)}I_n + [F_l(t, i) - F_{l-1}(t, i)]^*\mathcal{R}_i(t, X_{l-1}(t))$$
$$\times [F_l(t, i) - F_{l-1}(t, i)] + \sum_{k=1}^{r}[A_k(t, i) + B_k(t, i)F_l(t, i)]^*$$
$$\times (X_{l-1}(t, i) - X_l(t, i))[A_k(t, i) + B_k(t, i)F_l(t, i)]$$
$$+ \sum_{j=1, j\neq i}^{d} q_{ij}(X_{l-1}(t, j) - X_l(t, j))$$

for $l \geq 1$ and

$$\hat{\Delta}_l(t, i) \geq \frac{\varepsilon}{2}I_n + \left(F_0(t, i) - \tilde{F}_0(t, i)\right)^*\mathcal{R}_i(t, X_0(t))\left(F_0(t, i) - \tilde{F}_0(t, i)\right)$$

for $l = 0$.

Since $\hat{\Delta}_0(t, i) \geq 0$ and the zero state equilibrium of (4.68) for $l = 0$ is exponentially stable, it follows from (4.69) for $l = 0$ that $X_0(t, i) - X_1(t, i) \geq 0$, and further, by induction, we obtain that $\hat{\Delta}_l \geq 0$ for $l \geq 1$, which leads to $X_l(t, i) - X_{l+1}(t, i) \geq 0$; thus (c) is fulfilled.

From (a) and (c) we conclude that the sequences $\{X_l(t, i)\}_{l \geq 0}$, $i \in \mathcal{D}$ are convergent. More precisely we have the following theorem.

Theorem 21. *Assume that*

(i) *the system* $(\mathbf{A}, \mathbf{B}; Q)$ *is stabilizable, and*

(ii) *There exists* $\tilde{X}(t, i) \in \Gamma^\Sigma$, $(t, i) \in \mathcal{I} \times \mathcal{D}$. *Then for any choice of a stabilizing feedback gain* $\tilde{F}_0(t) = (\tilde{F}_0(t, 1), \tilde{F}_0(t, 2), \ldots, \tilde{F}_0(t, d))$, *the sequences* $\{X_l(t, i)\}_{l \geq 0}$, $i \in \mathcal{D}$, *constructed as solutions of (4.64) (the first terms* $X_0(t, i)$ *obtained by solving (4.54)) are convergent. If*

$$\tilde{X}(t, i) = \lim_{l\to\infty} X_l(t, i), \quad (t, i) \in \mathcal{I} \times \mathcal{D}, \tag{4.70}$$

then $\tilde{X}(t) = (\tilde{X}(t, 1), \tilde{X}(t, 2), \ldots, \tilde{X}(t, d))$ *is the maximal bounded solution of the system (4.1) verifying (4.8).*   □

Remark 6. (i) If condition (i) of Theorem 7 is fulfilled, the solution $\widetilde{X}(t)$ provided by (4.70) is just the stabilizing solution of the system (4.1).

(ii) Excepting the first step, when to obtain $X_0(t, i)$ we need to solve a system of linear inequalities of higher dimension, namely (4.54), to obtain the next terms of the sequences $\{X_l(t, i)\}_{l \geq 1}, i \in \mathcal{D}$, we need to solve a system of d uncoupled Lyapunov equations. We remark that to compute the gains $F_l(t, i)$ in (4.65) we need both the value of $X_l(t, i)$ and the value of $X_{l-1}(t, i)$.

(iii) Based on the uniqueness of the bounded solution of a Lyapunov equation, it follows that if the coefficients of the system (4.1) do not depend upon t, then the matrices X_l and F_l do not depend upon t. In this case (4.64) and (4.65) become

$$
\begin{aligned}
&\left[\widetilde{A}_0(i) + B_0(i)F_{l-1}(i)\right]^* X_l(i) \\
&\quad + X_l(i)\left[\widetilde{A}_0(i) + B_0(i)F_{l-1}(i)\right] + M_l(i) = 0, i \in \mathcal{D},
\end{aligned}
\tag{4.71}
$$

$$
\begin{aligned}
M_l(i) = M(i) &+ \frac{\varepsilon}{l+1} I_n + L(i)F_{l-1}(i) + F_{l-1}^*(i)L^*(i) + F_{l-1}^*(i)R(i)F_{l-1}(i), \\
&+ \sum_{k=1}^{r}[A_k(i) + B_k(i)F_{l-1}(i)]^* X_{l-1}(i)[A_k(i) + B_k(i)F_{l-1}(i)] \\
&+ \sum_{j=1, j \neq i}^{d} q_{ij} X_{l-1}(j), \quad l \geq 1,
\end{aligned}
$$

$$
\widetilde{A}_0(i) = A_0(i) + \frac{1}{2}q_{ii}I_n,
$$

$$
\begin{aligned}
F_l(i) = -&\left[R(i) + \sum_{k=1}^{r} B_k^*(i)X_{l-1}(i)B_k(i)\right]^{-1} \\
&\times \left[B_0^*(i)X_l(i) + \sum_{k=1}^{r} B_k^*(i)X_{l-1}(i)A_k(i) + L^*(i)\right], \quad l \geq 1,
\end{aligned}
\tag{4.72}
$$

while $X_0(i)$ is obtained solving the following system of LMIs:

$$
\begin{aligned}
&\left[A_0(i) + B_0(i)\widetilde{F}_0(i)\right]^* X_0(i) + X_0(i)[A_0(i) + B_0(i)\widetilde{F}_0(i)] \\
&\quad + \sum_{k=1}^{r}\left[A_k(i) + B_k(i)\widetilde{F}_0(i)\right]^* X_0(i)\left[A_k(i) + B_k(i)\widetilde{F}_0(i)\right] \\
&\quad + \sum_{j=1}^{d} q_{ij} X_0(j) + M(i) + \varepsilon I_n + L(i)\widetilde{F}_0(i) + \widetilde{F}_0^*(i)L^*(i) \\
&\quad + \widetilde{F}_0^*(i)R(i)\widetilde{F}_0(i) \leq 0, \quad i \in \mathcal{D},
\end{aligned}
\tag{4.73}
$$

and

$$F_0(i) = - \left[R(i) + \sum_{k=1}^{r} B_k^*(i) X_0(i) B_k(i) \right]^{-1}$$
$$\times \left[B_0^*(i) X_0(i) + \sum_{k=1}^{r} B_k^*(i) X_0(i) A_k(i) + L^*(i) \right].$$

(iv) In addition, from the uniqueness of the bounded solution of a Lyapunov equation, we deduce that if the coefficients of the system (4.1) are θ-periodic functions defined on $\mathbf{R}$, then the bounded solutions of (4.64) are θ-periodic functions too. Hence it is sufficient to compute the values of $X_l(t, i)$, $F_l(t, i)$ on the interval $[0, \theta]$. At each step l, the initial condition $X_l(0, i)$ is obtained by solving the linear equation

$$X_l(0, i) = \Phi_{l,i}^*(\theta, 0) X_l(0, i) \Phi_{l,i}(\theta, 0) + \int_0^{\theta} \Phi_{l,i}^*(s, 0) M_l(s, i) \Phi_{l,i}(s, 0) ds,$$

$\Phi_{l,i}(t, s)$ being the fundamental matrix solution of (4.68). For the first step, $X_0(t, i)$ is chosen as a periodic solution of the Lyapunov-type equation on $\mathcal{S}_n$:

$$\frac{d}{dt} X_0(t) + \mathcal{L}_{F_0}^*(t) X_0(t) + M_0(t) = 0,$$

where $M_0(t) = (M_0(t, 1), \ M_0(t, 2), \ldots, \ M_0(t, d))$,

$$M_0(t, i) = M(t, i) + \varepsilon I_n + L(t, i) \widetilde{F}_0(t, i) + \widetilde{F}_0^*(t, i) L^*(t, i)$$
$$+ \widetilde{F}_0^*(t, i) R(t, i) \widetilde{F}_0(t, i).$$

If $T_0(t, t_0)$ is the linear evolution operator defined by the linear differential equation on $\mathcal{S}_n^d$:

$$\frac{d}{dt} S(t) = \mathcal{L}_{\widetilde{F}_0}(t) S(t), \tag{4.74}$$

then the initial condition $X_0(0) = (X_0(0, 1), \ X_0(0, 2), \ldots, \ X_0(0, d))$ is given by

$$X_0(0) = \left[\widetilde{J} - T_0^*(\theta, 0) \right]^{-1} \int_0^{\theta} T_0^*(s, 0) M_0(s) ds,$$

where $\widetilde{J}$ is the identity operator on $\mathcal{S}_n^d$; $\widetilde{J} - T_0^*(\theta, 0)$ is invertible due to the exponential stability of the evolution defined by the differential equation (4.74).

In the final part of this section we present a procedure to compute the minimal semipositive solution $\widetilde{\widetilde{X}}(t)$.

First, we recall that the minimal solution $\widetilde{\widetilde{X}}(t)$ is obtained as

$$\widetilde{\widetilde{X}}(t) = \lim_{\tau \to \infty} X_\tau(t), \tag{4.75}$$

where $X_\tau(t) = (X_\tau(t, 1), X_\tau(t, 2), \ldots, X_\tau(t, d))$ is the solution of the system (4.44) with the terminal condition $X_\tau(\tau, i) = 0$, $i \in \mathcal{D}$ (see the proof of Theorem 14).

Let us consider the following systems of Itô differential equations:

$$dx_i(t) = \left[\tilde{A}_0(t, i)x_i(t) + B_0(t, i)u_i(t)\right]dt \tag{4.76}$$

$$+ \sum_{k=1}^{r} \left[A_k(t, i)x_i(t) + B_k(t, i)u_i(t)\right]dw_k(t),$$

$$y_i(t) = C_0(t, i)x_i(t), \quad i \in \mathcal{D},$$

where

$$\tilde{A}_0(t, i) = A_0(t, i) + \frac{1}{2}q_{ii}I_n.$$

For each $i \in \mathcal{D}$, we consider the Riccati-type differential equation

$$\frac{d}{dt}X_i(t) + \tilde{A}_0^*(t, i)X_i(t) + X_i(t)\tilde{A}_0(t, i) + \sum_{k=1}^{r} A_k^*(t, i)X_i(t)A_k(t, i)$$

$$- \left[X_i(t)B_0(t, i) + \sum_{k=1}^{r} A_k^*(t, i)X_i(t)B_k(t, i)\right]$$

$$\times \left[R(t, i) + \sum_{k=1}^{r} B_k^*(t, i)X_i(t)B_k(t, i)\right]^{-1}$$

$$\times \left[B_0^*(t, i)X_i(t) + \sum_{k=1}^{r} B_k^*(t, i)X_i(t)A_k(t, i)\right] + C_0^*(t, i)C_0(t, i) = 0. \tag{4.77}$$

If for each $i \in \mathcal{D}$, the system (4.76) is stochastically stabilizable and stochastically detectable or stochastically uniformly observable, then invoking Corollary 18 we obtain that the equation (4.77) has a bounded, stabilizing, and semipositive definite solution $X_i^0(t)$.

Taking $X_i^0(t)$ as a first step, we construct the sequences $\{X_i^l(t)\}_{l \geq 0}$, $i \in \mathcal{D}$, where for each l, $t \to X_i^l(t)$ is the unique bounded semipositive and stabilizing solution of the Riccati differential equation:

$$\frac{d}{dt}X_i^l(t) + \tilde{A}_0^*(t, i)X_i^l(t) + X_i^l(t)\tilde{A}_0(t, i) + \sum_{k=1}^{r} A_k^*(t, i)X_i^l(t)A_k(t, i)$$

$$- \left[X_i^l(t)B_0(t, i) + \sum_{k=1}^{r} A_k^*(t, i)X_i^l(t)B_k(t, i)\right]$$

$$\times \left[R(t, i) + \sum_{k=1}^{r} B_k^*(t, i)X_i^l(t)B_k(t, i)\right]^{-1}$$

$$\times \left[B_0^*(t, i)X_i^l(t) + \sum_{k=1}^{r} B_k^*(t, i)X_i^l(t)A_k(t, i)\right] + \tilde{M}_l(t, i) = 0, \tag{4.78}$$

where $\tilde{M}_l(t, i) = C_0^*(t, i)C_0(t, i) + \sum_{j=1, j \neq i}^{d} q_{ij}X_j^{l-1}(t)$.

Remark 7. Clearly, for each fixed $i \in \mathcal{D}$, the equation (4.78) is just the Riccati equation (4.49) associated with the following controlled system with multiplicative white noise:

$$dx_i(t) = \left[\widetilde{A}_0(t, i)x_i(t) + B_0(t, i)u(t)\right] dt \tag{4.79}$$

$$+ \sum_{k=1}^{r} [A_k(t, i)x_i(t) + B_k(t, i)u(t)]dw_k(t),$$

$$\widetilde{y}_l(t) = \widetilde{C}_l(t, i)x_i(t),$$

where

$$\widetilde{A}_0(t, i) = A_0(t, i) + \frac{1}{2}q_{ii}I_n,$$

$$\widetilde{C}_l(t, i) = \begin{pmatrix} C_0(t, i) \\ \widehat{C}_l(t, i) \end{pmatrix}, \widehat{C}_l(t, i) = \left[\sum_{j \neq i} q_{ij} X_j^{l-1}(t)\right]^{\frac{1}{2}}.$$

It is easy to see that if the system (4.76) is stochastically detectable, then the system (4.79) is stochastically detectable, and if the system (4.76) is stochastically uniformly observable, then (4.79) is stochastically uniformly observable too.

Proposition 22. *Assume that for each $i \in \mathcal{D}$,*
    (a) *the system (4.76) is stochastically stabilizable,*
    (b) *the system (4.76) is stochastically detectable or stochastically uniformly observable.*
Under these assumptions we have that
    (i) $X_i^{l+1}(t) \geq X_i^l(t) \geq 0$ *that* $l \geq 0, i \in \mathcal{D}, t \in \mathbf{R}_+$;
    (ii) $X_i^l(t) \leq \widehat{X}(t, i), (t, i) \in \mathbf{R}_+ \times \mathcal{D}, l \geq 0 \forall \widehat{X}(t) = (\widehat{X}(t, 1), \dots, \widehat{X}(t, d))$
semipositive and bounded solution of (4.44).

Proof. Combining Remark 7 with Corollary 18 we deduce that (4.78) has a stabilizing semipositive and bounded solution $X_i^l(t), l \geq 0, i \in \mathcal{D}$. By induction we obtain that $\check{M}_l(t, i) \geq 0$, which leads to $X_i^l(t) \geq 0$.

For each $l \geq 0, i \in \mathcal{D}$, consider the stabilizing feedback gain defined as follows:

$$F_i^l(t) = -\left[R(t, i) + \sum_{k=1}^{r} B_k^*(t, i)X_i^l(t)B_k(t, i)\right]^{-1} \tag{4.80}$$

$$\times \left[B_0^*(t, i)X_i^l(t, i) + \sum_{k=1}^{r} B_k^*(t, i)X_i^l(t)A_k(t, i)\right].$$

By direct calculation using (4.78) and (4.80) (for l replaced by $l+1$) we obtain

$$\frac{d}{dt}X_i^{l+1}(t) + \left[\widetilde{A}_0(t,i) + B_0(t,i)F_i^{l+1}(t,i)\right]^* X_i^{l+1}$$
$$+ X_i^{l+1}\left[\widetilde{A}_0(t,i) + B_0(t,i)F_i^{l+1}(t)\right]$$
$$+ \sum_{k=1}^{r}\left[A_k(t,i) + B_k(t,i)F_i^{l+1}(t)\right]^* X_i^{l+1}(t)\left[A_k(t,i) + B_k(t,i)F_i^{l+1}(t)\right]$$
$$+ \widetilde{M}_{l+1}(t,i) + \left(F_i^{l+1}(t)\right)^* R(t,i)F_i^{l+1}(t) = 0,$$

$$\frac{d}{dt}X_i^l(t) + \left[\widetilde{A}_0(t,i) + B_0(t,i)F_i^{l+1}(t)\right]^* X_i^l(t) + X_i^l(t)\left[\widetilde{A}_0(t,i) + B_0(t,i)F_i^{l+1}(t)\right]$$
$$+ \sum_{k=1}^{r}\left[A_k(t,i) + B_k(t,i)F_i^{l+1}(t)\right]^* X_i^l(t)\left[A_k(t,i) + B_k(t,i)F_i^{l+1}(t)\right]$$
$$+ \widetilde{M}_l(t,i) + (F_i^{l+1}(t))^* R(t,i)F_i^{l+1}(t),$$
$$- \left(F_i^{l+1}(t) - F_i^l(t)\right)^* \left(R(t,i) + \sum_{k=1}^{r} B_k^*(t,i)X_i^l(t,i)B_k(t,i)\right)$$
$$\times \left(F_i^{l+1}(t) - F_i^l(t)\right) = 0,$$

which leads to the fact that $t \to X_i^{l+1}(t) - X_i^l(t)$ is the bounded solution of the Lyapunov equation on S_n:

$$\frac{d}{dt}Y(t,i) + \left[\widetilde{A}_0(t,i) + B_0(t,i)F_i^{l+1}(t)\right]^* Y_i(t)$$
$$+ Y_i(t)\left[\widetilde{A}_0(t,i) + B_0(t,i)F_i^{l+1}(t)\right] + \sum_{k=1}^{r}\left[A_k(t,i) + B_k(t,i)F_i^{l+1}(t)\right]^*$$
$$\times Y_i(t)\left[A_k(t,i) + B_k(t,i)F_i^{l+1}(t)\right] + \widetilde{\Delta}_l(t,i) = 0, \tag{4.81}$$

where

$$\widetilde{\Delta}_l(t,i) = \sum_{j\neq i, j=1}^{d} q_{ij}\left[X_j^l(t) - X_j^{l-1}(t)\right] + \left(F_i^{l+1}(t) - F_i^l(t)\right)^*$$
$$\times \left[R(t,i) + \sum_{k=1}^{r} B_k^*(t,i)X_i^l(t)B_k(t,i)\right]\left(F_i^{l+1}(t) - F_i^l(t)\right).$$

Let $T_{l+1,i}(t,s)$ be the linear evolution operator on S_n defined by (4.81) with $\widetilde{\Delta}_l(t,i) = 0$.

Since $X_i^{l+1}(t)$ is the stabilizing solution of (4.78), we have $\|T_{l+1,i}(t,s)\| \leq \beta_{l+1,i}e^{-\alpha_{l+1,i}(t-s)}$ for some positive constants $\beta_{l+1,i}, \alpha_{l+1,i}$.

From the uniqueness of the bounded solution of the equation (4.81) we deduce that

$$X_i^{l+1}(t) - X_i^l(t) = \int_t^\infty T_{l+1,i}^*(s, t)\widetilde{\Delta}_l(s, i)\, ds.$$

Since $T_{l+1}^*(s, t)$ is a positive operator on $\mathcal{S}_n$, from the above equality we obtain that

$$\left(X_i^{l+1}(t) - X_i^l(t)\right) \geq 0$$

if $\widetilde{\Delta}_l(s, i) \geq 0$. This can be checked easily by induction.
    For $l = 0$ we have

$$\widetilde{\Delta}_0(s, i) = \sum_{j \neq i}^d q_{ij} X_j^0(s) + \left(F_i^1(s) - F_i^0(s)\right)^*$$

$$\times \left(R(s, i) + \sum_{k=1}^r B_k^*(s, i) X_i^0(t) B_k(s, i)\right) \left(F_i^1(s) - F_i^0(s)\right) \geq 0.$$

Thus assertion (i) in the statement is completely proved.
    To prove (ii) we recall that

$$X_i^l(t) = \lim_{\tau \to \infty} X_{\tau,i}^l(t) \tag{4.82}$$

(see the proof of Theorem 14), where $X_{\tau,i}^l(t)$ is the solution of the equation (4.78) with the terminal condition $X_{i,\tau}^l(\tau) = 0$. Let $\widehat{X}(t) = (\widehat{X}(t, 1)\, \widehat{X}(t, 2) \ldots \widehat{X}(t, d))$ be a bounded and semipositive solution of the system (4.44) and let $\widehat{F}(t) = (\widehat{F}(t, 1) \times \widehat{F}(t, 2) \ldots \widehat{F}(t, d))$ be the corresponding feedback gain, i.e., $\widehat{F}(t, i) = -\mathcal{R}_i^{-1}(t, \widehat{X}(t))\mathcal{P}_i(t, \widehat{X}(t))$, $i \in \mathcal{D}, t \geq 0$.
    By direct calculation we get:

$$\frac{d}{dt}\widehat{X}(t, i) + \left[\widetilde{A}_0(t, i) + B_0(t, i)\widehat{F}(t, i)\right]^*\widehat{X}(t, i)$$

$$+ \widehat{X}(t, i)\left[\widetilde{A}_0(t, i) + B_0(t, i)\widehat{F}(t, i)\right]$$

$$+ \sum_{k=1}^r \left[A_k(t, i) + B_k(t, i)\widehat{F}(t, i)\right]^*$$

$$\times \widehat{X}(t, i)\left[A_k(t, i) + B_k(t, i)\widehat{F}(t, i)\right] + \left[C_0(t, i) + D_0(t, i)\widehat{F}(t, i)\right]^*$$

$$\times \left[C_0(t, i) + D_0(t, i)\widehat{F}(t, i)\right] + \sum_{j=1, j \neq i}^d q_{ij}\widehat{X}(t, j) = 0$$

$$\frac{d}{dt}X_{\tau,i}^l(t) + \left[\tilde{A}_0(t,i) + B_0(t,i)\widehat{F}(t,i)\right]^* X_{\tau,i}^l(t)$$

$$+ X_{\tau,i}^l(t)\left[\tilde{A}_0(t,i) + B_0(t,i)\widehat{F}(t,i)\right] + \sum_{k=1}^{r}\left[A_k(t,i) + B_k(t,i)\widehat{F}(t,i)\right]^*$$

$$\times X_{i,\tau}^l(t)\left[A_k(t,i) + B_k(t,i)\widehat{F}(t,i)\right] + \sum_{j=1,j\neq i}^{d} q_{ij}X_{\tau,j}^{l-1}(t)$$

$$- \left[\widehat{F}(t,i) - F_{\tau,i}^l(t)\right]^*\left[R(t,i) + \sum_{k=1}^{r} B_k^*(t,i)X_{\tau,i}^l(t)B_k(t,i)\right]$$

$$\times \left[\widehat{F}(t,i) - F_{\tau,i}^l(t)\right] + C_0^*(t,i)C_0(t,i) + \widehat{F}^*(t,i)R(t,i)\widehat{F}(t,i) = 0,$$

where $F_{\tau,i}^l(t)$ is as in (4.80), with $X_i^l(t)$ replaced by $X_{\tau,i}^l(t)$.

We obtain in this way that $t \to \widehat{X}(t,i) - X_{\tau,i}^l(t)$ is the solution of the problem

$$\frac{d}{dt}Y_i(t) + \widehat{\mathcal{L}}_i^*(t)Y_i(t) + \widehat{\Delta}_l(t,i) = 0, \qquad (4.83)$$

$Y_i(\tau) = \widehat{X}(\tau,i) \geq 0$, where $\widehat{\mathcal{L}}_i^*(t)$ is the adjoint operator of the linear Lyapunov operator on $\mathcal{S}_n$ defined by

$$\mathcal{L}_i(t)Y = \left[\tilde{A}_0(t,i) + B_0(t,i)\widehat{F}(t,i)\right]Y + Y\left[\tilde{A}_0(t,i) + B_0(t,i)\widehat{F}(t,i)\right]^*$$

$$+ \sum_{k=1}^{r}\left[A_k(t,i) + B_k(t,i)\widehat{F}(t,i)\right]Y\left[A_k(t,i) + B_k(t,i)\widehat{F}(t,i)\right]^*$$

and

$$\widehat{\Delta}_l(t,i) = \sum_{j=1,j\neq i}^{d} q_{ij}\left(\widehat{X}(t,j) - X_{\tau,j}^{l-1}(t)\right) + \left(\widehat{F}(t,i) - F_{\tau,i}^l(t,i)\right)^*$$

$$\times \left[R(t,i) + \sum_{k=1}^{r} B_k^*(t,i)X_{\tau,i}^l(t)B_k(t,i)\right]\left(\widehat{F}(t,i) - F_{\tau,i}^l(t)\right).$$

If $\widehat{T}_i(t,s)$ is the linear evolution operator on $\mathcal{S}_n$ defined by the linear differential equation

$$\frac{d}{dt}Y(t) = \widehat{\mathcal{L}}_i(t)Y(t),$$

then from (4.83) we have the representation formula

$$\widehat{X}(t,i) - X_{\tau,i}^l(t) = \widehat{T}_i^*(\tau,t)\widehat{X}(\tau,i) + \int_t^{\tau}\widehat{T}_i^*(s,t)\widehat{\Delta}_l(s,i)\,ds, \ 0 \leq t \leq \tau.$$

Since $\widehat{T}_i^*(s,t)$ is a linear positive operator on $\mathcal{S}_n$, then from the above equality we deduce that $\widehat{X}(t,i) - X_{\tau,i}^l(t) \geq 0 \ \forall 0 \leq t \leq \tau, i \in \mathcal{D}$ if $\widehat{\Delta}_l(s,i) \geq 0$. This last condition may be checked by induction. To this end, we remark that if $l = 0$, we have

$$\widehat{\Delta}_0(s, i) = \sum_{j=1, j \neq i}^{d} q_{ij} \widehat{X}(s, j) + \left(\widehat{F}(s, i) - F^0_{\tau, i}(s) \right)^*$$

$$\times \left[R(s, i) + \sum_{k=1}^{r} B^*_k(s, i) X^0_{\tau, i}(s) B_k(s, i) \right] \left(\widehat{F}(s, i) - F^0_{\tau, i}(s) \right)$$

and it is obvious that $\widehat{\Delta}_0(s, i) \geq 0$, $0 \leq s \leq \tau < \infty$, $i \in \mathcal{D}$, which leads to $\widehat{X}(t, i) - X^0_{\tau, i}(t) \geq 0$; further, invoking (4.82) with $l = 0$ we conclude that $\widehat{X}(t, i) - X^0_i(t) \geq 0$ and the proof is complete.   □

Theorem 23. *Assume that:*

(i) $(\mathbf{A}, \mathbf{B}; Q)$ *is stabilizable;*

(ii) for each $i \in \mathcal{D}$*, the system* (4.76) *is either stochastically detectable or stochastically uniformly observable.*

Let be the sequences $\{X^l_i(t)\}_{l \geq 0}$*,* $i \in \mathcal{D}$*, where* $X^l_i(t)$ *is the unique bounded and stabilizing solution of the equation* (4.78). *Under the considered assumptions these sequences are convergent, and if we define* $\widetilde{\widetilde{X}}(t, i) = \lim_{l \to \infty} X^l_i(t)$*,* $(t, i) \in \mathbf{R}_+ \times \mathcal{D}$*, then* $\widetilde{\widetilde{X}}(t) = (\widetilde{\widetilde{X}}(t, 1) \ldots \widetilde{\widetilde{X}}(t, d))$ *is the minimal semipositive and bounded solution of the system* (4.44).

Proof. If $(\mathbf{A}, \mathbf{B}; Q)$ is stabilizable, then for each $i \in \mathcal{D}$, the system (4.76) is stochastically stabilizable. Therefore the assumptions of Proposition 22 are fulfilled and the sequences $\{X^l_i(t)\}_{l \geq 1}$, $i \in \mathcal{D}$ are well defined and monotonically increasing.

On the other hand, if assumption (i) is fulfilled, then applying Theorem 14 we obtain that the set of semipositive and bounded solutions of the system (4.44) is not empty. From Proposition 22(ii) we deduce that the sequences $\{X^l_i(t)\}_{l \geq 1}$, $i \in \mathcal{D}$, are bounded above. Then the functions $\widetilde{\widetilde{X}}(t, i)$ are well defined by $\widetilde{\widetilde{X}}(t, i) = \lim_{l \to \infty} X^l_i(t)$. By a standard method (based on the Lebesgue Theorem) we obtain that $\widetilde{\widetilde{X}}(t) = (\widetilde{\widetilde{X}}(t, 1) \ldots \widetilde{\widetilde{X}}(t, d))$ is a semipositive and bounded solution of (4.44). Applying Proposition 22(ii) again, we obtain that $\widetilde{\widetilde{X}}$ is the minimal semipositive and bounded solution of (4.44) and the proof is complete.   □

Remark 8. (i) In the particular case $A_k(t, i) = 0$, $B_k(t, i) = 0$, $k = 1, 2, \ldots, r$, and the system is in the time-invariant case, the iterative procedure proposed in the previous theorem was used in [1] to compute the stabilizing solution of a system of coupled algebraic Riccati equations associated with a linear system with Markovian jumping.

(ii) If for each $i \in \mathcal{D}$ the system (4.76) is stochastically uniformly observable, then the system $(C_0, A_0, \ldots, A_r; Q)$ is uniformly observable (see Proposition 9(iii) of Chapter 3), and in this case the solution $\widetilde{\widetilde{X}}(t)$ obtained in the previous theorem is just the stabilizing, bounded, and semipositive solution of the system (4.44).

(iii) At each step $l \geq 0$ the stabilizing solution $X^l_i(t)$ of (4.77) and (4.78), respectively, can be computed using the procedure provided by Theorem 21.

Numerical examples We shall illustrate the above iterative numerical procedures considering the linear time-invariant stochastic system of order $n = 2$, subjected to both multiplicative noise and Markovian jumps with $r = 1$ and $\mathcal{D} = \{1, 2\}$ having:

$$A_0(1) = \begin{bmatrix} -1 & 0 \\ 1 & -1 \end{bmatrix}, \quad A_0(2) = \begin{bmatrix} -1 & 1 \\ 0 & -1 \end{bmatrix},$$

$$A_1(1) = \begin{bmatrix} -1 & 1 \\ 0 & -2 \end{bmatrix}, \quad A_1(2) = \begin{bmatrix} -2 & 1 \\ 1 & -1 \end{bmatrix},$$

$$B_0(1) = \begin{bmatrix} 1 \\ -1 \end{bmatrix}, \quad B_0(2) = \begin{bmatrix} 1 \\ 1 \end{bmatrix},$$

$$L_0(1) = \begin{bmatrix} 1 \\ -1 \end{bmatrix}, \quad L_0(2) = \begin{bmatrix} 1 \\ 2 \end{bmatrix},$$

$$M_0(1) = \begin{bmatrix} 1 & 1 \\ 1 & 2 \end{bmatrix}, \quad M_0(2) = \begin{bmatrix} 1 & 1 \\ 1 & 4 \end{bmatrix},$$

$$R(1) = 1, \quad R(2) = 2.$$

Our purpose is to solve the SGRDE (4.1) corresponding to the above numerical values using the iterative procedure indicated in the statement of Theorem 21. Three distinct cases have been considered: the case when the system is subjected only to Markov jumps, the case when the system is subjected only to multiplicative white noise, and the case when the system is perturbed with both multiplicative white noise and Markovian jumps.

Case a. *The Markovian jumping case:* $A_1(i) = 0$, $B_1(i) = 0$, $i \in \mathcal{D}$. Using Proposition 3 in Chapter 3 we determined for the numerical values above that

$$\widetilde{F}_0(1) = [0.5923 \quad -0.7004], \quad \widetilde{F}_0(2) = [-0.0330 \quad 0.0653].$$

Then, solving (4.60), we obtained

$$X_0(1) = 10^3 \begin{bmatrix} 1.5519 & -0.0524 \\ -0.0524 & 1.7776 \end{bmatrix},$$

$$X_0(2) = 10^3 \begin{bmatrix} 1.1139 & 0.2680 \\ 0.2680 & 1.3970 \end{bmatrix}.$$

The solution of (4.1) for this case was determined solving (4.60) iteratively. For an imposed level of accuracy $\|X_{l+1}(i) - X_l(i)\| < 10^{-6}$ we obtained after 69 iterations:

$$X(1) = \begin{bmatrix} 30.7868 & 24.3960 \\ 24.3960 & 26.2218 \end{bmatrix}$$

$$X(2) = \begin{bmatrix} 21.5504 & -11.7226 \\ -11.7226 & 19.2254 \end{bmatrix}.$$

Case b. *The multiplicative white noise perturbations case:* $\mathcal{D} = \{1\}$, $A_i = A_i(1)$; $B_i = B_i(1)$, $i = 0, 1$.

In this case we obtained the initial values

$$\widetilde{F}_0 = [-0.4094 \quad 0.8482],$$

$$X_0 = \begin{bmatrix} 292.8945 & 163.9337 \\ 163.9337 & 140.9240 \end{bmatrix},$$

and after 202 iterations, the solution of (4.1):

$$X = \begin{bmatrix} 1.0782 & 1.0307 \\ 1.0307 & 0.5878 \end{bmatrix}.$$

Case c. The case when the system is subjected to both Markovian jumps and multiplicative white noise: In this situation we obtained the initial values:

$$\widetilde{F}_0(1) = [-0.3852 \quad 0.8594], \quad \widetilde{F}_0(2) = [-0.9000 \quad 0.5763],$$

$$X_0(1) = 10^8 \begin{bmatrix} 5.8005 & -4.5733 \\ -4.5733 & -3.7733 \end{bmatrix},$$

$$X_0(2) = 10^8 \begin{bmatrix} -0.7123 & -0.5110 \\ 0.5110 & -4.8453 \end{bmatrix}.$$

The solution of (4.1) was obtained after 133 iterations solving (4.60); thus we obtained

$$X(1) = \begin{bmatrix} 2.1893 & 2.0159 \\ 2.0159 & 2.0998 \end{bmatrix}, \quad X(2) = \begin{bmatrix} 0.7940 & -0.4088 \\ -0.4088 & 3.3714 \end{bmatrix}.$$

Notes and references

The Riccati equations of stochastic control were generally studied in connection with the linear quadratic problem either for controlled linear stochastic systems with state-dependent noise or for systems with Markov perturbations. For references concerning linear quadratic problems in the stochastic framework, see Chapter 5. Most of the results contained in this chapter were published for the first time in [30]. The iterative procedure to compute the stabilizing solution of SGRDE was also published in [31] and in [35]. Classes of nonlinear matrix differential equations which contain as particular cases Riccati differential equations arising in control problems for stochastic systems with multiplicative white noise have been studied in [23], [24], [28], [50], [51]. Iterative procedures for computation of the stabilizing solution of the algebraic Riccati equations associated with the linear stochastic systems with multiplicative white noise may be found in [57]. Iterative procedures to compute the stabilizing solution of systems of Riccati equations involved in the linear quadratic problem for stochastic systems with Markov parameters can be found, for example, in [1], [53]. Several aspects concerning the algebraic Riccati equations arising in the control of linear stochastic systems may be found in [22], [2] where rich lists of references dealing with symmetric and nonsymmetric Riccati equations may be found.

5

Linear Quadratic Control Problem for Linear Stochastic Systems

In this chapter as well as in the next two chapters one shows how the mathematical results derived in the previous chapters are involved in the design of stabilizing controllers with some imposed performances for a wide class of linear stochastic systems. The design problem of some stabilizing controls minimizing quadratic performance criteria is studied. The first two sections of this chapter deal with the so-called *linear quadratic optimization problem*. It will be seen that, depending on the class of admissible controls, the corresponding optimal control is obtained either with the stabilizing solution or with the minimal solution of a corresponding system of generalized Riccati differential equations. We also consider the case when the weights matrices do not have definite sign. Such situations may occur in a natural way in economy, ecology, and financial applications. A tracking problem is considered in Section 5.3.

In the last part of the chapter, the stochastic H^2 control problem is considered and solved in two significant cases: the full state access and the output feedback case, respectively.

5.1 Formulation of the linear quadratic problem

The linear quadratic optimization problem (LQOP) has received much attention in the control literature due to its wide area of applications. A more detailed overview of the main results obtained for stochastic and Markovian systems can be found in the "Notes and References" of this chapter. The main objective of the theoretical developments presented in the following consists in providing a unified approach to solving the LQOP for systems subjected both to multiplicative white noise and to Markovian jumping, the dynamics of which is described by the state-space equation:

$$dx(t) = \left[A_0(t, \eta(t))x(t) + B_0(t, \eta(t))u(t)\right]dt$$

$$+ \sum_{k=1}^{r} \left[A_k(t, \eta(t))x(t) + B_k(t, \eta(t))u(t)\right]dw_k(t), \qquad (5.1)$$

where $t \in \mathbf{R}_+$, with the state vector $x \in \mathbf{R}^n$ and with the control inputs $u \in \mathbf{R}^m$.

Let us consider the cost function

$$
J_1(t_0, x_0, u) = E \int_{t_0}^{\infty} \left[x_u^*(t) M(t, \eta(t)) x_u(t) + x_u^*(t) L(t, \eta(t)) u(t) \right.
$$
$$
\left. + u^*(t) L^*(t, \eta(t)) x_u(t) + u^*(t) R(t, \eta(t)) u(t) \right] dt, \quad (5.2)
$$

where $M(t, i) = M^*(t, i)$; $R(t, i) = R^*(t, i)$, $(t, i) \in \mathbf{R}_+ \times \mathcal{D}$; and $x_u(t)$ denotes the solution of the system (5.1) corresponding to the input $u(.)$ with the initial condition $(t_0, x_0) \in \mathbf{R}_+ \times \mathbf{R}^n$.

Two problems will be treated in the present chapter: the first one consists in determining the optimal state-feedback control:

$$
u(t) = F(t, \eta(t)) x(t), \quad (5.3)
$$

which stabilizes (5.1) and minimizes the cost function (5.2). The class of admissible controls for this problem is the set $\widetilde{\mathcal{U}}(t_0, x_0)$ of stochastic processes $u(t) \in L^2_{\eta, w}([t_0, T], \mathbf{R}^m)$ for all $T > t_0$, with the additional properties that $J_1(t_0, x_0, u)$ exists, and it is finite and $\lim_{t \to \infty} E|x_u(t)|^2 = 0$. The fact that $J_1(t_0, x_0, u)$ exists means that there exists

$$
\lim_{T \to \infty} E \int_{t_0}^{T} \left[x_u^*(t) M(t, \eta(t)) x_u(t) + x_u^*(t) L(t, \eta(t)) u(t) \right.
$$
$$
\left. + u^*(t) L^*(t, \eta(t)) x_u(t) + u^*(t) R(t, \eta(t)) u(t) \right] dt \in \mathbf{R}.
$$

An important feature specific to the systems subjected to multiplicative white noise is the one related to the *well-posedness* of the problem. Indeed, it will be shown that in contrast with the deterministic case, where the matrix

$$
\begin{bmatrix} M(t, i) & L(t, i) \\ L^*(t, i) & R(t, i) \end{bmatrix}
$$

must be positive semidefinite, in the stochastic case this condition is not necessary. In this chapter the optimization problem described by the controlled system (5.1), the cost functional (5.2), and the set of admissible controls $\widetilde{\mathcal{U}}(t_0, x_0)$ will be called the *first linear quadratic optimization problem* (LQOP1).

The second problem treated in the present chapter requires us to find the control of the form (5.3) such that the cost function

$$
J_2(t_0, x_0, u) = E \int_{t_0}^{\infty} |y_u(t)|^2 \, dt \quad (5.4)
$$

is minimized in the class $\mathcal{U}(t_0, x_0)$ of all stochastic processes

$$
u \in L^2_{\eta, w}([t_0, T], \mathbf{R}^m)
$$

for all $T > t_0$, $J_2(t_0, x_0, u) < \infty$, where

$$
y_u(t) = y_u(t, t_0, x_0) = C_0(t, \eta(t)) x_u(t) + D_0(t, \eta(t)) u(t) \quad (5.5)
$$

is an output in $\mathbf{R}^p$ This problem will be termed the *second linear quadratic optimization problem* (LQOP2).

In order to simplify the expressions involved in the solution of this problem we make the following assumption.

Assumption A (a) There exists $\rho > 0$ such that $D_0^*(t, i)D_0(t, i) \geq \rho I_m \ \forall(t, i) \in \mathbf{R}_+ \times \mathcal{D}$.

(b) $D_0^*(t, i)C_0(t, i) = 0 \forall(t, i) \in \mathbf{R}_+ \times \mathcal{D}$.

Remark 1. If the system (5.1) with the output (5.5) verifies (a), then without loss of generality, (b) is fulfilled. Indeed, if (a) is fulfilled, then by the change of control variables described by

$$u(t) = -\left[D_0^*(t, \eta(t))D_0(t, \eta(t))\right]^{-1}D_0^*(t, \eta(t))C_0(t, \eta(t))x(t) + \tilde{u}(t),$$

we may replace the given system (5.1)–(5.5) by the following modified system:

$$dx(t) = \left[\widehat{A}_0(t, \eta(t))x(t) + B_0(t, \eta(t))\tilde{u}(t)\right]dt$$

$$+ \sum_{k=1}^{r}\left[\widehat{A}_k(t, \eta(t))x(t) + B_k(t, \eta(t))\tilde{u}(t)\right]dw_k(t),$$

$$y(t) = \widehat{C}_0(t, \eta(t))x(t) + D_0(t, \eta(t))\tilde{u}(t),$$

where

$$\widehat{A}_k(t, i) = A_k(t, i) - B_k(t, i)R^{-1}(t, i)D_0^*(t, i)C_0(t, i), \quad k = 0, 1, \ldots, r,$$
$$\widehat{C}_0(t, i) = \left[I_p - D_0(t, i)R^{-1}(t, i)D_0^*(t, i)\right]C_0(t, i),$$
$$R(t, i) = D_0^*(t, i)D_0(t, i), \quad (t, i) \in \mathbf{R}_+ \times \mathcal{D}.$$

Clearly, this new system verifies both (a) and (b) of Assumption A.

5.2 Solution of the linear quadratic problems

In this section we shall present solutions of the optimization problems stated in Section 5.1. First, we recall several results which will be used repeatedly in subsequent developments.

For each quadruple $(t_0, \tau, x_0, i), 0 \leq t_0 < \tau < \infty, x_0 \in \mathbf{R}^n, i \in \mathcal{D}$, we consider the auxiliary cost functions $J(t_0, \tau, x_0, i, \cdot) : L^2_{\eta, w}([t_0, \tau], \mathbf{R}^m) \to \mathbf{R}$ by

$$J(t_0, \tau, x_0, i; u) = E\left[\int_{t_0}^{\tau}[x^*(t) \quad u^*(t)]\mathcal{M}(t, \eta(t))\begin{bmatrix} x(t) \\ u(t) \end{bmatrix}dt\,|\eta(t_0) = i\right],$$

(5.6)

where

$$\mathcal{M}(t, i) = \begin{pmatrix} M(t, i) & L(t, i) \\ L^*(t, i) & R(t, i) \end{pmatrix} \quad \text{and} \quad x(t) = x_u(t, t_0, x_0)$$

is the solution of the system (5.1) corresponding to the input $u(t)$ and having the initial condition (t_0, x_0).

Applying the Itô-type formula (Theorem 35 of Chapter 1) we obtain the following lemma.

Lemma 1. *If $t \to K(t, i) : \mathbf{R}_+ \to \mathcal{S}_n, i \in \mathcal{D}$, are C^1-functions, then we have*

$$J(t_0, \tau, x_0, i; u) = x_0^* K(t_0, i)x_0 - E[x^*(\tau)K(\tau, \eta(\tau))x(\tau)|\eta(t_0) = i]$$

$$+ E\left[\int_{t_0}^{\tau} [x^*(t) \quad u^*(t)]\mathcal{M}^K(t, \eta(t)) \begin{bmatrix} x(t) \\ u(t) \end{bmatrix} dt | \eta(t_0) = i \right],$$

$(\forall)\ 0 \leq t_0 < \tau < \infty,\ x_0 \in \mathbf{R}^n,\ i \in \mathcal{D},\ u \in L^2_{\eta, w}([t_0, \tau], \mathbf{R}^m)$, where

$$\mathcal{M}^K(t, i) = \begin{bmatrix} \mathcal{M}_{11}^K(t, i) & \mathcal{M}_{12}^K(t, i) \\ (\mathcal{M}_{12}^K(t, i))^* & \mathcal{M}_{22}^K(t, i) \end{bmatrix}$$

with

$$\mathcal{M}_{11}^K(t, i) = \frac{d}{dt}K(t, i) + A_0^*(t, i)K(t, i) + K(t, i)A_0(t, i)$$

$$+ \sum_{k=1}^{r} A_k^*(t, i)K(t, i)A_k(t, i) + \sum_{j=1}^{d} q_{ij}K(t, j) + M(t, i)$$

$$= \frac{d}{dt}K(t, i) + [\mathcal{L}^*(t)K(t)](i) + M(t, i),$$

$$\mathcal{M}_{12}^K(t, i) = K(t, i)B_0(t, i) + \sum_{k=1}^{r} A_k^*(t, i)K(t, i)B_k(t, i) + L(t, i)$$

$$= \mathcal{P}_i^*(t, K(t)),$$

$$\mathcal{M}_{22}^K(t, i) = R(t, i) + \sum_{k=1}^{r} B_k^*(t, i)K(t, i)B_k(t, i)$$

$$= \mathcal{R}_i(t, K(t)). \qquad \square$$

Corollary 2. *If $X(t) = (X(t, 1),\ X(t, 2), \ldots, X(t, d))$ is a solution of the system (4.1) defined on $[t_0, \tau]$, then we have*

$$J(t_0, \tau, x_0, i; u) = x_0^* X(t_0, i)x_0 - E[x^*(\tau)X(\tau, \eta(\tau))x(\tau)|\eta(t_0) = i]$$

$$+ E\left[\int_{t_0}^{\tau} \left(u(t) - F^X(t, \eta(t))x(t)\right)^* \left[R(t, \eta(t))\right.\right.$$

$$+ \sum_{k=1}^{r} B_k^*(t, \eta(t))X(t, \eta(t))B_k(t, \eta(t))\bigg](u(t)$$

$$- F^X(t, \eta(t))x(t))dt|\eta(t_0) = i \bigg] \qquad (5.7)$$

$\forall u \in L^2_{\eta, w}([t_0, \tau], \mathbf{R}^m),\ x_0 \in \mathbf{R}^n, i \in \mathcal{D}$, where

$$F^X(t, i) = -\mathcal{R}_i^{-1}(t, X(t))\mathcal{P}_i(t, X(t)), \qquad (5.8)$$

$(t, i) \in [t_0, \tau] \times \mathcal{D}$, and $x(t) = x_u(t, t_0, x_0)$. $\qquad \square$

5.2.1 Solution of LQOP1

In the following, we investigate the LQOP described by the cost function (5.2) and the system (5.1). As is shown in [4] and [14], while the cost functions of type (5.4) are always bounded below, the cost function J_1 may have values that approach $-\infty$. The same thing is expected to happen in the case of systems subjected both to multiplicative white noise and Markovian jumping.

For each $(t_0, x_0) \in \mathbf{R}_+ \times \mathbf{R}^n$ we denote

$$V(t_0, x_0) = \inf_{u \in \widetilde{\mathcal{U}}(t_0, x_0)} J_1(t_0, x_0, u),$$

the value function associated with the optimization problem.

Definition 1. *We say that the optimization problem described by the cost function (5.2) and the system (5.1) is* well posed *if* $-\infty < V(t_0, x_0) < \infty$ *for all* $(t_0, x_0) \in \mathbf{R}_+ \times \mathbf{R}^n$.

With the notations introduced in the previous chapter we have the following theorem.

Theorem 3. *Assume that*
   (i) *the system (5.1) is stochastically stabilizable;*
   (ii) *the set* Γ^Σ *defined in (4.6) is not empty.*
   Under the above conditions, the linear quadratic optimization problem described by the cost function (5.2) and the system (5.1) is well posed. Moreover,

$$V(t_0, x_0) = \sum_{i \in \mathcal{D}} \pi_i(t_0) x_0^* \widetilde{X}(t_0, i) x_0, \tag{5.9}$$

where $\pi_i(t_0) = \mathcal{P}(\eta(t_0) = i)$ *and* $\widetilde{X}(t) = (\widetilde{X}(t, 1) \dots \widetilde{X}(t, d))$ *is the maximal bounded solution of the system (4.1), which verifies*

$$\mathcal{R}_i(t, \widetilde{X}(t)) \geq \tilde{\rho} I_n > 0. \tag{5.10}$$

Proof. Let us remark that the assumption (i) implies $\widetilde{\mathcal{U}}(t_0, x_0) \neq \phi$ for all $t_0 \geq 0$ and $x_0 \in \mathbf{R}^n$. Based on Theorem 3 of Chapter 4 we deduce that the system (5.1) has a maximal solution $\widetilde{X}(t)$ which verifies (5.10). Applying Corollary 2 for $X(t, i)$ replaced by $\widetilde{X}(t, i)$, we get

$$J(t_0, \tau, x_0, i, u) = x_0^* \widetilde{X}(t_0, i) x_0 - E\left[x^*(\tau) \widetilde{X}(\tau, \eta(\tau)) x(\tau) | \eta(t_0) = i\right]$$
$$+ E\left[\int_{t_0}^{\tau} \left(u(t) - \widetilde{F}(t, \eta(t)) x(t)\right)^* \mathcal{R}_{\eta(t)}(t, \widetilde{X}(t))\right.$$
$$\left. \times \left(u(t) - \widetilde{F}(t, \eta(t)) x(t)\right) dt | \eta(t_0) = i\right] \tag{5.11}$$

for all $u \in \widetilde{\mathcal{U}}_m(t_0, x_0)$, $t_0 < \tau$, $x_0 \in \mathbf{R}^n$, $i \in \mathcal{D}$, where $\widetilde{F}(t, i)$ is defined as in (5.8).

Since $\widetilde{X}(t)$ is a bounded solution, it follows that there exists $\tilde{c} > 0$ such that $|\widetilde{X}(t, i)| \leq \tilde{c} \, \forall (t, i) \in \mathbf{R}^+ \times \mathcal{D}$. Then, from the inequality

$$|E[x^*(\tau)\widetilde{X}(\tau, \eta(\tau))x(\tau)|\eta(t_0) = i] \leq \tilde{c}E[|x(\tau)|^2|\eta(t_0) = i],$$

we obtain

$$\lim_{\tau \to \infty} E\big[x^*(\tau)\widetilde{X}(\tau, \eta(\tau))x(\tau)|\eta(t_0) = i\big] = 0.$$

Taking the limit in (5.11) we get

$$J_1(t_0, x_0, u) = \sum_{i \in \mathcal{D}} \pi_i(t_0)x_0^*\widetilde{X}(t_0, i)x_0 + \sum_{i \in \mathcal{D}} \pi_i(t_0)E$$
$$\times \Big[\int_{t_0}^{\infty} \big(u(t) - \widetilde{F}(t, \eta(t))x(t)\big)^* \mathcal{R}_{\eta(t)}(t, \widetilde{X}(t))(u(t)$$
$$- \widetilde{F}(t, \eta(t))x(t))dt|\eta(t_0) = i\Big] \tag{5.12}$$

for all $u \in \widetilde{\widetilde{\mathcal{U}}}(t_0, x_0)$, $x_0 \in \mathbf{R}^n$, $t_0 \in \mathbf{R}_+$. Combining (5.12) with (5.10) we obtain that $J_1(t_0, x_0, u) \geq \sum_{i \in \mathcal{D}} \pi_i(t_0)x_0^*\widetilde{X}(t_0, i)x_0 \; \forall u \in \widetilde{\widetilde{\mathcal{U}}}(t_0, x_0)$, which leads to

$$V(t_0, x_0) \geq \sum_{i \in \mathcal{D}} \pi_i(t_0)x_0^*\widetilde{X}(t_0, i)x_0.$$

This last inequality shows the well-posedness of the considered optimization problem. It remains to show that (5.9) holds.

To this end let us consider the following perturbed differential equations on $\mathcal{S}_n^d$:

$$\frac{d}{dt}X(t) + \mathcal{L}^*(t)X(t) - \mathcal{P}^*(t, X(t))\mathcal{R}^{-1}(t, X(t))\mathcal{P}(t, X(t)) + M(t) + \varepsilon_l J^d = 0, \tag{5.13}$$

where $\{\varepsilon_l\}_{l \geq 0}$ is a monotonically decreasing sequence with $\lim_{l \to \infty} \varepsilon_l = 0$.

Applying Theorem 7 of Chapter 4 (one uses the assumptions (i) and (ii)) we deduce that the equation (5.13) has a bounded and stabilizing solution $X_{\varepsilon_l}(t)$. Reasoning as in the proof of Theorem 3 of Chapter 4 we deduce that the sequence $\{X_{\varepsilon_l}(t)\}_{l \geq 0}$ is convergent and $\lim_{l \to \infty} X_{\varepsilon_l}(t) = \widetilde{X}(t)$, where $\widetilde{X}(t)$ is the maximal solution of the system (4.1) which verifies (5.10).

For each $l \geq 0$ we associate the cost function

$$J^{\varepsilon_l}(t_0, x_0, u) = E\Big[\int_{t_0}^{\infty} \{x^*(t)(M(t, \eta(t)) + \varepsilon_l I_n)x(t) + x^*(t)L(t, \eta(t))u(t)$$
$$+ u^*(t)L^*(t, \eta(t))x(t) + u^*(t)R(t, \eta(t))u(t)\}dt\Big],$$

$u \in \widetilde{\widetilde{\mathcal{U}}}(t_0, x_0)$. Clearly,

$$J^{\varepsilon_l}(t_0, x_0, u) = J_1(t_0, x_0, u) + \varepsilon_l E\Big[\int_{t_0}^{\infty} |x(t)|^2 dt\Big]. \tag{5.14}$$

Reasoning as in the first part of the proof we obtain the analogue of (5.12) for the perturbed cost function $J^{\varepsilon l}(t_0, x_0, u)$:

$$J^{\varepsilon l}(t_0, x_0, u) = \sum_{i \in \mathcal{D}} \pi_i(t_0) x_0^* X_{\varepsilon_l}(t_0, i) x_0$$

$$+ \sum_{i \in \mathcal{D}} \pi_i(t_0) E \left[\int_{t_0}^{\infty} (u(t) - F_{\varepsilon_l}(t, \eta(t)) x(t))^* \mathcal{R}_{\eta(t)}(t, X_{\varepsilon_l}(t))(u(t) \right.$$

$$\left. - F_{\varepsilon_l}(t, \eta(t)) x(t)) dt | \eta(t_0) = i \right] \qquad (5.15)$$

$\forall u \in \widetilde{\mathcal{U}}(t_0, x_0)$.

Let us consider the control

$$u_{\varepsilon_l}(t) = F_{\varepsilon_l}(t, \eta(t)) x_{\varepsilon_l}(t),$$

where

$$F_{\varepsilon_l}(t, i) = -\mathcal{R}_i^{-1}(t, X_{\varepsilon_l}(t)) \mathcal{P}_i(t, X_{\varepsilon_l}(t))$$

and $x_{\varepsilon_l}(t)$ is the solution of system (5.1) corresponding to the control $u_{\varepsilon_l}(t)$ and $x_{\varepsilon_l}(t_0) = x_0$.

Since $X_{\varepsilon_l}(t)$ is a stabilizing solution of the system (5.13), it follows that $u_{\varepsilon_l} \in \widetilde{\mathcal{U}}_m(t_0, x_0)$. Hence, from (5.15), with $u(t)$ replaced by $u_{\varepsilon_l}(t)$, we obtain

$$J^{\varepsilon l}(t_0, x_0, u_{\varepsilon_l}) = \sum_{i \in \mathcal{D}} \pi_i(t_0) x_0^* X_{\varepsilon_l}(t_0, i) x_0.$$

Therefore,

$$\sum_{i \in \mathcal{D}} \pi_i(t_0) x_0^* X_{\varepsilon_l}(t_0, i) x_0 = J^{\varepsilon l}(t_0, x_0, u_{\varepsilon_l}) \geq J_1(t_0, x_0, u_{\varepsilon_l})$$

$$\geq V(t_0, x_0) \geq \sum_{i \in \mathcal{D}} \pi_i(t_0) x_0^* \widetilde{X}(t_0, i) x_0,$$

and taking the limit for $l \to \infty$, we obtain that (5.9) holds and the proof is complete. □

Definition 2. A pair $(\tilde{x}(t), \tilde{u}(t))$, where $\tilde{u}(t) \in \widetilde{\mathcal{U}}(t_0, x_0)$ and $\tilde{x}(t) = x_{\tilde{u}}(t, t_0, x_0)$ is the solution of (5.1) corresponding to the input $\tilde{u}(t)$, is called optimal pair if $V(t_0, x_0) = J_1(t_0, x_0, \tilde{u})$. In this case the control $\tilde{u}(t)$ is termed the optimal control.

Corollary 4. Assume that the system (4.1) has a bounded and stabilizing solution, $\widetilde{X}(t) = (\widetilde{X}(t, 1) \ldots \widetilde{X}(t, d))$, which verifies (5.10). Set $\tilde{u}(t) = \widetilde{F}(t, \eta(t)) \tilde{x}(t)$, $\widetilde{F}(t, i) = -\mathcal{R}_i^{-1}(t, \widetilde{X}(t)) \mathcal{P}_i(t, \widetilde{X}(t))$, and let $\tilde{x}(t)$ be a solution of system (5.1) corresponding to the control $\tilde{u}$, $\tilde{x}(t_0) = x_0$. Under these assumptions $(\tilde{x}(t), \tilde{u}(t))$ is an optimal pair for the optimization problem described by (5.1)–(5.2).

Proof. From Corollary 8 of Chapter 4 it follows that the bounded and stabilizing solution of (4.1), if it exists, is just the maximal bounded solution $\widetilde{X}(t)$ which verifies (5.10). Now, the conclusion of this corollary follows in an obvious way, from (5.9), since $\tilde{u} \in \widetilde{\mathcal{U}}(t_0, x_0)$.   □

Theorem 5. *Assume that the assumptions of Theorem 3 hold. Then the linear quadratic optimization problem described by (5.1)–(5.2), has an optimal pair $(\hat{x}(t), \hat{u}(t))$ for some (t_0, x_0) if and only if*

$$\lim_{t \to \infty} x_0^* \big[T_{\tilde{F}}^*(t, t_0) J^d \big](i) x_0 = 0 \qquad (5.16)$$

$\forall i \in \mathcal{D}$, *where $T_{\tilde{F}}(t, t_0)$ is the linear evolution operator on $\mathcal{S}_n^d$ defined by the linear differential equation*

$$\frac{d}{dt} S(t) = \mathcal{L}_{\tilde{F}}(t) S(t); \qquad (5.17)$$

$\tilde{F}(t) = \big(\tilde{F}(t, 1) \ldots \tilde{F}(t, d) \big)$ *is associated by (5.8) to the maximal bounded solution of (4.1), which verifies (5.10).*

Proof. Let $(\hat{x}(t), \hat{u}(t))$ be an optimal pair. Using (5.12) we may write

$$V(t_0, x_0) = J_1(t_0, x_0, \hat{u}) = \sum_{i=1}^{d} \pi_i(t_0) x_0^* \widetilde{X}(t_0, i) x_0$$

$$+ E \left[\int_{t_0}^{\infty} \big(\hat{u}(t) - \tilde{F}(t, \eta(t)) \hat{x}(t) \big)^* \mathcal{R}_{\eta(t)} \big(t, \widetilde{X}(t) \big) \big(\hat{u}(t) \right.$$

$$\left. - \tilde{F}(t, \eta(t)) \hat{x}(t) \big) dt \right].$$

Taking into account the value of $V(t_0, x_0)$ given in Theorem 3, we get

$$E \left[\int_{t_0}^{\infty} \big(\hat{u}(t) - \tilde{F}(t, \eta(t)) \hat{x}(t) \big)^* \mathcal{R}_{\eta(t)} \big(t, \widetilde{X}(t) \big) \big(\hat{u}(t) - \tilde{F}(t, \eta(t)) \hat{x}(t) \big) dt \right] = 0,$$

which leads to

$$\hat{u}(t) - \tilde{F}(t, \eta(t)) \hat{x}(t) = 0, \qquad \text{a.e.}$$

By the uniqueness arguments we deduce that $\hat{x}(t)$ coincides a.s. with the solution $\tilde{x}(t)$ of the problem:

$$dx(t) = \big[A_0(t, \eta(t)) + B_0(t, \eta(t)) \tilde{F}(t, \eta(t)) \big] x(t) dt \qquad (5.18)$$

$$+ \sum_{k=1}^{r} \big[A_k(t, \eta(t)) + B_k(t, \eta(t)) \tilde{F}(t, \eta(t)) \big] x(t) dw_k(t),$$

$t \geq t_0, \; x(t_0) = x_0.$

Hence $\hat{u}(t)$ coincide a.s. with $\tilde{u}(t)$ given by $\tilde{u}(t) = \tilde{F}(t, \eta(t))\tilde{x}(t)$.

Let $\tilde{\Phi}(t, t_0)$ be the fundamental matrix solution of the stochastic differential equation (5.18), hence

$$\tilde{x}(t) = \tilde{\Phi}(t, t_0)x_0.$$

Since the optimal control $\tilde{u}(t) \in \tilde{\mathcal{U}}_m(t_0, x_0)$, it follows that

$$\lim_{t \to \infty} E\big[|\tilde{\Phi}(t, t_0)x_0|^2|\eta(t_0) = i\big] = 0, i \in \mathcal{D}.$$

Based on the representation formula given in Theorem 4 of Chapter 2, we obtain (5.16). The converse implication follows in a similar way. □

Corollary 6. *Suppose that the assumptions of Theorem 3 are fulfilled. Then the following are equivalent:*

(i) *For each* $(t_0, x_0) \in \mathbf{R}_+ \times \mathbf{R}^n$ *the optimization problem described by* (5.1)–(5.2) *has an optimal control* $u^{(t_0, x_0)}$, *that is,*

$$V(t_0, x_0) = J_1\big(t_0, x_0, u^{(t_0, x_0)}\big).$$

(ii)

$$\lim_{t \to \infty} \|T_{\tilde{F}}(t, t_0)\| = 0, \ \forall t_0 \geq 0, \tag{5.19}$$

and $T_{\tilde{F}}(t, t_0)$ *is the linear evolution operator defined by the differential equation* (5.17).

If (i) *or* (ii) *holds, then* $u^{(t_0, x_0)}(t) = \tilde{F}(t, \eta(t))\tilde{x}(t)$, *where* $\tilde{x}(t)$ *is the solution of* (5.18).

Proof. The proof follows immediately, taking into account that (5.16) is fulfilled for all $t_0 \geq 0, i \in \mathcal{D}, x_0 \in \mathbf{R}^n$, and

$$\|T_{\tilde{F}}^*(t, t_0)\| = |T_{\tilde{F}}^*(t, t_0)J^d| = \max_{i \in \mathcal{D}} \sup_{|x_0|=1} \{|x_0^*[T_{\tilde{F}}^*(t, t_0)J^d](i)x_0|\},$$

and the norms of the operators $T_{\tilde{F}}^*(t, t_0)$ and $T_{\tilde{F}}(t, t_0)$ are equivalent. □

Remark 2. The property of the evolution operator $T_{\tilde{F}}(t, t_0)$ stated in (5.19) shows that the maximal solution $\tilde{X}(t)$ of the system (4.1) has an additional property which consists in the attractively of the zero solution of the corresponding closed-loop system (5.18), that is,

$$\lim_{t \to \infty} E\big[|\tilde{\Phi}(t, t_0)x_0|^2|\eta(t_0) = i\big] = 0, i \in \mathcal{D}, t_0 \geq 0, x_0 \in \mathbf{R}^n.$$

It must be remarked that, in general, this property is not equivalent to the ESMS of the zero solution of the system (5.18), hence condition (5.19) does not imply that the maximal solution $\tilde{X}(t)$ coincides with the stabilizing solution of the system (4.1).

However, if the coefficients of the system (4.1) are θ-periodic functions, then (5.19) implies that the maximal solution $\tilde{X}(t)$ is just the stabilizing solution of the system (4.1).

This fact is stated in the following theorem.

Theorem 7. *Assume that the coefficients of the system (4.1) are θ-periodic functions and the assumptions of Theorem 3 are fulfilled. Then the following are equivalent:*

(i) *For all $(t_0, x_0) \in \mathbf{R}^+ \times \mathbf{R}^n$ there exists a control $u^{(t_0, x_0)} \in \widetilde{\mathcal{U}}_m(t_0, x_0)$ which verifies*

$$V(t_0, x_0) = J_1\left(t_0, x_0, u^{(t_0, x_0)}\right).$$

(ii) *The system of differential equations (4.1) has a stabilizing and bounded solution $\widetilde{X}(t)$ which verifies (5.10).*

Proof. From Corollary 6 we deduce that (i) is equivalent to (5.19). In particular,

$$\lim_{l \to \infty} \|T_{\widetilde{F}}(l\theta, 0)\| = 0. \tag{5.20}$$

Based on the identity $T_{\widetilde{F}}(t + \theta, t_0 + \theta) = T_{\widetilde{F}}(t, t_0) \ \forall t, t_0 \geq 0$, we may show by induction that $T_{\widetilde{F}}(l\theta, 0) = (T_{\widetilde{F}}(\theta, 0))^l$. Hence (5.20) is equivalent to

$$\lim_{l \to \infty} \|(T_{\widetilde{F}}(\theta, 0))^l\| = 0. \tag{5.21}$$

Since $T_{\widetilde{F}}(\theta, 0) : S_n^d \to S_n^d$ is a linear operator acting on a finite-dimensional Banach space, we obtain from (5.21) that all eigenvalues of $T_{\widetilde{F}}(\theta, 0)$ are located in the inside of the unit disk $|\lambda| < 1$. But $T_{\widetilde{F}}(\theta, 0)$ is the monodromy matrix of the equation (5.17); then, applying a well-known result concerning the uniform asymptotic stability of the zero state equilibrium of a linear differential equation with periodic coefficients (see [58]), we conclude that the zero solution of the equation (5.17) is exponentially stable. This means that the solution $\widetilde{X}(t)$ is just the stabilizing solution of the system (4.1), and thus the proof of the implication (i) $\Rightarrow$ (ii) is complete. The implication (ii) $\Rightarrow$ (i) follows from Corollary 4. $\qquad\square$

Corollary 8. *Assume the following.*

(a) *The system (5.1) and the cost function (5.2) are in the time-invariant case.*

(b) *$(\mathbf{A}, \mathbf{B}; Q)$ is stabilizable.*

(c) *The inequality $\mathcal{L}^*X - \mathcal{P}^*(X)\mathcal{R}^{-1}(X)\mathcal{P}(X) + M \geq 0$ has a solution*

$$\widehat{X} = \left(\widehat{X}(1), \ \widehat{X}(2), \ldots, \widehat{X}(d)\right),$$

which verifies the conditions $\mathcal{R}_i(\widehat{X}) > 0$, $i \in \mathcal{D}$. Then the following are equivalent:

(i) *For all $x_0 \in \mathbf{R}^n$ there exists an optimal control $u^{x_0} \in \widetilde{\mathcal{U}}_m(0, x_0)$, that is, $V(0, x_0) = J_1(0, x_0, u^{x_0})$.*

(ii) *The system of algebraic equations (4.31) has a stabilizing solution*

$$\widetilde{X} = \left(\widetilde{X}(1), \ \widetilde{X}(2), \ldots, \widetilde{X}(d)\right),$$

which verifies $\mathcal{R}_i(\widetilde{X}) > 0$, $i \in \mathcal{D}$.

(iii) *The system of linear matrix inequalities*

$$\begin{pmatrix} (\mathcal{L}^*X)(i) + M(i) & \mathcal{P}_i^*(X) \\ \mathcal{P}_i(X) & \mathcal{R}_i(X) \end{pmatrix} > 0, \quad i \in \mathcal{D},$$

has solutions in $\mathcal{S}_n^d$. *Under these conditions* $u^{x_0}(t) = \widetilde{F}(\eta(t))\tilde{x}(t)$, *where*

$$\widetilde{F}(i) = -R_i^{-1}(\widetilde{X})P_i(\widetilde{X}), \quad i \in \mathcal{D},$$

$\widetilde{X}$ *being the stabilizing solution of* (4.31) *and* $\tilde{x}(t)$ *being a solution of the corresponding closed-loop system* (5.18).

Proof. (i) $\Leftrightarrow$ (ii) follows from the previous theorem and (ii) $\Leftrightarrow$ (iii) follows from Theorem 9 of Chapter 4. $\square$

5.2.2 Solution of LQOP2

Since the cost functional (5.4) is a particular case of the cost functional (5.2), it follows that the solution of the optimization problem described by the controlled system (5.1), the cost functional (5.4), and the corresponding set of admissible controls $\widetilde{\mathcal{U}}(t_0, x_0)$ is obtained from the results derived in the previous section. The optimal control of this optimization problem is constructed with the stabilizing solution of SGRDE (4.44).

In this subsection we derive the solution of the optimization problem described by the controlled system (5.1), the cost functional (5.4), and the set of admissible controls $\mathcal{U}(t_0, x_0)$. Let $X(t)$ be a semipositive solution of the system (4.44) and let

$$F^X(t) = \left(F^X(t, 1) \; F^X(t, 2) \ldots F^X(t, d)\right)$$

be the corresponding feedback gain defined by (5.8). Set

$$u^X(t) = F^X(t, \eta(t))x^X(t), \quad t \geq 0,$$

where $x^X(t)$ is the solution of the system

$$dx(t) = \left[A_0(t, \eta(t)) + B_0(t, \eta(t))F^X(t, \eta(t))\right]x(t)dt \tag{5.22}$$

$$+ \sum_{k=1}^r \left[A_k(t, \eta(t)) + B_k(t, \eta(t))F^X(t, \eta(t))\right]x(t)dw_k(t),$$

$t \geq t_0, x(t_0) = x_0$.

Lemma 9. *For each bounded and semipositive solution* $X(t)$ *of the system* (4.44) *the control* $u^X(t)$ *belongs to* $\mathcal{U}(t_0, x_0)$, $t \geq 0$, $x_0 \in \mathbf{R}^n$.

Proof. Obviously the control $u^X(t) \in L_{\eta,w}^2([t_1, t_2], \mathbf{R}^m)$ for every compact interval $[t_1, t_2] \subset [t_0, \infty)$. Applying Corollary 2 for

$$M(t, i) = C_0^*(t, i)C_0(t, i),$$
$$L(t, i) = 0,$$
$$R(t, i) = D_0^*(t, i)D_0(t, i), \quad u(t) = u^X(t),$$

we obtain

$$E\left[\int_{t_0}^{\tau} |C_0(t, \eta(t))x^X(t) + D_0(t, \eta(t))u^X(t)|^2 dt | \eta(t_0) = i\right] \tag{5.23}$$

$$= x_0^* X(t_0, i)x_0 - E\left[x^*(\tau)X(\tau, \eta(\tau))x(\tau)|\eta(t_0) = i\right],$$

$\forall t_0 < \tau, x_0 \in \mathbf{R}^n, i \in \mathcal{D}, x(t) = x^X(t).$

Taking into account that $X(t)$ is a semipositive and bounded solution of the system (4.44) it follows that there exists a positive constant c, such that

$$E\left[\int_{t_0}^{\tau} |C_0(t, \eta(t))x(t) + D_0(t, \eta(t))u^X(t)|^2 dt | \eta(t_0) = i\right] \le x_0^* X(t_0, i)x_0 \le c|x_0|^2,$$

$\forall \tau \ge t_0, x_0 \in \mathbf{R}^n, i \in \mathcal{D}.$

Hence

$$E\left[\int_{t_0}^{\infty} |C_0(t, \eta(t))x(t) + D_0(t, \eta(t))u^X(t)|^2 dt | \eta(t_0) = i\right] \le x_0^* X(t_0, i)x_0,$$

which shows that $J_2(t_0, x_0, u^X)$ is well defined, and we have

$$J_2(t_0, x_0, u^X) \le \sum_{j \in \mathcal{D}} \pi_j(t_0)x_0^* X(t_0, j)x_0; \tag{5.24}$$

thus the proof is complete.   □

Theorem 10. *Assume that the system* $(\mathbf{A}, \mathbf{B}; Q)$ *is stabilizable. Then the optimization problem LQOP2 has a solution given by*

$$\widetilde{\widetilde{u}}(t) = \widetilde{\widetilde{F}}(t, \eta(t))\widetilde{\widetilde{x}}(t), \quad t \ge t_0,$$

where $\widetilde{\widetilde{F}}(t, i)$ *is defined as in (5.8) for X replaced by the minimal semipositive and bounded solution* $\widetilde{\widetilde{X}}(t)$ *of the system (4.44) and* $\widetilde{\widetilde{x}}$ *is the solution of the problem (5.22), where* $F^X(t, i)$ *is replaced by* $\widetilde{\widetilde{F}}(t, i)$. *Moreover the optimal value of the cost function is*

$$J_2(t_0, x_0, \widetilde{\widetilde{u}}) = \sum_{i=1}^{d} \pi_i(t_0)x_0^* \widetilde{\widetilde{X}}(t_0, i)x_0.$$

Proof. Let $X_\tau(t) = (X_\tau(t, 1) \dots X_\tau(t, d))$ be the solution of the system (4.44) which verifies the terminal condition $X_\tau(\tau, i) = 0$.

Based on Proposition 13 and Theorem 14 in Chapter 4 it follows that the solution $X_\tau(t)$ is defined for all $t \in [0, \tau]$ and

$$\lim_{\tau \to \infty} X_\tau(t) = \widetilde{\widetilde{X}}(t).$$

Applying Corollary 2 for $X(t, i)$ replaced by $X_\tau(t, i)$, we obtain

$$E\left[\int_{t_0}^{\tau} |C_0(t, \eta(t))x(t) + D_0(t, \eta(t))u(t)|^2 dt|\eta(t_0) = i\right]$$

$$= x_0^* X_\tau(t_0, i)x_0 + E\left[\int_{t_0}^{\tau} [u(t) - F_\tau(t, \eta(t))x(t)]^* \mathcal{R}_{\eta(t)}(t, X_\tau(t))\right.$$

$$\left. \times [u(t) - F_\tau(t, \eta(t))x(t)]dt|\eta(t_0) = i\right], \quad (5.25)$$

$\forall u \in L^2_{\eta, w}([t_0, \tau], \mathbf{R}^m)$.

Hence

$$E\left[\int_{t_0}^{\tau} |y_u(t)|^2 dt|\eta(t_0) = i\right] \geq x_0^* X_\tau(t_0, i)x_0 \quad (5.26)$$

and equality is possible if $u(t) = F_\tau(t, \eta(t))x_\tau(t), t \in [t_0, \tau], x_\tau(t)$ being the solution of the problem (5.22) for $F^X(t, i)$ replaced by $F_\tau(t, i) = -\mathcal{R}_i^{-1}(t, X_\tau(t))$ $\mathcal{P}_i(t, X_\tau(t))$. From (5.26) for $u(t) = \widetilde{\widetilde{u}}(t)$, we obtain easily that

$$J_2(t_0, x_0, \widetilde{\widetilde{u}}) \geq \sum_{i \in \mathcal{D}} \pi_i(t_0)x_0^* \widetilde{\widetilde{X}}(t_0, i)x_0. \quad (5.27)$$

Combining (5.24) with (5.27) we get

$$J_2(t_0, x_0, \widetilde{\widetilde{u}}) = \sum_{i \in \mathcal{D}} \pi_i(t_0)x_0^* \widetilde{\widetilde{X}}(t_0, i)x_0.$$

Let $u \in \mathcal{U}(t_0, x_0)$ be arbitrary. Applying (5.25) to the restriction of u to the interval $[t_0, \tau]$ and taking the limit for $\tau \to \infty$, we obtain

$$E\left[\int_{t_0}^{\infty} |y_u(t)|^2 dt|\eta(t_0) = i\right]$$

$$= x_0^* \widetilde{\widetilde{X}}(t_0, i)x_0 + E\left[\int_{t_0}^{\infty} \left(u(t) - \widetilde{\widetilde{F}}(t, \eta(t))x(t)\right)^* \mathcal{R}_{\eta(t)}\left(t, \widetilde{\widetilde{X}}(t)\right)\right.$$

$$\left. \times \left(u(t) - \widetilde{\widetilde{F}}(t, \eta(t))x(t)\right)dt|\eta(t_0) = i\right],$$

which leads to

$$J_2(t, x_0, u) = \sum_{i \in \mathcal{D}} \pi_i(t_0)x_0^* \widetilde{\widetilde{X}}(t_0, i)x_0$$

$$+ \sum_{i \in \mathcal{D}} \pi_i(t_0)E\left[\int_{t_0}^{\infty} \left(u(t) - \widetilde{\widetilde{F}}(t, \eta(t))x(t)\right)^* \mathcal{R}_{\eta(t)}\left(t, \widetilde{\widetilde{X}}(t)\right)\right.$$

$$\left. \times \left(u(t) - \widetilde{\widetilde{F}}(t, \eta(t))x(t)\right)dt|\eta(t_0) = i\right]$$

$\forall u \in \mathcal{U}(t_0, x_0)$, which completes the proof. $\square$

Remark 3. From (5.23) and (5.26) for $u(t) = \widetilde{\widetilde{u}}(t)$, we obtain

$$\lim_{\tau \to \infty} E\left[\widetilde{\widetilde{x}}^*(\tau)\widetilde{\widetilde{X}}(\tau, \eta(\tau))\widetilde{\widetilde{x}}(\tau)|\eta(t_0) = i\right] = 0,$$

which is the single item of information concerning the behavior of the optimal trajectory of the system for $t \to \infty$.

Theorem 11. *Assume that the assumptions in Theorem 17 of Chapter 4 are fulfilled. Under these conditions the solutions of the optimization problems LQOP1 and LQOP2 described by the cost function (5.4) and the controlled system (5.1) coincide, and they are given by*

$$\tilde{u}(t) = \widetilde{F}(t, \eta(t))\tilde{x}(t), \tag{5.28}$$

where $\widetilde{F}(t, i)$ is defined as in (5.8), with $X(t)$ replaced by the stabilizing and bounded solution $\widetilde{X}(t)$ of the system (4.44), and $\tilde{x}(t)$ is the solution of the problem (5.22), with $F^X(t, i)$ replaced by $\widetilde{F}(t, i)$. Moreover the optimal value of the cost function is given by

$$J_2(t_0, x_0, \tilde{u}) = \sum_{i \in \mathcal{D}} \pi_i(t_0)x_0^* \widetilde{X}(t_0, i)x_0.$$

Proof. Under the considered assumptions, the system (4.44) has a unique bounded and semipositive solution, and that solution is a stabilizing one. Therefore the control $\tilde{u}(t)$ given by (5.28) coincides with $\widetilde{\widetilde{u}}(t)$ and hence the conclusion of the theorem follows immediately. ☐

Remark 4. Since $\widetilde{\widetilde{\mathcal{U}}}(t_0, x_0) \subseteq \mathcal{U}(t_0, x_0)$ it follows that

$$J_2(t_0, x_0, \tilde{u}) = \min_{u \in \widetilde{\widetilde{\mathcal{U}}}(t_0.x_0)} J_2(t_0, x_0, u) \geq \min_{u \in \mathcal{U}(t_0.x_0)} J_2(t_0, x_0, u)$$

$$= J_2\left(t_0, x_0, \widetilde{\widetilde{u}}\right). \tag{5.29}$$

On the other hand, from Theorem 11 and Corollary 4 it follows that if the system (5.1) is stochastic stabilizable and the system

$$dx(t) = A_0(t, \eta(t))x(t)dt + \sum_{k=1}^{r} A_k(t, \eta(t))x(t)dw_k(t),$$

$$y(t) = C_0(t, \eta(t))x(t)$$

is either stochastic detectable or stochastic uniformly observable, then in (5.29) we have equality, and additionally $\tilde{u} = \widetilde{\widetilde{u}}$ (a.s.).

The next numerical example shows that in the absence of the properties of detectability and observability in (5.29), the equality does not always take place.

Numerical example. Consider the system (5.1) in the particular case $n = 2, r = 1$, $d = 1$, $m = 1$. In this case the system becomes

$$dx(t) = (A_0 x(t) + B_0 u(t)) \, dt + (A_1 x(t) + B_1 u(t)) \, dw_1(t), \qquad (5.30)$$

$$x = \begin{bmatrix} x_1 \\ x_2 \end{bmatrix} \in \mathbf{R}^2, \quad u(t) \in \mathbf{R},$$

and the coefficient matrices are those from the numerical example at the end of Section 4.4. The cost functional is

$$J_2(0, x_0, u) = E \left[\int_0^\infty (x_1^2(t) + u^2(t)) dt \right]. \qquad (5.31)$$

From Corollary 4 one obtains that the solution of the optimization problem described by the system (5.30), the cost functional (5.31), and the set of admissible controls $\widetilde{\mathcal{U}}(0, x_0)$ is constructed with the stabilizing solution of the SGRAE (4.50), and the optimal value is given by

$$J_2(0, x_0, \tilde{u}) = \begin{bmatrix} x_{10} & x_{20} \end{bmatrix} \begin{bmatrix} 8 & -21 \\ -21 & 63 \end{bmatrix} \begin{bmatrix} x_{10} \\ x_{20} \end{bmatrix}, \qquad (5.32)$$

where $x_0 = \begin{bmatrix} x_{10} & x_{20} \end{bmatrix}^T$. On the other hand, from Theorem 10 it follows that the solution of the optimization problem described by the system (5.30), the cost function (5.31), and the set of admissible controls $\mathcal{U}(0, x_0)$ is constructed with the minimal solution of the SGRAE (4.50). The optimal value is

$$J_2(0, x_0, \widetilde{\tilde{u}}) = \begin{bmatrix} x_{10} & x_{20} \end{bmatrix} \begin{bmatrix} 1 & 0 \\ 0 & 0 \end{bmatrix} \begin{bmatrix} x_{10} \\ x_{20} \end{bmatrix}. \qquad (5.33)$$

From (5.32) and (5.33) one sees that $J_2(0, x_0, \tilde{u}) \neq J_2(0, x_0, \widetilde{\tilde{u}})$.

5.3 The tracking problem

Consider the stochastic system (5.1) with the output (5.5) together with Assumption A (a) and (b) stated at the beginning of this chapter. Then, if $t \to r(t) = (r(t, 1), r(t, 2), \ldots, r(t, d)) : \mathbf{R}_+ \to (\mathbf{R}^p)^d$ is a continuous and bounded function, the *tracking problem* consists in finding a control $\hat{u}(\cdot) \in \widetilde{\mathcal{U}}_m(t_0, x_0)$ which minimizes the cost function

$$\hat{J}(u) = \overline{\lim_{T \to \infty}} \frac{1}{T - t_0} E \left[\int_{t_0}^T |y_u(t, t_0, x_0) - r(t, \eta(t))|^2 dt \right] \qquad (5.34)$$

in the class of all stochastic processes $\widetilde{\mathcal{U}}_m(t_0, x_0)$, where $\widetilde{\mathcal{U}}_m(t_0, x_0)$ is the set of all stochastic processes $u : [t_0, \infty) \times \Omega \to \mathbf{R}^m$ with the properties $u \in L^2_{\eta, w}([t_0, T], \mathbf{R}^m)$ for all $T > t_0$ and $\sup E |x_u(t, t_0, x_0)|^2 < \infty, t \geq t_0$.

For each $(t_0, \tau, x_0, i) \in \mathbf{R}_+ \times \mathbf{R}_+ \times \mathbf{R}^n \times \mathcal{D}$ with $0 \leq t_0 < \tau$, we consider the auxiliary cost functions

$$\mathcal{W}(t_0, \tau, x_0, i, u) = E\left[\int_{t_0}^\tau |y_u(t, t_0, x_0) - r(t, \eta(t))|^2 dt | \eta(t_0) = i\right],$$

for all $u \in L^2_{\eta,w}([t_0, \tau], \mathbf{R}^m)$. Based on Itô-type formula given in Theorem 35 of Chapter 1 we obtain the following lemma.

Lemma 12. *Let* $t \to K(t, i) : \mathbf{R}_+ \to \mathcal{S}_n, t \to g(t, i) : \mathbf{R}_+ \to \mathbf{R}^n, t \to h(t, i) : \mathbf{R}_+ \to \mathbf{R}, i \in \mathcal{D}$ *be* C^1-*functions, and let*

$$v(t, x, i) = x^* K(t, i)x + 2g^*(t, i)x + h(t, i).$$

Then:

$$\mathcal{W}(t_0, \tau, x_0, i, u) = v(t_0, x_0, i) - E\left[v(\tau, x(\tau), \eta(\tau)) | \eta(t_0) = i\right]$$

$$+ E\left[\int_{t_0}^\tau \{(x^*(t) \quad u^*(t))\mathcal{M}^K(t, \eta(t))\binom{x(t)}{u(t)}\right.$$

$$+ 2\left[\frac{\partial}{\partial t}g^*(t, \eta(t)) + g^*(t, \eta(t))A_0(t, \eta(t))\right.$$

$$+ \sum_{j=1}^d q_{\eta(t)j}g^*(t, j) - r^*(t, \eta(t))C_0(t, \eta(t))\right]x(t)$$

$$+ 2\left[g^*(t, \eta(t))B_0(t, \eta(t)) - r^*(t, \eta(t))D_0(t, \eta(t))\right]u(t)$$

$$+ r^*(t, \eta(t))r(t, \eta(t)) + \frac{\partial}{\partial t}h(t, \eta(t))$$

$$+ \sum_{j=1}^d q_{\eta(t)j}h(t, j)\}dt | \eta(t_0) = i\right],$$

for all $t_0, 0 \leq t_0 < \tau, x_0 \in \mathbf{R}^n, i \in \mathcal{D}, u \in L^2_{\eta,w}([t_0, \tau], \mathbf{R}^m)$, *where* $x(t) = x_u(t, t_0, x_0)$, $\mathcal{M}^K(t, i)$ *being as in Lemma* 1, *with* $M(t, i) = C_0^*(t, i)C_0(t, i)$, $L(t, i) = 0, R(t, i) = D_0^*(t, i)D_0(t, i)$. $\square$

Let $\widetilde{X}(t)$ be the stabilizing and bounded solution of the system (4.44). Set $\widetilde{F}(t) = (\widetilde{F}(t, 1), \widetilde{F}(t, 2), \ldots, \widetilde{F}(t, d)), \widetilde{F}(t, i) = -\mathcal{R}_i^{-1}(t, \widetilde{X}(t))\mathcal{P}_i(t, \widetilde{X}(t))$ be the stabilizing feedback gain. This means that the zero state equilibrium of the corresponding closed-loop system (5.18) is ESMS. Then, by Corollary 27 in Chapter 2, the zero solution of the differential equation with Markovian jumping

$$\frac{d}{dt}x(t) = \left[A_0(t, \eta(t)) + B_0(t, \eta(t))\widetilde{F}(t, \eta(t))\right]x(t)$$

is ESMS.

Now, applying Theorem 25 of Chapter 2 we deduce that the zero state equilibrium of the linear differential equation on $(\mathbf{R}^n)^d$:

$$\frac{d}{dt} y_i(t) = \left[A_0(t, i) + B_0(t, i)\widetilde{F}(t, i)\right] y_i(t) + \sum_{j=1}^{d} q_{ji} y_j(t), \quad i \in \mathcal{D},$$

is exponentially stable.

Let $\tilde{g}(t) = \left(\tilde{g}(t, 1) \ \tilde{g}(t, 2) \ldots \tilde{g}(t, d)\right)$ (see Corollary 26 of Chapter 2) be the unique bounded solution on $\mathbf{R}_+$ of the affine differential equations

$$\frac{d}{dt} y_i(t) + \left[A_0(t, i) + B_0(t, i)\widetilde{F}(t, i)\right]^* y_i(t) \tag{5.35}$$

$$+ \sum_{j=1}^{d} q_{ij} y_j(t) - \left[C_0(t, i) + D_0(t, i)\widetilde{F}(t, i)\right]^* r(t, i)$$

$$= 0,$$

$i \in \mathcal{D}$. From the previous lemma we have the following corollary.

Corollary 13. *Assume that the system* (4.1) *has a bounded and stabilizing solution* $\widetilde{X}(t)$. *Let* $\tilde{g}(t)$ *be the unique bounded solution of the equations* (5.35) *and* $h(t, i)$ *be arbitrary* C^1-*functions as in the previous lemma. If* $v(t, x, i) = x^* \widetilde{X}(t, i)x + 2\tilde{g}^*(t, i)x + h(t, i)$, *we have*

$$\mathcal{W}(t_0, \tau, x_0, i, u) = v(t_0, x_0, i) - E[v(\tau, x(\tau), \eta(\tau))|\eta(t_0) = i]$$

$$+ E\left[\int_{t_0}^{\tau} \{[u(t) - \widetilde{F}(t, \eta(t))x(t)]^* \mathcal{R}_{\eta(t)}(t, \widetilde{X}(t))[u(t) - \widetilde{F}(t, \eta(t))x(t)] \right.$$

$$+ 2[\tilde{g}^*(t, \eta(t))B_0(t, \eta(t)) - r^*(t, \eta(t))D_0(t, \eta(t))][u(t) - \widetilde{F}(t, \eta(t))x(t)]$$

$$\left. + \frac{\partial}{\partial t}h(t, \eta(t)) + \sum_{j=1}^{d} q_{\eta(t)j}h(t, j) + r^*(t, \eta(t))r(t, \eta(t))\}dt \Big| \eta(t_0) = i \right],$$

$$\tag{5.36}$$

for all t_0, $0 \le t_0 < \tau$, $x_0 \in \mathbf{R}$, $i \in \mathcal{D}$, $u \in L^2_{\eta, w}([t_0, \tau], \mathbf{R}^m)$, $x(t) = x_u(t, t_0, x_0)$. $\square$

Remark 5. If $\widetilde{X}$ is a bounded and stabilizing solution of the system (4.44) then we may write

$$\frac{d}{dt}\widetilde{X}(t) + \mathcal{L}^*_{\widetilde{F}}(t)\widetilde{X}(t) + \left[C_0(t) + D_0(t)\widetilde{F}(t)\right]^* \left[C_0(t) + D_0(t)\widetilde{F}(t)\right] = 0,$$

which shows that the stabilizing and bounded solution of the system (4.44), if it exists, is always semipositive. Therefore, the condition $\mathcal{R}_i(t, \widetilde{X}(t)) \ge \rho I_m > 0$ is fulfilled.

For each $\tau > 0$ set $h_\tau(t) = (h_\tau(t, 1) \ldots h_\tau(t, d))^*$, the solution of the system of affine differential equations

$$\frac{d}{dt} h(t) + Qh(t) + \tilde{m}(t) = 0,$$

with the terminal condition $h_\tau(\tau) = 0$, where

$$\tilde{m}(t) = \left(\tilde{m}_1(t)\, \tilde{m}_2(t) \ldots \tilde{m}_d(t)\right)^*$$

$$\tilde{m}_j(t) = r^*(t, j)r(t, j) - \left[\tilde{g}^*(t, j)B_0(t, j) - r^*(t, j)D_0(t, j)\right]$$
$$\times R_j^{-1}(t, \tilde{X}(t))\left[B_0^*(t, j)\tilde{g}(t, j) - D_0^*(t, j)r(t, j)\right], \qquad (5.37)$$

$i \in \mathcal{D}, t \geq 0, \ Q = \{q_{ij}\}_{i,j\in\mathcal{D}}$. Let $v_\tau(t, x, i)$ be defined by

$$v_\tau(t, x, i) = x^*\tilde{X}(t, i)x + 2\tilde{g}^*(t, i)x + h_\tau(t, i).$$

From Corollary 13 we get

$$W(t_0, \tau, x_0, i, u) = v_\tau(t_0, x_0, i) - E\left[v_\tau(\tau, x(\tau), \eta(\tau))|\eta(t_0) = i\right]$$
$$+ E\left[\int_{t_0}^\tau \left(u(t) - \tilde{F}(t, \eta(t))x(t) - \psi(t, \eta(t))\right)^* R_{\eta(t)}\left(t, \tilde{X}(t)\right)\left(u(t)\right.\right.$$
$$\left.\left. - \tilde{F}(t, \eta(t))x(t) - \psi(t, \eta(t))\right)dt \,|\eta(t_0) = i\right] \qquad (5.38)$$

for all $t_0, \ 0 \leq t_0 < \tau, x_0 \in \mathbf{R}^n, i \in \mathcal{D}, u \in L^2_{\eta,w}([t_0, \tau], \mathbf{R}^m), x(t) = x_u(t, t_0, x_0)$, where

$$\psi(t, i) = -R_i^{-1}(t, \tilde{X}(t))[B_0^*(t, i)\tilde{g}(t, i) - D_0^*(t, i)r(t, i)]. \qquad (5.39)$$

Now we are able to prove the main result of this section.

Theorem 14. *Assume that the system of differential equations (4.44) has a bounded and stabilizing solution $\tilde{X}(t)$. Let $\tilde{g}(t) = (\tilde{g}(t, 1), \ \tilde{g}(t, 2), \ldots, \tilde{g}(t, d))$ be the unique bounded on $\mathbf{R}_+$ solution of the equations (5.35) and $\psi(t, i)$ defined by (5.39). Under these conditions we have*

$$\min_{u\in\tilde{\mathcal{U}}_m(t_0,x_0)} \hat{J}(u) = \hat{J}(\bar{u}) = \lim_{t\to\infty} \frac{1}{T}\int_0^T \sum_{i=1}^d \sum_{j=1}^d \pi_i(t_0)\tilde{p}_{ij}\tilde{m}_j(t)dt,$$

for all $i \in \mathcal{D}, t_0 \geq 0, x_0 \in \mathbf{R}^n$, where $\bar{u}(t) = \tilde{F}(t, \eta(t))\bar{x}(t) + \psi(t, \eta(t)), \ \bar{x}(t)$ being the solution of the problem

$$dx(t) = \left[\left(A_0(t, \eta(t)) + B_0(t, \eta(t))\tilde{F}(t, \eta(t))\right)x(t) + B_0(t, \eta(t))\psi(t, \eta(t))\right]dt$$
$$+ \sum_{k=1}^r \left[\left(A_k(t, \eta(t)) + B_k(t, \eta(t))\tilde{F}(t, \eta(t))\right)x(t)\right. \qquad (5.40)$$
$$\left. + B_k(t, \eta(t))\psi(t, \eta(t))\right]dw_k(t),$$

$t \geq t_0, \bar{x}(t_0) = x_0$ *and*

$$\tilde{P} = (\tilde{p}_{ij})_{i,j\in\mathcal{D}} = \lim_{t\to\infty} P(t) = \lim_{t\to\infty} e^{Qt}.$$

Proof. Applying Theorem 32 of Chapter 2 to the system (5.40) we deduce that $\sup_{t \geq t_0} E|\bar{x}(t)|^2 < \infty$ and therefore $\bar{u}(t)$ belongs to $\tilde{\mathcal{U}}_m(t_0, x_0)$. It is easy to see that for each $u \in \tilde{\mathcal{U}}_m(t_0, x_0)$ we have

$$\hat{J}(u) = \limsup_{T \to \infty} \frac{1}{T - t_0} \sum_{i=1}^{d} \pi_i(t_0) \mathcal{W}(t_0, T, x_0, i, u).$$

Then from (5.38) we have for $u \in \tilde{\mathcal{U}}_m(t_0, x_0)$

$$\hat{J}(u) \geq \limsup_{T \to \infty} \frac{1}{T - t_0} \sum_{i=1}^{d} \pi_i(t_0) \{ v_T(t_0, x_0, i) - E[v_T(T, x(T), \eta(T)) | \eta(t_0) = i] \}$$

$$= \limsup_{T \to \infty} \frac{1}{T - t_0} \sum_{i=1}^{d} \pi_i(t_0) h_T(t_0, i) = \hat{J}(\bar{u}).$$

But

$$h_T(t) = \int_t^T e^{Q(s-t)} \tilde{m}(s)ds = \int_t^T P(s-t)\tilde{m}(s)ds.$$

Therefore

$$h_T(t_0) = \int_{t_0}^T \left[P(s-t_0) - \tilde{P} \right] \tilde{m}(s)ds + \int_{t_0}^T \tilde{P}\tilde{m}(s)ds.$$

Since $\lim_{t \to \infty} P(t) = \tilde{P}$ and $\tilde{m}(t)$ is a continuous and bounded function we have

$$\lim_{T \to \infty} \frac{1}{T - t_0} \int_{t_0}^T \left(P(s-t_0) - \tilde{P} \right) \tilde{m}(s)ds = 0.$$

Hence

$$\limsup_{T \to \infty} \frac{1}{T - t_0} \sum_{i=1}^{d} \pi_i(t_0) h_T(t_0, i)$$

$$= \overline{\lim}_{T \to \infty} \frac{1}{T - t_0} \int_{t_0}^T \sum_{i=1}^{d} \sum_{j=1}^{d} \pi_i(t_0) \tilde{p}_{ij} \tilde{m}_j(t)dt$$

$$= \overline{\lim}_{T \to \infty} \frac{1}{T} \int_0^T \sum_{i=1}^{d} \sum_{j=1}^{d} \pi_i(t_0) \tilde{p}_{ij} \tilde{m}_j(t)dt.$$

The last equality follows since $\sum_{i=1}^{d} \sum_{j=1}^{d} \pi_i(t_0) \tilde{p}_{ij} \tilde{m}_j(t)$ is a bounded function on $\mathbf{R}_+$. Thus the proof is complete. $\qquad \square$

Remark 6. Concerning the feasibility aspects of the control $\bar{u}(t) = \widetilde{F}(t, \eta(t))\bar{x}(t) + \psi(t, \eta(t))$, which is the solution of the above tracking problem, we distinguish two important situations:

(i) If the system (5.1), (5.5) is in the time-invariant case and the signal $r(t)$ satisfies $r(t, i) = r(i), (t, i) \in \mathbf{R}_+ \times \mathcal{D}$, then the stabilizing solution of the system (4.44) is constant and solves the system of algebraic equations. This solution may be computed applying the iterative procedure described in Section 4.6.

By uniqueness arguments it follows that the bounded solution of the system (5.35) is constant, and it solves the system of linear equations

$$\left[A_0(i) + B_0(i)\widetilde{F}(i)\right]^* \tilde{g}(i) + \sum_{j=1}^d q_{ij}\tilde{g}(j) - \left[C_0(i) + D_0(i)\widetilde{F}(i)\right]^* r(i) = 0, \ i \in \mathcal{D}.$$

(ii) If the coefficients of the system (5.1), (5.5) are θ-periodic functions, then the stabilizing solution of the system (4.44) is a θ-periodic function, and it can be computed with the iterative procedure given in Section 4.6. From the uniqueness arguments the bounded solution of the system (5.35) is a θ-periodic function, and its initial conditions can be obtained by solving a linear system of algebraic equations.

(iii) Under the assumptions of Theorem 14 it follows that the optimal value of the tracking problem does not depend upon x_0.

5.4 Stochastic H^2 controllers

In this section we assume that the controlled system (5.1) is also subjected to an additive white noise perturbation. For this perturbed system we shall introduce a norm extending the well-known H^2 norm from the deterministic framework.

The optimization problem that we address in this section consists in finding a stabilizing output feedback controller which minimizes the H^2 norm of the resulting system.

In the following we shall focus our attention only on the time-invariant case.

5.4.1 Stochastic H^2 norms

Consider the linear stochastic system **G** described by

$$dx(t) = A_0(\eta(t))x(t)dt + \sum_{k=1}^r A_k(\eta(t))x(t)dw_k(t) \qquad (5.41)$$

$$+ B_v(\eta(t))dv(t),$$
$$z(t) = C(\eta(t))x(t)$$

with $x \in \mathbf{R}^n$, $z \in \mathbf{R}^p$, $A_k(i) \in \mathbf{R}^{n \times n}$, $k = 0, \ldots, r$, $B_v(i) \in \mathbf{R}^{n \times m_v}$, $C(i) \in \mathbf{R}^{p \times n}$, $i \in \mathcal{D}$, $w_k(t), t \geq 0$ being a scalar Wiener process and $v(t), t \geq 0$, being an m_v-dimensional Wiener process. As in the previous sections $w(t) = (w_1(t), \ldots, w_r(t))^*$ and $\eta(t)$ are a standard Wiener process and a Markov process, respectively, with the properties in Section 1.8; $v(t)$, $t \geq 0$, is an m_v-dimensional standard Wiener process independent of the pair $(w(t), \eta(t))$, $t \geq 0$. Throughout this section, $\mathcal{F}_t$, $\mathcal{G}_t$, $\mathcal{H}_t$ are the σ-algebras defined in Chapter 1 related to the processes $w(t)$ and $\eta(t)$, and $\widehat{\mathcal{H}}_t$ is the smallest σ-algebra containing $\mathcal{H}_t$ and the σ-algebra generated by $v(s)$, $0 \leq s \leq t$. Denoting by $\Phi(t, s)$ the fundamental matrix solution of the system

$$dx(t) = A_0(\eta(t))x(t)dt + \sum_{k=1}^{r} A_k(\eta(t))x(t)dw_k(t), \qquad (5.42)$$

according to (1.29) the solutions of (5.41) have the following representation:

$$x(t) = \Phi(t, 0)x_0 + \Phi(t, 0) \int_0^t \Phi^{-1}(s, 0) B_v(\eta(s))dv(s). \qquad (5.43)$$

In particular, the solution of (5.41) with zero initial conditions is

$$x_0(t) = \Phi(t, 0) \int_0^t \Phi^{-1}(s, 0) B_v(\eta(s))dv(s). \qquad (5.44)$$

We prove the following lemma.

Lemma 15. *For each $\tau > 0$ and $j \in \mathcal{D}$ we have*

$$E\left[x_0(\tau)x_0^*(\tau)\chi_{\eta(\tau)=j}\right] = E\left[\int_0^{\tau} \Phi(\tau, s)B_v(\eta(s))B_v^*(\eta(s))\Phi^*(\tau, s)\chi_{\eta(\tau)=j}ds\right]. \qquad (5.45)$$

Proof. Set

$$\Psi(s) = \Phi^{-1}(s, 0) B_v(\eta(s)). \qquad (5.46)$$

It is obvious that the components of Ψ belong to $L^{2p}_{\eta,w}[0, \tau]$ for all integers $p \geq 1$, and in particular for $p = 2$.

We prove that

$$E\left[\Phi(\tau, 0) \int_0^{\tau} \Psi(t)dv(t)\left(\Phi(\tau, 0) \int_0^{\tau} \Psi(t)dv(t)\right)^* \chi_{\eta(\tau)=j}\right]$$
$$= E \int_0^{\tau} \Phi(\tau, 0)\Psi(t)\Psi^*(t)\Phi^*(\tau, 0)\chi_{\eta(\tau)=j}dt. \qquad (5.47)$$

To this end we prove (5.47) for the case when the elements of Ψ are step functions in $L^4_{\eta,w}[0, \tau]$. Indeed, let

$$\Psi(t) = \sum_{i=0}^{k-1} \Psi(t_i)\chi_{[t_i, t_{i+1})}, \quad 0 = t_0 < t_1 < \cdots < t_{k-1} < t_k = \tau,$$

$\Psi(t_i)$ being $\mathcal{H}_{t_i}$ measurables, $0 \le i \le k$, $E |\Psi(t_i)|^4 < \infty$. We have

$$E \left[\Phi(\tau, 0) \int_0^\tau \Psi(t) dv(t) \left(\Phi(\tau, 0) \int_0^\tau \Psi(t) dv(t) \right)^* \chi_{\eta(\tau)=j} \mid \mathcal{H}_\tau \right]$$

$$= E \left[\Phi(\tau, 0) \sum_{i,l} \Psi(t_i)(v(t_{i+1}) - v(t_i))(v(t_{l+1}) - v(t_l))^* \Psi^*(t_l) \right.$$

$$\left. \times \Phi^*(\tau, 0) \chi_{\eta(\tau)=j} \mid \mathcal{H}_\tau \right] \tag{5.48}$$

$$= \Phi(\tau, 0) \sum_{i,l} \Psi(t_i) E \left[(v(t_{i+1}) - v(t_i))(v(t_{l+1}) - v(t_l))^* \mid \mathcal{H}_\tau \right]$$

$$\times \Psi^*(t_l) \Phi^*(\tau, 0) \chi_{\eta(\tau)=j}$$

$$= \Phi(\tau, 0) \left(\sum_{i=0}^{k-1} \Psi(t_i) \Psi^*(t_i)(t_{i+1} - t_i) \right) \Phi^*(\tau, 0) \chi_{\eta(\tau)=j}.$$

The last equality above has been obtained by taking into account that the σ-algebra generated by $\{v(t) - v(s), t, s \in [0, \tau]\}$ is independent of $\mathcal{H}_\tau$ and therefore

$$E \left[(v(t_{i+1}) - v(t_i))(v(t_{l+1}) - v(t_l))^* \mid \mathcal{H}_\tau \right]$$
$$= E \left[(v(t_{i+1}) - v(t_i))(v(t_{l+1}) - v(t_l))^* \right] = \delta_{i,l}(t_{i+1} - t_i) I_{m_v},$$

where $\delta_{i,l}$ are the Kronecker coefficients. Hence, by taking expectation in (5.48), one concludes that (5.45) holds if the elements of Ψ are step functions in $L_{\eta,w}^4([0, \tau])$. Now, based on Remark 9 of Chapter 1, take a sequence $\{\Psi_k(t)\}_{k=0,1,...}$ of step functions in $L_{\eta,w}^4([0, \tau])$ such that

$$\lim_{k \to \infty} E \int_0^\tau |\Psi_k(t) - \Psi(t)|^4 dt = 0. \tag{5.49}$$

Writing (5.47) for each Ψ_k, one obtains

$$E \left[\left(\Phi(\tau, 0) \int_0^\tau \Psi_k(t) dv(t) \right) \left(\Phi(\tau, 0) \int_0^\tau \Psi_k(t) dv(t) \right)^* \chi_{\eta(\tau)=j} \right]$$

$$= E \int_0^\tau \Phi(\tau, 0) \Psi_k(t) \Psi_k^*(t) \Phi^*(\tau, 0) \chi_{\eta(\tau)=j} dt. \tag{5.50}$$

Using Theorem 27 of Chapter 1 and (5.49) above, it follows that

$$\lim_{k \to \infty} E \left[\left(\Phi(\tau, 0) \int_0^\tau \Psi_k(t) dv(t) \right) \left(\Phi(\tau, 0) \int_0^\tau \Psi_k(t) dv(t) \right)^* \chi_{\eta(\tau)=j} \right]$$

$$= E \left[\left(\Phi(\tau, 0) \int_0^\tau \Psi(t) dv(t) \right) \left(\Phi(\tau, 0) \int_0^\tau \Psi(t) dv(t) \right)^* \chi_{\eta(\tau)=j} \right]$$

and

$$\lim_{k \to \infty} E \int_0^\tau \Phi(\tau, 0)\Psi_k(t)\Psi_k^*(t)\Phi^*(\tau, 0)\chi_{\eta(\tau)=j} dt$$

$$= E \int_0^\tau \Phi(\tau, 0)\Psi(t)\Psi^*(t)\Phi^*(\tau, 0)\chi_{\eta(\tau)=j} dt.$$

Combining the last two equalities with (5.50), one obtains (5.47). By replacing $\Psi(t)$ in (5.47) with (5.46), (5.45) directly follows since $\Phi(\tau, 0)\Phi^{-1}(s, 0) = \Phi(\tau, s)$ a.s., and thus the proof is complete.   □

Remark 7. If we consider the particular case when $A_k(i) = 0$, $1 \le k \le r$, $i \in \mathcal{D}$, the proof of the above lemma does not become simpler. This is due to the fact that in the representation formula (5.44) we cannot write

$$x_0(\tau) = \int_0^\tau \dot{\Phi}(\tau, s) B_v(\eta(s)) dv(s), \tag{5.51}$$

since the expression under the integral is random, and it is measurable with respect to $\mathcal{H}_\tau$. On the other hand, the integral in (5.51) is well defined if the function under the integral is measurable with respect to

$$\widehat{\mathcal{H}}_s = \mathcal{H}_s \vee \sigma(v(t), 0 \le t \le s)$$

for all $s < \tau$.

Let us introduce the following notations

$$\pi_i(t) = P\{\eta(t) = i\}, \tag{5.52}$$

$$\widetilde{P} = \lim_{t \to \infty} P(t) \quad \text{with elements } \tilde{p}_{ij}, \tag{5.53}$$

$$\pi_i = P(\eta(0) = i) = \pi_i(0), \tag{5.54}$$

$$\pi_{i\infty} = \sum_{j=1}^d \pi_j \tilde{p}_{ji}. \tag{5.55}$$

It is obvious that

$$\pi_i(t) = \sum_{j=1}^d \pi_j p_{ji}(t)$$

and hence

$$\lim_{t \to \infty} \pi_i(t) = \pi_{i\infty}.$$

Set

$$\widehat{B}_v(s, i) = \pi_i(s) B_v(i) B_v^*(i), \tag{5.56}$$

$$\widehat{B}_v(i) = \pi_{i\infty} B_v(i) B_v^*(i). \tag{5.57}$$

It is clear that

$$\lim_{s \to \infty} \widehat{B}_v(s, i) = \widehat{B}_v(i) \text{ for all } i \in \mathcal{D}. \tag{5.58}$$

With these notations we prove the following lemma.

Lemma 16. *With $x_0(t)$ defined by (5.44), we have*

$$E\left[x_0(\tau)x_0^*(\tau)\chi_{\eta(\tau)=j}\right] = \int_0^\tau \left(e^{\mathcal{L}(\tau-s)}\widehat{B}_v(s)\right)(j)ds,$$

where $\widehat{B}_v(s) = \left(\widehat{B}_v(s,1), \ldots, \widehat{B}_v(s,d)\right)$ with $\widehat{B}_v(s,i)$ given by (5.56) and $\mathcal{L}$ is the Lyapunov operator defined by the system $(A_0, A_1, \ldots, A_r; Q)$.

Proof. Based on Lemma 15 we may write successively:

$$E\left[x_0(\tau)x_0^*(\tau)\chi_{\eta(\tau)=j}\right]$$

$$= \int_0^\tau E\left[\Phi(\tau,s)B_v(\eta(s))B_v^*(\eta(s))\Phi^*(\tau,s)\chi_{\eta(\tau)=j}\right]ds$$

$$= \int_0^\tau \sum_{i=1}^d \pi_i(s)E\left[\Phi(\tau,s)B_v(\eta(s))B_v^*(\eta(s))\Phi^*(\tau,s)\chi_{\eta(\tau)=j} \mid \eta(s)=i\right]ds$$

$$= \int_0^\tau \sum_{i=1}^d E\left[\Phi(\tau,s)\widehat{B}_v(\eta(s))\Phi^*(\tau,s)\chi_{\eta(\tau)=j} \mid \eta(s)=i\right]ds$$

$$= \int_0^\tau \left(T(\tau,s)\widehat{B}_v(s)\right)(j)ds.$$

For the last equality above we used the representation formula (2.24) of the evolution operator $T(t,s)$.

The conclusion follows since in the time-invariant case, $T(t,s) = e^{\mathcal{L}(t-s)}$ (see Remark 3 of Chapter 2). □

Lemma 17. *Assume that the system $(A_0, A_1, \ldots, A_r; Q)$ is stable. Then we have*

$$\lim_{\tau\to\infty} E\left[x_0(\tau)x_0^*(\tau)\chi_{\eta(\tau)=j}\right] = \widehat{P}_c(j), \qquad (5.59)$$

where $\widehat{P}_c = \left(\widehat{P}_c(1), \ldots, \widehat{P}_c(d)\right)$ is the unique semipositive solution of the Lyapunov-like equation $\mathcal{L}P + \widehat{B}_v = 0$ with $\widehat{B}_v = \left(\widehat{B}_v(1), \ldots, \widehat{B}_v(d)\right)$, $\widehat{B}_v$ being defined by (5.57).

Proof. Based on Lemma 16 we have

$$E\left[x_0(\tau)x_0^*(\tau)\chi_{\eta(\tau)=j}\right] = \int_0^\tau \left(e^{\mathcal{L}(\tau-s)}\widehat{B}_v(s)\right)(j)ds$$

$$= \int_0^\tau \left(e^{\mathcal{L}(\tau-s)}\left(\widehat{B}_v(s) - \widehat{B}_v\right)\right)(j)ds + \int_0^\tau \left(e^{\mathcal{L}(\tau-s)}\widehat{B}_v\right)(j)ds.$$

By a simple change of integration variable we get

$$E\left[x_0(\tau)x_0^*(\tau)\chi_{\eta(\tau)=j}\right] = \int_0^\tau \left(e^{\mathcal{L}(\tau-s)}\left(\widehat{B}_v(s) - \widehat{B}_v\right)\right)(j)ds$$

$$+ \int_0^\tau \left(e^{\mathcal{L}s}\widehat{B}_v\right)(j)ds. \qquad (5.60)$$

Since the system $(A_0, A_1, \ldots, A_r; Q)$ is stable, there exist $\beta \geq 1$, $\alpha > 0$ such that $\|e^{\mathcal{L}s}\| \leq \beta e^{-\alpha s}$ for all $s \geq 0$. Further, we have

$$\left| \int_0^\tau \left(e^{\mathcal{L}(\tau-s)} \left(\widehat{B}_v(s) - \widehat{B}_v \right) \right) (j) ds \right| \leq \left| \int_0^\tau \left(e^{\mathcal{L}(\tau-s)} \left(\widehat{B}_v(s) - \widehat{B}_v \right) \right) ds \right|$$

$$\leq \beta \int_0^\tau e^{-\alpha(\tau-s)} \left| \widehat{B}_v(s) - \widehat{B}_v \right| ds.$$

Taking $\tau \to \infty$, one obtains from (5.58) by standard arguments

$$\lim_{\tau \to \infty} \beta \int_0^\tau e^{-\alpha(\tau-s)} \left| \widehat{B}_v(s) - \widehat{B}_v \right| ds = 0,$$

which leads to

$$\lim_{\tau \to \infty} \int_0^\tau \left(e^{\mathcal{L}(\tau-s)} \left(\widehat{B}_v(s) - \widehat{B}_v \right) \right) (j) ds = 0,$$

and hence from (5.60) we get

$$\lim_{\tau \to \infty} E\left[x_0(\tau) x_0^*(\tau) \chi_{\eta(\tau)=j} \right] = \int_0^\infty \left(e^{\mathcal{L}s} \widehat{B}_v \right) (j) ds = \widehat{P}_c(j).$$

The last equality follows from the proof of Theorem 15 of Chapter 2. Thus the proof is complete.   □

Remark 8. From the representation formulae (5.43) and (5.44) and from Lemma 17 it follows that if the system $(A_0, A_1, \ldots, A_r; Q)$ is stable, then

$$\lim_{t \to \infty} E\left[x(t) x^*(t) \chi_{\eta(t)=j} \right] = \lim_{t \to \infty} E\left[x_0(t) x_0^*(t) \chi_{\eta(t)=j} \right] = \widehat{P}_c(j)$$

for all $j \in \mathcal{D}$ and for any solution $x(t)$ of the system (5.41).

Theorem 18. *Assume that the system* $(A_0, A_1, \ldots, A_r; Q)$ *is stable. Then*

$$\lim_{t \to \infty} E|z(t)|^2 = \sum_{j=1}^d Tr\left(C(j) \widehat{P}_c(j) C^*(j) \right)$$

$$= \sum_{j=1}^d \pi_{j\infty} Tr\left(B_v^*(j) \widehat{P}_o(j) B_v(j) \right),$$

where $\widehat{P}_o = \left(\widehat{P}_o(1), \ldots, \widehat{P}_o(d) \right)$ *is the unique positive semidefinite solution of the equation*

$$\mathcal{L}^* P_o + \widetilde{C} = 0$$

with $\widetilde{C} = \left(\widetilde{C}(1), \ldots, \widetilde{C}(d) \right)$, $\widetilde{C}(j) = C^*(j) C(j)$, $j \in \mathcal{D}$.

Proof. First we shall prove the result in the statement for $z_0 = C(\eta(t))x_0(t)$. To this end we have

$$\lim_{t\to\infty} E|z_0(t)|^2 = \lim_{t\to\infty} Tr\, E\left[z_0(t)z_0^*(t)\right]$$

$$= \lim_{t\to\infty} Tr E\left[C(\eta(t))x_0(t)x_0^*(t)C^*(\eta(t))\right]$$

$$= \lim_{t\to\infty} \sum_{j=1}^{d} Tr\, E\left[C(\eta(t))x_0(t)x_0^*(t)\chi_{\eta(t)=j}C^*(\eta(t))\right]$$

$$= \lim_{t\to\infty} \sum_{j=1}^{d} Tr\, C(j)E\left[x_0(t)x_0^*(t)\chi_{\eta(t)=j}\right]C^*(j).$$

Then, based on Lemma 17, we get

$$\lim_{t\to\infty} E|z_0(t)|^2 = \sum_{j=1}^{d} Tr\left(C(j)\widehat{P}_c(j)C^*(j)\right). \qquad (5.61)$$

Taking into account the definition of the inner product in $\mathcal{S}_n^d$ and the representation formulae of $\widehat{P}_c$ and $\widehat{P}_o$, we have

$$\sum_{j=1}^{d} Tr\left(C(j)\widehat{P}_c(j)C^*(j)\right) = \sum_{j=1}^{d} Tr\left(\widehat{P}_c(j)C^*(j)C(j)\right)$$

$$= \langle \widehat{P}_c, \widetilde{C}\rangle = \int_0^\infty \langle e^{\mathcal{L}t}\widehat{B}_v, \widetilde{C}\rangle dt$$

$$= \int_0^\infty \langle \widehat{B}_v, e^{\mathcal{L}^*t}\widetilde{C}\rangle dt = \langle \widehat{B}_v, \widehat{P}_o\rangle$$

$$= \sum_{j=1}^{d} Tr\left(\widehat{B}_v(j)\widehat{P}_o(j)\right)$$

$$= \sum_{j=1}^{d} \pi_{j\infty} Tr\left(B_v^*(j)\widehat{P}_o(j)B_v(j)\right).$$

Finally we remark that, based on the representation formula (5.43), it follows that for any output $z(t)$ we have

$$\lim_{t\to\infty} E|z(t)|^2 = \lim_{t\to\infty} E|z_0(t)|^2,$$

and the proof is complete.   □

For the system **G** defined by (5.41), under the assumption of Theorem 18 we introduce the following norm.

Definition 3. *We call the H^2 norm of the system* (5.41):

$$\|G\|_2 = \left[\lim_{t \to \infty} E|z(t)|^2 \right]^{\frac{1}{2}}. \tag{5.62}$$

Remark 9. The result in Theorem 18 shows that the right-hand side of (5.62) is well defined, and a characterization of the H^2 norm can be given in terms of the controllability and observability Gramians $\widehat{P}_c$ and $\widehat{P}_o$, respectively, which extends to the case of stochastic systems of type (5.41) the well-known results from the deterministic setting.

Further we prove Theorem 19.

Theorem 19. *Under the assumption of Theorem 18 we have*

$$\lim_{T \to \infty} \frac{1}{T} E \left[\int_0^T |z(s)|^2 ds \mid \eta(0) = i \right] = \sum_{j=1}^d Tr \left(B_v^*(j) \widehat{P}_o(j) B_v(j) \right) \tilde{p}_{ij}. \tag{5.63}$$

Proof. Applying the Itô-type formula (Theorem 35 of Chapter 1) for the system (5.41) and for the function $v(x, i) = x^* \widehat{P}_0(i)x$, $x \in \mathbf{R}^n$, $i \in \mathcal{D}$, one obtains

$$E \left[\int_0^T |y(s)|^2 ds \mid \eta(0) = i \right] \tag{5.64}$$

$$= E \left[\int_0^T Tr \left(B_v^*(\eta(s)) \widehat{P}_o(\eta(s)) B_v(\eta(s)) \right) ds \mid \eta(0) = i \right]$$
$$+ x_0^* \widehat{P}_o(i) x_0 - E \left[x^*(T) \widehat{P}_o(\eta(T)) x(T) \mid \eta(0) = i \right].$$

But

$$E \left[\int_0^T Tr \left(B_v^*(\eta(s)) \widehat{P}_o(\eta(s)) B_v(\eta(s)) \right) ds \mid \eta(0) = i \right]$$

$$= E \left[\int_0^T \sum_{j=1}^d Tr \left(B_v^*(j) \widehat{P}_o(j) B_v(j) \chi_{\eta(s)=j} \right) ds \mid \eta(0) = i \right] \tag{5.65}$$

$$= \sum_{j=1}^d Tr \left(B_v^*(j) \widehat{P}_o(j) B_v(j) \right) \int_0^T E \left[\chi_{\eta(s)=j} \mid \eta(0) = i \right] ds$$

$$= \sum_{j=1}^d Tr \left(B_v^*(j) \widehat{P}_o(j) B_v(j) \right) \int_0^T p_{ij}(s) ds.$$

Since $\lim_{s\to\infty} p_{ij}(s) = \tilde{p}_{ij}$ we obtain from (5.65) that

$$
\lim_{T\to\infty} \frac{1}{T} E\left[\int_0^T Tr\left(B_v^*(\eta(s))\widehat{P}_o(\eta(s))B_v(\eta(s))\right) ds \mid \eta(0) = i\right]
$$

$$
= \sum_{j=1}^d Tr\left(B_v^*(j)\widehat{P}_o(j)B_v(j)\right) \lim_{T\to\infty} \frac{1}{T} \int_0^T p_{ij}(s)ds \qquad (5.66)
$$

$$
= \sum_{j=1}^d Tr\left(B_v^*(j)\widehat{P}_o(j)B_v(j)\right) \tilde{p}_{ij}.
$$

Based on Lemma 17 it follows that

$$
\lim_{T\to\infty} \frac{1}{T}\left\{x_0^*\widehat{P}_o(i)x_0 - E\left[x^*(T)^*\widehat{P}_o(\eta(T))x(T) \mid \eta(0) = i\right]\right\}
$$

$$
= \lim_{T\to\infty} \frac{1}{T}\left\{x_0^*\widehat{P}_o(i)x_0 - \sum_{j=1}^d Tr\left(\widehat{P}_o(j)\right) E\left[x(T)x^*(T)\chi_{\eta(T)=j} \mid \eta(0) = i\right]\right\}
$$

$$
= 0. \qquad (5.67)
$$

Finally, from (5.64) combined with (5.66) and (5.67), we get (5.63), and the proof is complete. □

Evidently, the next result holds.

Corollary 20. *Under the assumption of Theorem* 18 *the following hold:*

$$
\lim_{T\to\infty} \frac{1}{T} \int_0^T |z(t)|^2 dt = \lim_{T\to\infty} E|z(T)|^2 = \|\mathbf{G}\|_2^2. \qquad □
$$

Theorem 21. *Assume that the system* $(A_0, A_1, \ldots, A_r; Q)$ *is stable. Then*

$$
\lim_{T\to\infty} \frac{1}{T} \sum_{i=1}^d E\left[\int_0^T |z(s)|^2 ds \mid \eta(0) = i\right]
$$

$$
= \sum_{j=1}^d \delta_j Tr\left(B_v^*(j)\widehat{P}_o(j)B_v(j)\right)
$$

$$
= \sum_{j=1}^d Tr\left(C(j)\widetilde{P}_c(j)C^*(j)\right),
$$

where

$$
\delta_j = \sum_{i=1}^d \tilde{p}_{ij}
$$

and

$$
\widetilde{P}_c = \left(\widetilde{P}_c(1), \ldots, \widetilde{P}_c(d)\right)
$$

is the unique positive semidefinite solution of the equation $\mathcal{L}P + \widehat{M} = 0$, *with* $\widehat{M}(i) = \delta_i B_v(i)B_v^*(i)$.

Proof. From Theorem 19 we have

$$\lim_{T\to\infty} \frac{1}{T} \sum_{i=1}^{d} \left[\int_0^T |z(s)|^2 ds \mid \eta(0) = i \right]$$

$$= \sum_{i,j=1}^{d} Tr\left(B_v^*(j)\widehat{P}_o(j)B_v(j)\right) \tilde{p}_{ij}$$

$$= \sum_{j=1}^{d} \delta_j Tr\left(B_v^*(j)\widehat{P}_o(j)B_v(j)\right)$$

$$= \langle \widehat{P}_o, \widehat{M} \rangle = \int_0^\infty \left\langle e^{\mathcal{L}^* t}\widetilde{C}, \widehat{M} \right\rangle dt$$

$$= \int_0^\infty \left\langle \widetilde{C}, e^{\mathcal{L}t}\widehat{M} \right\rangle dt = \langle \widetilde{C}, \widetilde{P}_c \rangle$$

$$= \sum_{j=1}^{d} Tr\left(C(j)\widetilde{P}_c(j)C^*(j)\right),$$

and hence the proof is complete.   □

Using the result in the above theorem, one can introduce a new norm for the system **G** given by Theorem 21.

Definition 4. *If the zero solution of the system* (5.41) *in the absence of the additive noise* $v(t)$ *is ESMS, then define*

$$|||\mathbf{G}|||_2^2 = \lim_{T\to\infty} \frac{1}{T} \sum_{i=1}^{d} E\left[\int_0^T |z(s)|^2 ds \mid \eta(0) = i \right].$$

Remark 10. (i) Based on the results in Theorems 18 and 21 it follows that while $\|\mathbf{G}\|_2$ depends on the initial repartition $\pi = (\pi_1, \ldots, \pi_d)$ of the process $\eta(t)$, the norm $|||\mathbf{G}|||_2$ does not depend on the initial repartition of $\eta(t)$.

(ii) In the particular case when the system (5.41) is subjected only to white noise perturbations, the two norms defined above coincide. The difference between them is due to the Markov jump perturbations.

(iii) It is obvious that

$$\|\mathbf{G}\|_2 \le |||\mathbf{G}|||_2.$$

5.4.2 Stochastic H^2 optimal control: the state full access case

In this subsection we shall state and solve the design problem of a stabilizing controller that minimizes the H^2 norm of a controlled system whose states are accessible for measurement.

Consider the system **G** described by

$$dx(t) = [A_0(\eta(t))x(t) + B_0(\eta(t))u(t)] \, dt$$

$$+ \sum_{k=1}^{r} [A_k(\eta(t))x(t) + B_k(\eta(t))u(t)] \, dw_k(t) \qquad (5.68)$$

$$+ B_v(\eta(t))dv(t),$$

$$z(t) = C(\eta(t))x(t) + D(\eta(t))u(t),$$

where $x \in \mathbf{R}^n$ is the state vector, $u \in \mathbf{R}^m$ denotes the vector of control variables, $z \in \mathbf{R}^p$ is the regulated output, and $A_k(i), B_k(i), 0 \le k \le r, C(i), D(i), B_v(i), i \in \mathcal{D}$ are constant matrices of appropriate dimensions with real elements. The stochastic processes $\{w(t)\}_{t \ge 0} = (w_1(t), \ldots, w_r(t))^*, \{\eta(t)\}_{t \ge 0}$ and $\{v(t)\}_{t \ge 0}$ have the properties stated in the preceding subsection.

Consider the following family of controllers **G**$_c$ described by

$$\dot{x}_c(t) = A_c(\eta(t))x_c(t) + B_c(\eta(t))u_c(t), \qquad (5.69)$$

$$y_c(t) = C_c(\eta(t))x_c(t) + D_c(\eta(t))u_c(t),$$

where $x_c \in \mathbf{R}^{n_c}, u_c \in \mathbf{R}^n, y_c \in \mathbf{R}^m$. Let us remark that the controller **G**$_c$ of form (5.69) is completely determined by the set of parameters $(n_c, A_c(i), B_c(i), C_c(i), D_c(i), i \in \mathcal{D})$ where $n_c \ge 0$ denotes the controller order. In the particular case $n_c = 0$ the controller (5.69) reduces to

$$y_c(t) = D_c(\eta(t))u_c(t),$$

which shows that the zero order (state-feedback) controllers are included in the set of controllers (5.69).

The resulting system **G**$_{cl}$ obtained by coupling a controller of form (5.69) to the system (5.68) by taking $u_c(t) = x(t)$ and $u(t) = y_c(t)$ is

$$dx_{cl}(t) = A_{0cl}(\eta(t))x_{cl}(t)dt + \sum_{k=1}^{r} A_{kcl}(\eta(t))x_{cl}(t)dw_k(t)$$

$$+ B_{vcl}(\eta(t))dv(t), \qquad (5.70)$$

$$y_{cl}(t) = C_{cl}(\eta(t))x_{cl}(t),$$

where

$$x_{cl} = \begin{bmatrix} x \\ x_{cl} \end{bmatrix};$$

$$A_{0cl}(i) = \begin{bmatrix} A_0(i) + B_0(i)D_c(i) & B_0(i)C_c(i) \\ B_c(i) & A_c(i) \end{bmatrix};$$

$$A_{kcl}(i) = \begin{bmatrix} A_k(i) + B_k(i)D_c(i) & B_k(i)C_c(i) \\ 0 & 0 \end{bmatrix};$$

$$B_{vcl}(i) = \begin{bmatrix} B_v(i) \\ 0 \end{bmatrix};$$

$$C_{cl}(i) = [C(i) + D(i)D_c(i) \quad D(i)C_c(i)].$$

Definition 5. *A controller* $\mathbf{G}_c$ *of form* (5.69) *is called* stabilizing *for the system* (5.68) *if the zero solution of the closed-loop system* (5.70) *(in the absence of the noise v) is* ESMS.

By $\mathcal{K}_s(\mathbf{G})$ we denote the set of all stabilizing controllers $\mathbf{G}_c$ of the form (5.69). Then two optimization problems will be formulated and solved as follows.

(OP1) Find a stabilizing controller of the form (5.69) minimizing $\|\mathbf{G}_{cl}\|_2$.

(OP2) Find a stabilizing controller of the form (5.69) minimizing $\||\mathbf{G}_{cl}|\|_2$.

For the sake of simplicity we shall unify the notation, writing $\|.\|_{2,\ell}$, $\ell = 1, 2$, where $\|.\|_{2,1}$ stands for $\|.\|_2$ and $\|.\|_{2,2}$ stands for $\||.|\|_2$. Thus, from Theorems 18 and 21 we have

$$\|\mathbf{G}_{cl}\|_{2,\ell}^2 = \sum_{i=1}^{d} \varepsilon_i Tr\left(B_{vcl}^*(i)\widehat{P}_{ocl}(i)B_{vcl}(i)\right), \tag{5.71}$$

where

$$\varepsilon_i = \pi_{i\infty} \quad \text{for } \ell = 1, \tag{5.72}$$
$$\varepsilon_i = \delta_i \quad \text{for } \ell = 2,$$

and $\widehat{P}_{ocl}(i) = \left(\widehat{P}_{ocl}(1), \ldots, \widehat{P}_{ocl}(d)\right)$ is the unique positive semidefinite solution of the Lyapunov-type equation on $\mathcal{S}_{n+n_c}^d$, with n_c denoting the order of the controller:

$$A_{0cl}^*(i)\widehat{P}_{ocl}(i) + \widehat{P}_{ocl}(i)A_{0cl}(i) + \sum_{k=1}^{r} A_{kcl}^*(i)\widehat{P}_{ocl}(i)A_{kcl}(i)$$

$$+ \sum_{j=1}^{d} q_{ij}\widehat{P}_{ocl}(j) + C_{cl}^*(i)C_{cl}(i) = 0, \ i \in \mathcal{D}. \tag{5.73}$$

One can associate with the system (5.68) the following SGRAEs:

$$A_0^*(i)X(i) + X(i)A_0(i) + \sum_{k=1}^{r} A_k^*(i)X(i)A_k(i)$$

$$+ \sum_{j=1}^{d} q_{ij}X(j) - \left[X(i)B_0(i) + \sum_{k=1}^{r} A_k^*(i)X(i)B_k(i) + C^*(i)D(i)\right]$$

$$\times \left[D^*(i)D(i) + \sum_{k=1}^{r} B_k^*(i)X(i)B_k(i)\right]^{-1}$$

$$\times \left[B_0^*(i)X(i) + \sum_{k=1}^{r} B_k^*(i)X(i)A_k(i) + D^*(i)C(i)\right] + C^*(i)C(i) = 0, \tag{5.74}$$

$i \in \mathcal{D}$, which can be written in compact form as

$$\mathcal{L}X - \mathcal{P}^*(X)\mathcal{R}^{-1}(X)\mathcal{P}(X) + \widetilde{C} = 0,$$

where $\mathcal{L}$ is the Lyapunov operator defined by the system $(A_0, A_1, \ldots, A_r; Q)$ and

$$\mathcal{P}(X) = (\mathcal{P}_1(X), \ldots, \mathcal{P}_d(X))$$

with

$$\mathcal{P}_i(X) = B_0^*(i)X(i) + \sum_{k=1}^{r} B_k^*(i)X(i)A_k(i) + D^*(i)C(i)$$

and

$$\mathcal{R}(X) = (\mathcal{R}_1(X), \ldots, \mathcal{R}_d(X))$$

with

$$\mathcal{R}_i(X) = D^*(i)D(i) + \sum_{k=1}^{r} B_k^*(i)X(i)B_k(i).$$

Denote by

$$\mathcal{N}(X) = (\mathcal{N}_1(X), \ldots, \mathcal{N}_d(X)) \in \mathcal{S}_{n+m}^d$$

the *generalized dissipation matrix*, where

$$\mathcal{N}_i(X) = \begin{bmatrix} (\mathcal{L}^*X)(i) + \tilde{C}(i) & \mathcal{P}_i^*(X) \\ \mathcal{P}_i(X) & \mathcal{R}_i(X) \end{bmatrix}.$$

Assume that the following conditions are fulfilled.

H1. The system $(\mathbf{A}, \mathbf{B}; Q)$ is stabilizable, where as usual, $\mathbf{A} = (A_0, A_1, \ldots, A_r)$, $\mathbf{B} = (B_0, B_1, \ldots, B_r)$.

H2. There exists $\widehat{X} = (\widehat{X}(1), \ldots, \widehat{X}(d))$ such that $\mathcal{N}(\widehat{X}) > 0$.

Applying Theorem 9 of Chapter 4, we deduce that the SGRAE (5.74) has a stabilizing solution $\tilde{X}$. Now defining the gains

$$\tilde{F}(i) = -\mathcal{R}_i^{-1}(\tilde{X})\,\mathcal{P}_i(\tilde{X}), \quad i \in \mathcal{D}, \tag{5.75}$$

it results that the control

$$u = \tilde{F}(\eta(t))x(t)$$

stabilizes the system (5.68) in the absence of the additive noise $v(t)$.

The corresponding closed-loop system $\tilde{\mathbf{G}}_{cl}$ is

$$dx_{cl}(t) = \left[A_0(\eta(t)) + B_0(\eta(t))\tilde{F}(\eta(t))\right]x(t)dt$$

$$+ \sum_{k=1}^{r}\left[A_k(\eta(t)) + B_k(\eta(t))\tilde{F}(\eta(t))\right]x(t)dw_k(t) + B_v(\eta(t))dv(t), \tag{5.76}$$

$$z(t) = \left[C(\eta(t)) + D(\eta(t))\tilde{F}(\eta(t))\right]x(t).$$

Then the following result is valid.

Proposition 22. *Under the assumptions* **H1** *and* **H2** *we have*

$$\left\|\tilde{\mathbf{G}}_{cl}\right\|_{2,\ell}^2 = \sum_{j=1}^{d}\varepsilon_j Tr\left(B_v^*(j)\tilde{X}(j)B_v(j)\right).$$

Proof. By direct algebraic manipulations (see also Lemma 1(ii) of Chapter 4) we obtain that the SGRAE (5.74) verified by $\widetilde{X}$ can be written in a Lyapunov form as follows:

$$\left[A_0(i) + B_0(i)\widetilde{F}(i)\right]^* \widetilde{X}(i) + \widetilde{X}(i)\left[A_0(i) + B_0(i)\widetilde{F}(i)\right]$$

$$+ \sum_{k=1}^{r} \left[A_k(i) + B_k(i)\widetilde{F}(i)\right]^* \widetilde{X}(i)\left[A_0(i) + B_0(i)\widetilde{F}(i)\right]$$

$$+ \sum_{j=1}^{r} q_{ij}\widetilde{X}(j) + \left[C(i) + D(i)\widetilde{F}(i)\right]^* \left[C(i) + D(i)\widetilde{F}(i)\right] = 0,$$

which shows that the observability Gramian $\widehat{P}_{ocl}$ associated with the closed-loop system (5.76) coincides with the stabilizing solution $\widetilde{X}$ of the SGRAE (5.74). The conclusion in the statement follows from Theorems 18 and 21. □

The main result of this subsection is the following theorem.

Theorem 23. *Assume that* **H1** *and* **H2** *are fulfilled. Under these conditions we have*

$$\min_{G_c \in \mathcal{K}_s(\mathbf{G})} \|\mathbf{G}_{cl}\|_{2,\ell} = \left[\sum_{j=1}^{d} \varepsilon_j Tr\left(B_v^*(j)\widetilde{X}(j)B_v(j)\right) \right]^{\frac{1}{2}},$$

and the optimal control is

$$u(t) = \widetilde{F}(\eta(t))x(t),$$

where $\widetilde{X}$ is the stabilizing solution of SGRAE (5.74), $\widetilde{F} = \left(\widetilde{F}(1), \ldots, \widetilde{F}(d)\right)$, is the stabilizing feedback gain defined by (5.75) and ε_i are as defined in (5.72).

Proof. Let $\mathbf{G}_c \in \mathcal{K}_s(\mathbf{G})$ and $\mathbf{G}_{cl}$ be the corresponding closed-loop system and $\widehat{P}_{ocl}(i)$ denote the observability Gramian. Let

$$\begin{bmatrix} U_{11}(i) & U_{12}(i) \\ U_{12}^*(i) & U_{22}(i) \end{bmatrix}$$

be a partition of $\widehat{P}_{ocl}(i)$ conformably with the partition of the state matrix of the resulting system. Partitioning (5.73) according with the partition of $\widehat{P}_{ocl}(i)$, we get

$$(A_0(i) + B_0(i)D_c(i))^*U_{11}(i) + B_c^*(i)U_{12}^*(i)$$
$$+ U_{11}(i)(A_0(i) + B_0(i)D_c(i)) + U_{12}(i)B_c(i)$$
$$+ \sum_{k=1}^{r} (A_k(i) + B_k(i)D_c(i))^*U_{11}(i)(A_k(i) + B_k(i)D_c(i))$$
$$+ \sum_{j=1}^{d} q_{ij}U_{11}(j) + (C(i) + D(i)D_c(i))^*(C(i) + D(i)D_c(i)) = 0, \qquad (5.77)$$

$$(A_0(i) + B_0(i)D_c(i))^*U_{12}(i) + B_c^*(i)U_{22}(i) + U_{11}(i)B_0(i)C_c(i)$$

$$+ U_{12}(i)A_c(i) + \sum_{k=1}^{r}(A_k(i) + B_k(i)D_c(i))^*U_{11}(i)B_k(i)C_c(i)$$

$$+ \sum_{j=1}^{d}q_{ij}U_{12}(j) + (C(i) + D(i)D_c(i))^*D(i)C_c(i) = 0, \qquad (5.78)$$

$$C_c^*(i)B_0^*(i)U_{12}(i) + A_c^*(i)U_{22}(i) + U_{12}^*B_0(i)C_c(i)$$

$$+ U_{22}(i)A_c(i) + \sum_{k=1}^{r}C_c^*(i)B_k^*(i)U_{11}(i)B_k(i)C_c(i)$$

$$+ \sum_{j=1}^{d}q_{ij}U_{22}(j) + C_c^*(i)D^*(i)D(i)C_c(i) = 0. \qquad (5.79)$$

Using Lemma 1(i) of Chapter 4, SGRAE (5.74) for the stabilizing solution $\widetilde{X}$ can be written as follows:

$$(A_0(i) + B_0(i)D_c(i))^*\widetilde{X}(i) + \widetilde{X}(i)(A_0(i) + B_0(i)D_c(i))$$

$$+ \sum_{k=1}^{r}(A_k(i) + B_k(i)D_c(i))^*\widetilde{X}(i)(A_k(i) + B_k(i)D_c(i))$$

$$+ \sum_{j=1}^{d}q_{ij}\widetilde{X}_j + (C(i) + D(i)D_c(i))^*(C(i) + D(i)D_c(i))$$

$$- \left(D_c(i) - \widetilde{F}(i)\right)^*\mathcal{R}\left(\widetilde{X}\right)\left(D_c(i) - \widetilde{F}(i)\right) = 0. \qquad (5.80)$$

Denoting by

$$\widetilde{U}_{11}(i) = U_{11}(i) - \widetilde{X}(i)$$

and subtracting (5.80) from (5.77), one easily obtains that the triplet $\left(\widetilde{U}_{11}(i), U_{12}(i), U_{22}(i)\right)$ solves the following system of algebraic equations:

$$(A_0(i) + B_0(i)D_c(i))^*\widetilde{U}_{11}(i) + \widetilde{U}_{11}(i)(A_0(i) + B_0(i)D_c(i))$$

$$+ B_c^*(i)U_{12}^*(i) + U_{12}(i)B_c(i) + \sum_{k=1}^{r}(A_k(i) + B_k(i)D_c(i))^*$$

$$\times \widetilde{U}_{11}(i)(A_k(i) + B_k(i)D_c(i)) + \sum_{j=1}^{d}q_{ij}\widetilde{U}_{11}(j)$$

$$+ \left(D_c(i) - \widetilde{F}(i)\right)^*\mathcal{R}\left(\widetilde{X}\right)\left(D_c(i) - \widetilde{F}(i)\right) = 0, \qquad (5.81)$$

$$(A_0(i) + B_0(i)D_c(i))^*U_{12}(i) + B_c^*(i)U_{22}(i) + \tilde{U}_{11}(i)B_0(i)C_c(i)$$

$$+ U_{12}(i)A_c(i) + \sum_{k=1}^{r}(A_k(i) + B_k(i)D_c(i))^*\tilde{U}_{11}(i)B_k(i)C_c(i)$$

$$+ \sum_{j=1}^{d}q_{ij}U_{12}(j) + \left(D_c(i) - \tilde{F}(i)\right)^* \mathcal{R}\left(\tilde{X}\right)C_c(i) = 0, \qquad (5.82)$$

$$C_c^*(i)B_0^*(i)U_{12}(i) + A_c^*(i)U_{22}(i) + U_{12}^*(i)B_0(i)C_c(i)$$

$$+ U_{22}(i)A_c(i) + \sum_{k=1}^{r}C_c^*(i)B_k^*(i)\tilde{U}_{11}(i)B_k(i)C_c(i)$$

$$+ \sum_{j=1}^{d}q_{ij}U_{22}(j) + C_c^*(i)\mathcal{R}\left(\tilde{X}\right)C_c(i) = 0. \qquad (5.83)$$

Setting

$$\tilde{U}(i) = \begin{bmatrix} \tilde{U}_{11}(i) & U_{12}(i) \\ U_{12}^*(i) & U_{22}(i) \end{bmatrix},$$

equations (5.81)–(5.83) can be written in compact form as follows:

$$A_{0cl}^*(i)\tilde{U}(i) + \tilde{U}(i)A_{0cl}(i) + \sum_{k=1}^{r}A_{kcl}^*(i)\tilde{U}(i)A_{kcl}(i)$$

$$+ \sum_{j=1}^{d}q_{ij}\tilde{U}(i) + \Theta^*(i)\mathcal{R}\left(\tilde{X}\right)\Theta(i) = 0,$$

where

$$\Theta(i) = \begin{bmatrix} D_c(i) - \tilde{F}(i) & C_c(i) \end{bmatrix}.$$

Since the system $(A_{0cl}, A_{1cl}, \ldots, A_{rcl}; Q)$ is stable, it follows that $\tilde{U}(i) \geq 0$. Further, we have

$$\|\mathbf{G}_{cl}\|_{2,\ell}^2 = \sum_{i=1}^{d}\varepsilon_i Tr\left(B_{vcl}^*(i)\hat{P}_{0cl}(i)B_{vcl}(i)\right)$$

$$= \sum_{i=1}^{d}\varepsilon_i Tr\left(B_v^*(i)\tilde{X}(i)B_v(i)\right) + \sum_{i=1}^{d}\varepsilon_i Tr\left(B_{vcl}^*(i)\tilde{U}(i)B_{vcl}(i)\right),$$

Since $\tilde{U}(i)$ is positive semidefinite it follows that

$$\|\mathbf{G}_{cl}\|_{2,\ell}^2 \geq \sum_{i=1}^{d}\varepsilon_i Tr\left(B_v^*(i)\tilde{X}(i)B_v(i)\right)$$

for all stabilizing controllers $\mathbf{G}_c$. Using Proposition 22 the conclusion in the statement immediately follows. $\qquad \square$

Remark 11. From Theorem 23 it follows that both optimization problems (OP1) and (OP2) have the same optimal solution given by the controllers with the set of parameters $n_c = 0$, $A_c(i) = 0$, $B_c(i) = 0$, $C_c(i) = 0$, $D_c(i) = \tilde{F}(i)$, $i \in \mathcal{D}$.

The theoretical results derived in this subsection are illustrated by the following numerical example.

Consider the stochastic linear system subjected both to Markovian jumps and to multiplicative noise of form (5.1) with $n = 2$, $\mathcal{D} = \{1, 2\}$, and $r = 1$, where

$$A_0(1) = \begin{bmatrix} -1 & 0 \\ 1 & -1 \end{bmatrix}, A_0(2) = \begin{bmatrix} -1 & 1 \\ 0 & -1 \end{bmatrix},$$

$$A_1(1) = \begin{bmatrix} -1 & 1 \\ 0 & -2 \end{bmatrix}, A_1(2) = \begin{bmatrix} -2 & 1 \\ 1 & -1 \end{bmatrix},$$

$$B_0(1) = \begin{bmatrix} 1 \\ -1 \end{bmatrix}, B_0(2) = \begin{bmatrix} 1 \\ 1 \end{bmatrix},$$

$$B_1(1) = \begin{bmatrix} -1 \\ 1 \end{bmatrix}, B_1(2) = \begin{bmatrix} -2 \\ 1 \end{bmatrix},$$

$$B_v(1) = \begin{bmatrix} 1 \\ -2 \end{bmatrix}, B_v(2) = \begin{bmatrix} 1 \\ 3 \end{bmatrix},$$

$$C(1) = [1 \ 3], \quad C(2) = [2 \ -1],$$

$$D(1) = 1, \quad D(2) = 3,$$

$$Q = \begin{bmatrix} -1 & 1 \\ 1 & -1 \end{bmatrix},$$

and the initial distribution (0.5 0.5). Applying the iterative algorithm presented above for a precision of 10^{-6}, after 205 iterations the following solution has been obtained:

$$F(1) = [-0.2863 \quad -1.5672],$$
$$F(2) = [-0.8547 \quad 0.2353],$$

providing the optimal H^2 norm of the resulting system, which equals 4.4028.

5.4.3 Stochastic H^2 optimal control: the output feedback control

Consider the system **G** described by

$$dx(t) = [A_0(\eta(t))x(t) + B_0(\eta(t))u(t)]dt$$
$$+ \sum_{k=1}^{r} [A_k(\eta(t))x(t) + B_k(\eta(t))u(t)]dw_k(t)$$
$$+ B_v(\eta(t))dv(t), \tag{5.84}$$

$$dy(t) = C_0(\eta(t))x(t)dt + \sum_{k=1}^{r} C_k(\eta(t))x(t)dw_k(t)$$
$$+ D_v(\eta(t))dv(t),$$
$$z(t) = C(\eta(t))x(t) + D(\eta(t))u(t),$$

where $x \in \mathbf{R}^n$ denotes the state, $u \in \mathbf{R}^m$ is the control variable, $y \in \mathbf{R}^p$ is the measured output, and $z \in \mathbf{R}^s$ denotes the regulated output; $\eta(t)$, $w(t)$, $v(t)$, $t \geq 0$ are stochastic processes with the properties given in the previous subsection.

Associate with the system (5.84) the following class of controllers $\mathbf{G}_c$ of the form

$$dx_c(t) = A_c(\eta(t))x_c(t)dt + \sum_{k=1}^{r} A_{kc}(\eta(t))\, x_c(t)dw_k(t) \qquad (5.85)$$
$$+\, B_c(\eta(t))dy(t),$$
$$u(t) = C_c(\eta(t))x_c(t).$$

By coupling $\mathbf{G}_c$ to $\mathbf{G}$ one obtains the resulting system $\mathbf{G}_{cl}$ with the state equations

$$dx_{cl}(t) = A_{0cl}(\eta(t))x_{cl}(t)dt + \sum_{k=1}^{r} A_{kcl}(\eta(t))x_{cl}(t)dw_k(t)$$
$$+\, B_{vcl}(\eta(t))dv(t), \qquad (5.86)$$
$$z(t) = C_{cl}(\eta(t))x_{cl}(t),$$

where

$$x_{cl} = \begin{bmatrix} x \\ x_c \end{bmatrix},$$

$$A_{0cl}(i) = \begin{bmatrix} A_0(i) & B_0(i)C_c(i) \\ B_c(i)C_0(i) & A_c(i) \end{bmatrix},$$

$$A_{kcl}(i) = \begin{bmatrix} A_k(i) & B_k(i)C_c(i) \\ B_c(i)C_k(i) & A_{kc}(i) \end{bmatrix}, \; k = 1, \ldots, r,$$

$$B_{vcl}(i) = \begin{bmatrix} B_v(i) \\ B_c(i)D_v(i) \end{bmatrix},$$

$$C_{cl}(i) = [C(i) \quad D(i)C_c(i)], \quad i \in \mathcal{D}.$$

Definition 6. *The controller* $\mathbf{G}_c$ *is said to be the* stabilizing controller *of* $\mathbf{G}$ *if the zero solution of the closed-loop system* (5.86) *in the absence of the white noise* $v(t)$ *is ESMS. The set of all stabilizing controllers will be denoted by* $\mathcal{K}(\mathbf{G})$.

A controller in $\mathcal{K}(\mathbf{G})$ is determined by the set of the following parameters: $n_c \geq 1$, $A_c(i) \in \mathbf{R}^{n_c \times n_c}$, $B_c(i) \in \mathbf{R}^{n_c \times p}$, $C_c(i) \in \mathbf{R}^{m \times n_c}$. The controller order n_c is not a priori fixed. For a stabilizing controller $\mathbf{G}_c$, consider the norms $\|\mathbf{G}_{cl}\|_2$ and $\||\mathbf{G}_{cl}\||_2$ corresponding to the closed-loop system. Then two optimization problems will be formulated and solved in the following.

(OP1') Find a stabilizing controller minimizing $\|\mathbf{G}_{cl}\|_2$.
(OP2') Find a stabilizing controller minimizing $\||\mathbf{G}_{cl}\||_2$.

It is expected that the solutions of the two problems formulated above will be different. In the particular case when the whole state vector is available for measurements, the solutions of (OP1') and (OP2') coincide, and they are given by a stabilizing state feedback.

Consider the associated SGRAE:

$$A_0(i)Y(i) + Y(i)A_0^*(i) + \sum_{k=1}^{r} A_k(i)Y(i)A_k^*(i) + \sum_{j=1}^{d} q_{ji}Y(j)$$

$$- \left[Y(i)C_0^*(i) + \sum_{k=1}^{r} A_k(i)Y(i)C_k^*(i) + \varepsilon_i B_v(i)D_v^*(i) \right]$$

$$\times \left[\varepsilon_i D_v(i)D_v^*(i) + \sum_{k=1}^{r} C_k(i)Y(i)C_k^*(i) \right]^{-1}$$

$$\times \left[C_0(i)Y(i) + \sum_{k=1}^{r} C_k^*(i)Y(i)A_k^*(i) + \varepsilon_i D_v(i)B_v^*(i) \right]$$

$$+ \varepsilon_i B_v(i)B_v^*(i) = 0, \quad i \in \mathcal{D}, \tag{5.87}$$

where ε_i have been introduced in the previous section. Recall that

$$\tilde{Y} = \left(\tilde{Y}(1), \ldots, \tilde{Y}(d) \right) \in \mathcal{S}_n^d,$$

is a stabilizing solution of (5.87) if the system

$$\left(A_0 + \tilde{K}C_0, A_1 + \tilde{K}C_1, \ldots, A_r + \tilde{K}C_r; Q \right)$$

is stable, where

$$\tilde{K}(i) = - \left[\tilde{Y}(i)C_0^*(i) + \sum_{k=1}^{r} A_k(i)\tilde{Y}(i)C_k^*(i) + \varepsilon_i B_v(i)D_v^*(i) \right]$$

$$\times \left[\varepsilon_i D_v(i)D_v^*(i) + \sum_{k=1}^{r} C_k(i)Y(i)C_k^*(i) \right]^{-1}, \quad i \in \mathcal{D}. \tag{5.88}$$

A necessary and sufficient condition which guarantees the existence of the stabilizing solution of (5.87) is proved by Theorem 18. To this end we introduce the corresponding generalized dissipation matrix:

$$\tilde{N}(Y) = \left(\tilde{N}_1(Y), \ldots, \tilde{N}_d(Y) \right),$$

where

$$\tilde{N}_i(Y) = \begin{bmatrix} (\mathcal{L}Y)(i) + \varepsilon_i B_v(i)B_v^*(i) & \tilde{P}_i(Y) \\ \tilde{P}_i^*(Y) & \tilde{R}_i(Y) \end{bmatrix} \tag{5.89}$$

with

$$\tilde{P}_i(Y) = Y(i)C_0^*(i) + \sum_{k=1}^{r} A_k(i)\tilde{Y}(i)C_k^*(i) + \varepsilon_i B_v(i)D_v^*(i)$$

and

$$\tilde{R}_i(Y) = \varepsilon_i D_v(i)D_v^*(i) + \sum_{k=1}^{r} C_k(i)Y(i)C_k^*(i), \quad i \in \mathcal{D},$$

for all $Y = (Y(1), \ldots, Y(d)) \in \mathcal{S}_n^d$. From Theorem 19 of Chapter 4 it follows that the SGRAE (5.87) has a stabilizing solution if and only if the triplet $(\mathbf{C}, \mathbf{A}; Q)$ is detectable and there exists $\widehat{Y} \in \mathcal{S}_n^d$ such that $\widetilde{\mathcal{N}}(\widehat{Y}) > 0$. Further, if $\mathbf{G}_{cl}$ is the closed-loop system obtained by coupling a stabilizing controller of the set $\mathcal{K}(\mathbf{G})$ to the system (5.84), then according to Theorems 18 and 21 we have

$$\|G_{cl}\|_{2,\ell} = \sum_{i=1}^{d} \varepsilon_i Tr \left(B_{vcl}^*(i) \widehat{P}_{ocl}(i) B_{vcl}(i) \right), \qquad (5.90)$$

where

$$\widehat{P}_{ocl} = \left(\widehat{P}_{ocl}(1), \dots, \widehat{P}_{ocl}(d) \right)$$

is the observability Gramian of the closed-loop system and it verifies the Lyapunov-type system:

$$A_{0cl}^*(i) \widehat{P}_{ocl}(i) + \widehat{P}_{ocl}(i) A_{0cl} + \sum_{k=1}^{r} A_{kcl}^*(i) \widehat{P}_{ocl}(i) A_{kcl}$$

$$+ \sum_{i=1}^{d} q_{ij} \widehat{P}_{ocl}(i) + C_{cl}^* C_{cl} = 0. \qquad (5.91)$$

Since $(A_{0cl}, A_{1cl}, \dots, A_{rcl})$ is stable, the system (5.91) has a unique positive semidefinite solution $\widehat{P}_{ocl}(i)$.

Let $\widetilde{X} = \left(\widetilde{X}(1), \dots, \widetilde{X}(d) \right)$ be the stabilizing solution of SGRAE (5.74). Denote by

$$U(i) = \widehat{P}_{ocl}(i) - \begin{bmatrix} \widetilde{X}(i) & 0 \\ 0 & 0 \end{bmatrix}, \quad i \in \mathcal{D}.$$

By direct calculation one obtains as in the proof of Theorem 23 that

$$U = (U(1), \dots, U(d)) \in \mathcal{S}_{n+n_c}^d$$

is the solution of the Lyapunov-type equation

$$A_{0cl}^*(i) U(i) + U(i) A_{0cl}(i) + \sum_{k=1}^{r} A_{kcl}^*(i) U(i) A_{kcl}(i)$$

$$+ \sum_{j=1}^{d} q_{ij} U(j) + \widehat{C}_{cl}^*(i) \widehat{C}_{cl}(i) = 0, \quad i \in \mathcal{D}, \qquad (5.92)$$

where

$$\widehat{C}_{cl}(i) = \begin{bmatrix} -\Pi(i) \widetilde{F}(i) & \Pi(i) C_c(i) \end{bmatrix}$$

with

$$\Pi(i) = \left(D^*(i) D(i) + \sum_{k=1}^{r} B_k^*(i) \widetilde{X}(i) B_k(i) \right)^{\frac{1}{2}}.$$

Since the system $(A_{0cl}, A_{1cl}, \dots, A_{kcl}; Q)$ is stable, it follows that the unique solution of (5.92) is semipositive. As in the proof of Theorem 23, the equality (5.90) can be written as

$$\|\mathbf{G}_{cl}\|_{2,\ell}^2 = \sum_{i=1}^{d} \varepsilon_i Tr \left(B_v^*(i) \widetilde{X}(i) B_v(i) \right) \tag{5.93}$$

$$+ \sum_{i=1}^{d} \varepsilon_i Tr \left(B_{vcl}^*(i) U(i) B_{vcl}(i) \right).$$

On the other hand, since

$$U = (U(1), \ldots, U(d))$$

is the observability Gramian associated with the triplet

$$\left(\widehat{C}_{cl}, (A_{0cl}, \ldots, A_{rcl}) ; Q \right),$$

then according to the results in Theorems 18 and 21, we get

$$\sum_{i=1}^{d} \varepsilon_i Tr \left(B_{vcl}^*(i) U(i) B_{vcl}(i) \right) = \sum_{i=1}^{d} Tr \left(\widehat{C}_{cl}(i) \widehat{P}_{ccl}(i) \widehat{C}_{cl}^*(i) \right), \tag{5.94}$$

where

$$\widehat{P}_{ccl} = \left(\widehat{P}_{ccl}(1), \ldots, \widehat{P}_{ccl}(d) \right)$$

is the unique solution of the Lyapunov equation on $\mathcal{S}_{n+n_c}^d$:

$$A_{0cl}(i) \widehat{P}_{ccl}(i) + \widehat{P}_{ccl}(i) A_{0cl}^*(i) + \sum_{k=1}^{r} A_{kcl}(i) \widehat{P}_{ccl}(i) A_{kcl}^*(i)$$

$$+ \sum_{j=1}^{d} q_{ji} \widehat{P}_{ccl}(j) + \varepsilon_i B_{vcl}(i) B_{vcl}^*(i) = 0. \tag{5.95}$$

From (5.93) and (5.94) one obtains:

$$\|\mathbf{G}_{cl}\|_{2,\ell}^2 = \sum_{i=1}^{d} \varepsilon_i Tr \left(B_v^*(i) \widetilde{X}(i) B_v(i) \right) \tag{5.96}$$

$$+ \sum_{i=1}^{d} Tr \left(\widehat{C}_{cl}(i) \widehat{P}_{ccl}(i) \widehat{C}_{cl}^*(i) \right).$$

Let

$$\widetilde{Y} = \left(\widetilde{Y}(1), \ldots, \widetilde{Y}(d) \right)$$

be the stabilizing solution of SGRAE (5.87) and define

$$V(i) = \widehat{P}_{ccl}(i) - \begin{bmatrix} \widetilde{Y}(i) & 0 \\ 0 & 0 \end{bmatrix}.$$

Let

$$\begin{bmatrix} Y_{11}(i) & Y_{12}(i) \\ Y_{12}^*(i) & Y_{22}(i) \end{bmatrix}$$

be the partition of $\widehat{P}_{ccl}(i)$ according to the partition of the state matrix of the closed-loop matrix of the closed-loop system. It is easy to see that (5.95) can be partitioned as follows:

$$A_0(i)Y_{11}(i) + Y_{11}(i)A_0^*(i) + B_0(i)C_c(i)Y_{12}^*(i) + Y_{12}(i)C_c^*(i)B_0^*(i)$$

$$+ \sum_{k=1}^{r} \left(A_k(i)Y_{11}(i)A_k^*(i) + B_k(i)C_c(i)Y_{12}^*A_k^* \right.$$

$$+ A_k(i)Y_{12}C_c^*(i)B_k^*(i) + B_k(i)C_c(i)Y_{22}(i)C_c^*(i)B_k^*(i) \Big)$$

$$+ \sum_{j=1}^{d} q_{ji}Y_{11}(j) + \varepsilon_i B_v(i)B_v^*(i) = 0,$$

$$A_0(i)Y_{12}(i) + B_0(i)C_c(i)Y_{22}(i) + Y_{11}(i)C_0^*(i)B_c^*(i) + Y_{12}(i)A_c^*(i)$$

$$+ \sum_{k=1}^{r} \left(A_k(i)Y_{11}(i)C_k^*(i)B_c^*(i) + B_k(i)C_c(i)Y_{12}^*(i)C_k^*(i)B_c^*(i) \right.$$

$$+ A_k(i)Y_{12}(i)A_{kc}^*(i) + B_k(i)C_c(i)Y_{22}(i)A_{kc}^*(i) \Big) \qquad (5.97)$$

$$+ \sum_{j=1}^{d} q_{ji}Y_{12}(j) + \varepsilon_i B_v(i)D_v^*(i)B_c^*(i) = 0,$$

$$B_c(i)C_0(i)Y_{12}(i) + A_c(i)Y_{22}(i) + Y_{12}^*(i)C_0^*(i)B_c^*(i)$$

$$+ Y_{22}(i)A_c^*(i) + \sum_{k=1}^{r} \left(B_c(i)C_k(i)Y_{11}(i)C_k^*(i)B_c^*(i) \right.$$

$$+ A_{kc}(i)Y_{12}^*(i)C_k^*(i)B_c^*(i) + B_c(i)C_k(i)Y_{12}(i)A_{kc}^*(i) + A_{kc}(i)Y_{22}(i)A_{kc}^*(i) \Big)$$

$$+ \sum_{j=1}^{d} q_{ji}Y_{22}(j) + \varepsilon_i B_c(i)D_v(i)D_v^*(i)B_c^*(i) = 0.$$

By direct calculations based on (5.97) and (5.87), we deduce that $V = (V(1), \ldots, V(d))$ is a solution of the following Lyapunov-type equation on $\mathcal{S}_{n+n_c}^d$:

$$A_{0cl}(i)V(i) + V(i)A_{0cl}^*(i) + \sum_{k=1}^{r} A_{kcl}(i)V(i)A_{kcl}^*(i)$$

$$+ \sum_{j=1}^{d} q_{ji}V(j) + \widehat{B}_{vcl}(i)\widehat{B}_{vcl}^*(i) = 0, \quad i \in \mathcal{D}, \qquad (5.98)$$

where

$$\widehat{B}_{vcl}(i) = \begin{bmatrix} -\widetilde{K}(i) \\ B_c(i) \end{bmatrix} \widehat{\Pi}(i)$$

with

$$\widehat{\Pi}(i) = \left(\varepsilon_i D_v(i)D_v^*(i) + \sum_{k=1}^{r} C_k(i)\widetilde{Y}(i)C_k^*(i) \right)^{\frac{1}{2}}, \quad i \in \mathcal{D}.$$

Since the system

$$(A_{0cl}, \ldots, A_{rcl}; Q)$$

is stable, the equation (5.98) has a unique solution $V(i) \geq 0$. Furthermore (5.96) can be rewritten in the form

$$\|G_{cl}\|_{2,\ell}^2 = \sum_{i=1}^{d} \varepsilon_i Tr\left(B_v^*(i)\widetilde{X}(i)B_v(i)\right)$$

$$+ \sum_{i=1}^{d} Tr\left(\Pi(i)\widetilde{F}(i)\widetilde{Y}(i)\widetilde{F}^*(i)\Pi(i)\right) \qquad (5.99)$$

$$+ \sum_{i=1}^{d} Tr\left(\widehat{C}_{cl}(i)V(i)\widehat{C}_{cl}^*(i)\right).$$

Now we are able to prove the main result of this subsection.

Theorem 24. *Assume the following.*
  (i) *The triplet* $(\mathbf{A}, \mathbf{B}; Q)$ *is stabilizable and* $(\mathbf{C}, \mathbf{A}; Q)$ *is detectable.*
  (ii) *There exists* $\widehat{X} \in S_n^d$ *verifying*

$$\mathcal{N}\left(\widehat{X}\right) > 0,$$

where $\mathcal{N}$ *denotes the generalized dissipation matrix.*
  (iii) *There exists* $\widehat{Y} \in S_n^d$ *verifying*

$$\widetilde{\mathcal{N}}\left(\widehat{Y}\right) > 0,$$

where $\widetilde{\mathcal{N}}$ *is defined by* (5.89).
  Under the above conditions we have

$$\min_{G_c \in \mathcal{K}(G)} \|G_{cl}\|_{2,\ell}^2 = \sum_{i=1}^{d} \varepsilon_i Tr\left(B_v^*(i)\widetilde{X}(i)B_v(i)\right)$$

$$+ \sum_{i=1}^{d} Tr\left(\Pi(i)\widetilde{F}(i)\widetilde{Y}(i)\widetilde{F}^*(i)\Pi(i)\right),$$

and this minimum is attained by the optimal controller

$$dx_c(t) = \widetilde{A}_{0c}(\eta(t))x_c(t)dt$$

$$+ \sum_{k=1}^{r} \widetilde{A}_{kc}(\eta(t))x_c(t)dw_k(t) \qquad (5.100)$$

$$+ \widetilde{B}_c(\eta(t))dy(t),$$

$$u(t) = \widetilde{C}_c(\eta(t))x_c(t),$$

with

$$\widetilde{A}_{kc}(i) = A_k(i) + \widetilde{K}(i)C_k(i) + B_k(i)\widetilde{F}(i), \quad k = 0, \ldots, r,$$
$$\widetilde{B}_c(i) = -\widetilde{K}(i),$$
$$\widetilde{C}_c(i) = \widetilde{F}(i), \quad i \in \mathcal{D},$$

where $\widetilde{K}(i)$ *and* $\widetilde{F}(i)$ *are defined by* (5.88) *and* (5.75), *respectively.*

Proof. From (5.99) and from the positivity of the solution V of (5.98), it follows that

$$\|\mathbf{G}_{cl}\|_{2,\ell}^2 \geq \sum_{i=1}^{d} \varepsilon_i Tr\left(B_v^*(i)\widetilde{X}(i)B_v(i)\right) \tag{5.101}$$

$$+ \sum_{i=1}^{d} Tr\left(\Pi(i)\widetilde{F}(i)\widetilde{Y}(i)\widetilde{F}^*(i)\Pi(i)\right)$$

for all stabilizing controllers $\mathbf{G}_c \in \mathcal{K}(\mathbf{G})$. We show now that the controller given by (5.100) belongs to the class of stabilizing controllers $\mathcal{K}(\mathbf{G})$, and for this controller (5.101) becomes equality. The closed-loop system corresponding to the controller (5.100) is

$$
\begin{aligned}
dx(t) &= \left(A_0(\eta(t))x(t) + B_0(\eta(t))\widetilde{F}(\eta(t))x_c(t)\right)dt \\
&\quad + \sum_{k=1}^{r}\left(A_k(\eta(t))x(t) + B_k(\eta(t))\widetilde{F}(\eta(t))x_c(t)\right)dw_k(t) \\
&\quad + B_v(\eta(t))dv(t), \\
dx_c(t) &= \left(-\widetilde{K}(\eta(t))C_0(\eta(t))x(t) \right. & (5.102) \\
&\quad \left. + \left(A_0(\eta(t)) + B_0(\eta(t))\widetilde{F}(\eta(t)) + \widetilde{K}(\eta(t))C_0(\eta(t))\right)x_c\right)dt \\
&\quad + \sum_{k=1}^{r}\left(-\widetilde{K}(\eta(t))C_k(\eta(t))x(t) + \left(A_k(\eta(t)) + B_k(\eta(t))\widetilde{F}(\eta(t))\right.\right. \\
&\qquad\qquad \left.\left. + \widetilde{K}(\eta(t))C_k(\eta(t))\right)x_c(t)\right) \\
&\quad \times dw_k(t) - \widetilde{K}(\eta(t))D_v(\eta(t))dv(t), \\
z(t) &= C(\eta(t))x(t) + D(\eta(t))\widetilde{F}(\eta(t))x_c(t).
\end{aligned}
$$

If $\begin{bmatrix} x^*(t) & x_c^*(t) \end{bmatrix}^*$ is a solution of (5.102) in the absence of the additive noise $v(t)$, define

$$\xi(t) = x(t) - x_c(t), \ t \geq 0.$$

Then, by direct computations, it follows that the stochastic process $[x^*(t) \ \xi^*(t)]^*$ verifies the system

$$
\begin{aligned}
dx(t) &= \left(\left(A_0(\eta(t)) + B_0(\eta(t))\widetilde{F}(\eta(t))\right)x(t) \right. \\
&\qquad \left. - B_0(\eta(t))\widetilde{F}(\eta(t))\xi(t)\right)dt \\
&\quad + \sum_{k=1}^{r}\left(\left(A_k(\eta(t)) + B_k(\eta(t))\widetilde{F}(\eta(t))\right)x(t) \right. \\
&\qquad\qquad \left. - B_k(\eta(t))\widetilde{F}(\eta(t))\xi(t)\right)dw_k(t), & (5.103) \\
d\xi(t) &= \left(A_0(\eta(t)) + \widetilde{K}(\eta(t))C_0(\eta(t))\right)\xi(t)dt \\
&\quad + \sum_{k=1}^{r}\left(A_k(\eta(t)) + \widetilde{K}(\eta(t))C_k(\eta(t))\right)\xi(t)dw_k(t).
\end{aligned}
$$

Since $\widetilde{Y}$ is the stabilizing solution of SGRAE (5.87), from the second equation (5.103) one obtains

$$E\left[|\xi(t)|^2 \mid \eta(0) = i\right] \le \beta e^{-\alpha t} |\xi(0)|^2, \quad t \ge 0, \quad i \in \mathcal{D}, \qquad (5.104)$$

for some $\alpha > 0$ and $\beta \ge 1$. Further, the first equation (5.103) can be rewritten as follows:

$$dx(t) = \left(\left(A_0(\eta(t)) + B_0(\eta(t))\widetilde{F}(\eta(t))\right)x(t) + f_0(t)\right)dt$$
$$+ \sum_{k=1}^{r}\left(\left(A_k(\eta(t)) + B_k(\eta(t))\widetilde{F}(\eta(t))\right)x(t) + f_k(t)\right)dw_k(t)$$

with

$$f_k(t) = -B_k(\eta(t))\widetilde{F}(\eta(t))\xi(t), \quad t \ge 0, \quad k = 0, 1, \ldots, r.$$

Applying Theorem 32(i) of Chapter 2, one deduces that there exist $\tilde{\beta} \ge 1$ and $\tilde{\alpha} > 0$ such that

$$E\left[|x(t)|^2 \mid \eta(0) = i\right] \le \tilde{\beta} e^{-\tilde{\alpha} t}\left(|x(0)|^2 + |\xi(0)|^2\right). \qquad (5.105)$$

From (5.104) and (5.105) we get

$$E\left[|x_c(t)|^2 \mid \eta(0) = i\right] \le \hat{\beta} e^{-\hat{\alpha} t}\left(|x(0)|^2 + |\xi(0)|^2\right),$$

where $\hat{\alpha} = \min(\alpha, \tilde{\alpha})$, $\hat{\beta} = \max(\beta, \tilde{\beta})$, and therefore we conclude that the controller (5.100) is a stabilizing controller. On the other hand, we may write with this controller:

$$\sum_{i=1}^{d} Tr\left(\widetilde{C}_{cl}(i)V(i)\widetilde{C}_{cl}^*(i)\right) \qquad (5.106)$$

$$= \sum_{i=1}^{d} Tr\left(\Pi(i)\widetilde{F}(i)\left(V_{11}(i) - V_{12}(i) - V_{12}^*(i) + V_{22}(i)\right)\widetilde{F}^*(i)\Pi(i)\right),$$

where

$$\begin{bmatrix} V_{11}(i) & V_{12}(i) \\ V_{12}^*(i) & V_{22}(i) \end{bmatrix}$$

is the partition of the solution $V(i)$ of equation (5.98) corresponding to the controller (5.100).

Partitioning the equation (5.98) we obtain the following system:

$$A_0(i)V_{11}(i) + B_0(i)\widetilde{F}(i)V_{12}^*(i) + V_{11}(i)A_0^*(i) + V_{12}(i)\widetilde{F}(i)B_0^*(i)$$
$$+ \sum_{k=1}^{r}\left(A_k(i)V_{11}(i)A_k^*(i) + B_k(i)\widetilde{F}(i)V_{12}^*(i)A_k^*(i)\right.$$
$$+ A_k(i)V_{12}(i)\widetilde{F}^*(i)B_k^*(i) + B_k(i)\widetilde{F}(i)V_{22}(i)\widetilde{F}^*(i)B_k^*(i)\right)$$
$$+ \sum_{j=1}^{d} q_{ji}V_{11}(j) + \widetilde{K}(i)\widehat{\Pi}^2(i)\widetilde{K}^*(i) = 0,$$

$$A_0(i)V_{12}(i) + B_0(i)\widetilde{F}(i)V_{22}(i) - V_{11}(i)C_0^*(i)\widetilde{K}^*(i) + V_{12}(i)\widetilde{A}_{0c}^*(i)$$

$$+ \sum_{k=1}^{r}\left(-A_k(i)V_{11}(i)C_k^*(i)\widetilde{K}^*(i) - B_k(i)\widetilde{F}(i)V_{12}^*(i)C_k^*(i)\widetilde{K}^*(i)\right.$$

$$\left. + A_k(i)V_{12}(i)\widetilde{A}_{kc}^*(i) + B_k(i)\widetilde{F}(i)V_{22}(i)\widetilde{A}_{kc}^*(i)\right)$$

$$+ \sum_{j=1}^{d} q_{ji} V_{12}(j) + \widetilde{K}(i)\widehat{\Pi}^2(i)\widetilde{K}^*(i) = 0,$$

$$- \widetilde{K}(i)C_0(i)V_{12}(i) + \widetilde{A}_{0c}(i)V_{22}(i) - V_{12}^*(i)C_0^*\widetilde{K}^*(i) + V_{22}(i)\widetilde{A}_{0c}^*(i)$$

$$+ \sum_{k=1}^{r}\left(\widetilde{K}(i)C_k(i)V_{11}(i)C_k^*(i)\widetilde{K}^*(i) - \widetilde{A}_{kc}(i)V_{12}^*(i)C_k^*(i)\widetilde{K}^*(i)\right.$$

$$\left. - \widetilde{K}(i)C_k(i)V_{12}(i)\widetilde{A}_{kc}^*(i) + \widetilde{A}_{kc}(i)V_{22}(i)\widetilde{A}_{kc}^*(i)\right)$$

$$+ \sum_{j=1}^{d} q_{ji} V_{22}(j) + \widetilde{K}(i)\widehat{\Pi}^2(i)\widetilde{K}^*(i) = 0.$$

(5.107)

By summing the first and the third equations of (5.107) and by then subtracting the second equation (5.107) and its transpose, one obtains that

$$W(i) = V_{11}(i) - V_{12}(i) - V_{12}^*(i) + V_{22}(i)$$

verifies the equation

$$\left(A_0(i) + \widetilde{K}(i)C_0(i)\right)W(i) + W(i)\left(A_0(i) + \widetilde{K}(i)C_0(i)\right)^*$$

$$+ \sum_{k=1}^{r}\left(A_k(i) + \widetilde{K}(i)C_k(i)\right)W(i)\left(A_k(i) + \widetilde{K}(i)C_k(i)\right)^*$$

$$+ \sum_{j=1}^{d} q_{ji} W(j) = 0.$$

Since the system

$$\left(A_0 + \widetilde{K}C_0, A_1 + \widetilde{K}C_1, \ldots, A_r + \widetilde{K}C_r; Q\right)$$

is stable, the above equation has a unique solution from which we deduce that $W(i) = 0$, $i \in \mathcal{D}$. Based on (5.105) this shows that

$$\sum_{i=1}^{d} Tr\left(\widetilde{C}_{cl}(i)V(i)\widetilde{C}_{cl}^*(i)\right) = 0,$$

and therefore

$$\left\|\widetilde{G}_{cl}\right\|_{2,\ell}^{2} = \sum_{i=1}^{d} \varepsilon_i Tr\left(B_v^*(i)\widetilde{X}(i)B_v(i)\right)$$

$$+ \sum_{i=1}^{d} Tr\left(\Pi(i)\widetilde{F}(i)\widetilde{Y}(i)\widetilde{F}^*(i)\Pi(i)\right),$$

where $\widetilde{G}_{cl}$ is the closed-loop system corresponding to the controller (5.100) and thus the proof is complete. $\square$

Remark 12. In the particular case when $\mathcal{D} = \{1\}$, $A_k = 0$, $B_k = 0$, $C_k = 0$, $k = 1, 2, \ldots, r$, the controller (5.100) reduces to the well-known Kalman–Bucy filter which solves the classic H^2 optimization problem. Therefore, it is natural that in the general framework considered here, the solution of the H^2 optimization problem has a form similar to the Kalman–Bucy filter. Unfortunately, in the general case, when the nominal plant is corrupted with multiplicative white noise, the solution of the H^2 optimization problem is a stochastic system with multiplicative noise, which leads to implementation difficulties. This fact leads us to consider an H^2 optimization problem in the class of controllers with $A_{kc}(i) = 0$, $k = 1, \ldots, r$, which still remains an open problem.

At the end of this section we focus our attention on the strictly Markovian case, namely $d > 1$, $A_k(i) = 0$, $B_k(i) = 0$, $C_k(i) = 0$, $A_{kc}(i) = 0$, $1 \leq k \leq r$, $i \in \mathcal{D}$. Therefore, the controlled system is in this case:

$$dx(t) = (A_0(\eta(t))x(t) + B_0(\eta(t))u(t)) \, dt + B_v(\eta(t))dv(t),$$
$$dy(t) = C_0(\eta(t))x(t)dt + D_v(\eta(t))dv(t), \qquad (5.108)$$
$$z(t) = C(\eta(t))x(t) + D(\eta(t))u(t).$$

In this particular case Theorem 24 leads to the following corollary.

Corollary 25. *Assume the following.*
  (i) *The triplet $(A_0, B_0; Q)$ is stabilizable and $(C_0, A_0; Q)$ is detectable.*
  (ii) *There exists $\widehat{X} = \left(\widehat{X}(1), \ldots, \widehat{X}(d)\right) \in \mathcal{S}_n^d$ satisfying the LMI*

$$\mathcal{N} := \begin{bmatrix} \mathcal{L}^*\left(\widehat{X}\right)(i) + C^*(i)C(i) & \widehat{X}(i)B_0(i) + C^*(i)D(i) \\ B_0^*(i)\widehat{X}(i) + D^*(i)C(i) & D^*(i)D(i) \end{bmatrix} > 0,$$

where

$$\mathcal{L}^*\left(\widehat{X}\right)(i) = A_0^*(i)\widehat{X}(i) + \widehat{X}(i)A_0(i) + \sum_{j=1}^{d} q_{ij}\widehat{X}(j).$$

  (iii) *There exists $\widehat{Y} = \left(\widehat{Y}(1), \ldots, \widehat{Y}(d)\right) \in \mathcal{S}_n^d$ satisfying the LMI*

$$\widetilde{\mathcal{N}} := \begin{bmatrix} \mathcal{L}\left(\widehat{Y}\right)(i) + \varepsilon_i B_v(i)B_v^*(i) & \widehat{Y}(i)C_0(i) + \varepsilon_i B_v(i)D_v^*(i) \\ C_0^*(i)\widehat{Y}(i) + \varepsilon_i D_v(i)B_v^*(i) & \varepsilon_i D_v(i)D_v^*(i) \end{bmatrix} > 0,$$

where ε_i are either $\pi_{i\infty}$ or δ_i introduced in Section 5.4.1. Then the controller

$$dx_c(t) = \widetilde{A}_c(\eta(t))x_c(t)dt + \widetilde{B}_c(\eta(t))dy(t), \qquad (5.109)$$
$$u(t) = \widetilde{C}_c(\eta(t))x_c(t),$$

with

$$\tilde{A}_c(i) = A_0(i) + B_0(i)\tilde{F}(i) + \tilde{K}(i)C_0(i),$$
$$\tilde{B}_c(i) = -\tilde{K}(i),$$
$$\tilde{C}_c(i) = \tilde{F}(i)$$

stabilizes the system (5.108) and

$$\|\tilde{\mathbf{G}}_{cl}\|_{2,\ell}^2 = \sum_{i=1}^{d} \varepsilon_i Tr\left(B_v^*(i)\tilde{X}(i)B_v(i)\right)$$

$$+ \sum_{i=1}^{d} Tr\left((D^*(i)D(i))^{\frac{1}{2}}\tilde{F}(i)\tilde{Y}(i)\tilde{F}^*(i)\left(D^*(i)D(i)\right)^{\frac{1}{2}}\right)$$

$$= \min_{\mathbf{G}_c \in \mathcal{K}(\mathbf{G})} \|\mathbf{G}_{cl}\|_{2,\ell}^2,$$

where $\tilde{\mathbf{G}}_{cl}$ *is the closed-loop system obtained by coupling the controller (5.109) to the system (5.108);* $\tilde{X}$ *and* $\tilde{Y}$ *are the stabilizing solutions of the Riccati-type equations:*

$$A_0^*(i)X(i) + X(i)A_0(i) + \sum_{j=1}^{d} q_{ij}X(j) - \left(X(i)B_0(i) + C^*(i)D(i)\right)$$
$$\times \left(D^*(i)D(i)\right)^{-1}\left(B_0^*(i)X(i) + D^*(i)C(i)\right) + C^*(i)C(i) = 0;$$

$$A_0(i)Y(i) + Y(i)A_0^*(i) + \sum_{j=1}^{d} q_{ji}Y(j) - \left(Y(i)C_0(i) + \varepsilon_i B_v(i)D_v^*(i)\right)$$
$$\times \left(\varepsilon_i D_v(i)D_v^*(i)\right)^{-1}\left(C_0^*(i)Y(i) + \varepsilon_i D_v(i)B_v^*(i)\right) + \varepsilon_i B_v(i)B_v^*(i) = 0;$$

and $\tilde{F}$ *and* $\tilde{K}$ *are given by*

$$\tilde{F}(i) = -\left(D^*(i)D(i)\right)^{-1}\left(B_0^*(i)\tilde{X}(i) + D^*(i)C(i)\right),$$
$$\tilde{K}(i) = -\left(\tilde{Y}(i)C_0(i) + \varepsilon_i B_v(i)D_v^*(i)\right)\left(\varepsilon_i D_v(i)D_v^*(i)\right)^{-1}. \qquad \square$$

In order to illustrate the above results we shall present a numerical example. Consider a helicopter dynamics having the state-space equations

$$\dot{x}(t) = A(\eta)x(t) + B(\eta(t))u(t) + Ew(t),$$
$$z(t) = C_1x(t) + D_1u(t),$$
$$y(t) = C_2x(t) + D_2w(t),$$

where $\eta(t)$ indicates the airspeed and the state variables are the horizontal velocity x_1, the vertical velocity x_2, the pitch rate x_3, and the pitch angle x_4. The matrices in the above state-space representation have the form (see [25])

$$A(i) = \begin{bmatrix} -0.0366 & 0.0271 & 0.0188 & -0.4555 \\ 0.0482 & -1.01 & 0.0024 & -4.0208 \\ 0.1002 & a_{32}(i) & -0.707 & a_{34}(i) \\ 0 & 0 & 1 & 0 \end{bmatrix},$$

$$B(i) = \begin{bmatrix} 0.4422 & 0.1761 \\ b_{21}(i) & -7.5922 \\ -5.5200 & 4.4900 \\ 0 & 0 \end{bmatrix},$$

$$E = [I_{4\times4} \quad 0_{4\times1}], \quad C_1 = \begin{bmatrix} I_{4\times4} \\ 0_{2\times4} \end{bmatrix}, \quad D_1 = \begin{bmatrix} 0_{4\times2} \\ I_{2\times2} \end{bmatrix},$$
$$C = [0 \quad 1 \quad 0 \quad 0], \quad D_2 = [0 \quad 0 \quad 0 \quad 0 \quad 1], \quad i = 1,2,3,$$

where $a_{32}(\cdot)$, $a_{34}(\cdot)$, and $b_{21}(\cdot)$ are given in Table 5.1 as a function of the airspeed. The behavior of $\eta(t)$ is modelled as a Markov chain with three states corresponding to the three values of the airspeed: 135, 60, and 170 knots.

Airspeed (knots)	a_{32}	a_{34}	b_{21}
135	0.3681	1.4200	3.5446
60	0.0664	0.1198	0.9775
170	0.5047	2.5460	5.1120

Table 5.1.

The following three transition matrices have been considered:

$$Q_1 = \begin{bmatrix} -0.0907 & 0.0671 & 0.0236 \\ 0.0671 & -0.0671 & 0 \\ 0.0236 & 0 & -0.0236 \end{bmatrix},$$

$$Q_2 = \begin{bmatrix} -0.0171 & 0.0007 & 0.0164 \\ 0.0013 & -0.0013 & 0 \\ 0.0986 & 0 & -0.0986 \end{bmatrix},$$

$$Q_3 = \begin{bmatrix} -0.0450 & 0.0002 & 0.0448 \\ 0.0171 & -0.0171 & 0 \\ 0.0894 & 0 & -0.0894 \end{bmatrix}.$$

The initial assumed distributions are (0.333 0.333 0.333), (0.6 0.3 0.1), and (0.6 0.1 0.3), respectively. The optimal H^2 corresponding norms obtained using the method described in this section are presented in Table 5.2.

Q	Optimal H^2 norms computed by the method in the present paper
Q_1	$\\|\mathbf{G}_{cl}\\|_{2.1} = 4.6735;\quad \\|\mathbf{G}_{cl}\\|_{2.2} = 8.0988$
Q_2	$\\|\mathbf{G}_{cl}\\|_{2.1} = 4.5196;\quad \\|\mathbf{G}_{cl}\\|_{2.2} = 7.8264$
Q_3	$\\|\mathbf{G}_{cl}\\|_{2.1} = 4.8113;\quad \\|\mathbf{G}_{cl}\\|_{2.2} = 8.3333$

Table 5.2.

Here only the optimal H^2 controller for the case $Q = Q_1$ is given. Its realization is the following:

$$A_c(1) = \begin{bmatrix} -0.4431 & 0.3328 & 0.4106 & 0.0327 \\ -3.4133 & -10.3798 & 4.8501 & 6.3131 \\ 5.3252 & 5.2657 & -6.8663 & -9.4439 \\ 0 & 1.7630 & 1 & 0 \end{bmatrix}, \quad B_c(1) = \begin{bmatrix} -0.1509 \\ 3.0100 \\ -1.1841 \\ -1.7630 \end{bmatrix},$$

$$C_c(1) = \begin{bmatrix} -0.9282 & 0.0139 & 0.9616 & 1.3881 \\ 0.0226 & 0.8442 & -0.1896 & -0.7131 \end{bmatrix},$$

$$A_c(2) = \begin{bmatrix} -0.4133 & 0.4164 & 0.3727 & -0.0675 \\ -2.0379 & -9.7852 & 3.6641 & 4.2692 \\ 5.8528 & 3.3426 & -7.5378 & -10.9517 \\ 0 & 1.3828 & 1 & 0 \end{bmatrix}, \quad B_c(2) = \begin{bmatrix} -0.1727 \\ 2.6160 \\ -0.4174 \\ -1.3828 \end{bmatrix},$$

$$C_c(2) = \begin{bmatrix} -0.9144 & 0.1586 & 0.9440 & 1.2483 \\ 0.1570 & 0.8317 & -0.3607 & -0.9312 \end{bmatrix},$$

$$A_c(3) = \begin{bmatrix} -0.4517 & 0.2545 & 0.4437 & 0.1318 \\ -4.3958 & -11.1936 & 5.5719 & 7.2984 \\ 5.0354 & 6.8942 & -6.4680 & -7.9318 \\ 0 & 2.2062 & 1 & 0 \end{bmatrix}, \quad B_c(3) = \begin{bmatrix} -0.1030 \\ 3.4319 \\ -2.2534 \\ -2.2062 \end{bmatrix},$$

$$C_c(3) = \begin{bmatrix} -0.9240 & -0.0573 & 0.9882 & 1.5154 \\ -0.0368 & 0.8507 & -0.0682 & -0.4705 \end{bmatrix}.$$

Let us finally remark that no ill-conditioned computations occurred when the iterative procedure described in this section was applied.

Notes and references

The results presented in this chapter are mainly based on the papers [30], [31], [94]. The linear quadratic problem in the stochastic case has been investigated starting with [117]. For stochastic linear systems with multiplicative noise we mention [77], [8], [67], [116], [14], [88], [3], [4], and for the infinite-dimensional case we cite [18]–[20] and [111]. In the case of stochastic systems subjected to Markovian perturbations, the linear quadratic problem has been addressed in [86], [70], [89]. As concerns the H^2 control problem for stochastic systems with multiplicative white noise, we cite [19] and [39], and for systems with Markovian jump we mention [15], [25], where suboptimal solutions of the same order as the order of the nominal system are considered.

6

Stochastic Version of the Bounded Real Lemma and Applications

The main goal of this chapter is to investigate the robustness properties of a stable linear stochastic system with respect to various classes of uncertainties.

A crucial role in determining a lower bound of robustness radius will be played by the norm of a linear bounded operator associated with the given plant. This operator will be called the *input–output operator* and it will be introduced in Section 6.1. In the next section a stochastic version of the so-called *Bounded Real Lemma* will be proved. This result provides an estimation of the norm of the input–output operator in terms of feasibility of some linear matrix inequalities (LMIs) or in terms of existence of stabilizing solutions of a generalized algebraic Riccati-type equation.

Further, the stochastic version of the so-called *Small Gain Theorem* will be proved. This result will be used to derive a lower bound of robustness with respect to linear structural uncertainties. Then we shall investigate the stability robustness with respect to a wide class of nonlinear uncertainties.

As in the previous chapters a unitary approach will be used for systems subjected both to multiplicative white noise disturbances and to Markovian switching. In order to simplify the developments in this chapter we restrict our attention to the systems in the time-invariant case.

6.1 Input–output operators

Consider the linear system described by

$$dx(t) = [A_0(\eta(t))x(t) + B_0(\eta(t))u(t)]dt$$
$$+ \sum_{k=1}^{r}[A_k(\eta(t))x(t) + B_k(\eta(t))u(t)]dw_k(t), \qquad (6.1)$$
$$y(t) = C(\eta(t))x(t) + D(\eta(t))u(t),$$

with the state $x(t) \in \mathbf{R}^n$, the input $u(t) \in \mathbf{R}^m$, and the output $y(t) \in \mathbf{R}^p$. $A_k(i)$, $B_k(i)$, $k = 0, 1, \ldots, r$, $C(i)$, $D(i)$, $i \in \mathcal{D}$, are constant matrices of appropriate dimensions. The stochastic processes $\eta(t), t \geq 0$, $w(t) = (w_1(t), \ldots, w_r(t))_{t \geq 0}$

have the properties given in Chapter 1. If $u(t)$, $t \geq 0$, is a stochastic process having the components in $L^2_{\eta,w}[0, \infty)$, $x_u(t)$, $t \geq 0$ stands for the solution of (6.1) with the initial condition $x_u(0) = 0$. According to the results derived in Section 1.12, the components of the process $x_u(t)$, $t \geq 0$, are in $L^2_{\eta,w}[0, \tau]$ $\forall \tau > 0$. Moreover, if the system $(A_0, \ldots, A_r; Q)$ is stable, then based on Theorem 32 of Chapter 2, with $f_k(t) = B_k(\eta(t))u(t)$, it follows that $x_u(.)$ is in $L^2_{\eta,w}([0, \infty), \mathbf{R}^n)$. On the other hand, by uniqueness arguments one easily obtains that the map $u \mapsto x_u(.)$ is linear. Therefore, if the system $(A_0, \ldots, A_r; Q)$ is stable, we may consider the operator $\mathcal{T}$ defined on the space of stochastic processes $L^2_{\eta,w}([0, \infty), \mathbf{R}^m)$ with values in $L^2_{\eta,w}([0, \infty), \mathbf{R}^p)$, as follows:

$$(\mathcal{T}u)(t) = y_u(t),$$

where

$$y_u(t) = C(\eta(t))x_u(t) + D(\eta(t))u(t). \tag{6.2}$$

From Theorem 19 of Chapter 1 it follows that $L^2_{\eta,w}([0, \infty), \mathbf{R}^\ell)$ is a closed subspace of the Hilbert space $L^2([0, \infty), \mathbf{R}^\ell)$. Therefore,

$$L^2_{\eta,w}([0, \infty), \mathbf{R}^\ell)$$

is a real Hilbert space with the usual inner product:

$$\langle u, v \rangle = E \int_0^\infty u^*(t)v(t)dt = \int_0^\infty E u^*(t)v(t)dt.$$

The norm induced by this inner product will be denoted by $\| \cdot \|$.

Obviously

$$\|z\| = \left(E \int_0^\infty |z(t)|^2 \, dt \right)^{\frac{1}{2}} = \left(\sum_{j=1}^d \pi_j E \left[\int_0^\infty |z(t)|^2 \, dt \mid \eta(0) = j \right] \right)^{\frac{1}{2}} \tag{6.3}$$

for all $z \in L^2_{\eta,w}([0, \infty), \mathbf{R}^\ell)$, where $\pi_i = P\{\eta(0) = i\}$. Again invoking Theorem 32 of Chapter 2, it immediately follows that there exists $c > 0$ not depending on u such that

$$\|x_u\|^2 = \sum_{j=1}^d \pi_j E \left[\int_0^\infty |x_u(t)|^2 \, dt \mid \eta(0) = j \right]$$

$$\leq c \sum_{j=1}^d \pi_j E \left[\int_0^\infty |u(t)|^2 dt \mid \eta(0) = j \right] = c\|u\|^2.$$

This allows us to conclude that the operator $\mathcal{T}$ defined by (6.2) is linear and bounded. The operator $\mathcal{T}$ introduced above will be termed the *input–output operator* associated with the system (6.1), and the system (6.1) will be a state-space realization of the operator T. As in the deterministic case the state-space realization of the input–output operator is not unique. The set of operators

$$T : L^2_{\eta,w}([0, \infty), \mathbf{R}^m) \rightarrow L^2_{\eta,w}([0, \infty), \mathbf{R}^p)$$

which admits state-space realizations is a subspace of the Banach space

$$\mathcal{L}\left(L^2_{\eta,w}([0, \infty), \mathbf{R}^m), L^2_{\eta,w}([0, \infty), \mathbf{R}^p)\right).$$

Indeed, one can easily check that if

$$T_\ell : L^2_{\eta,w}([0, \infty), \mathbf{R}^m) \rightarrow L^2_{\eta,w}([0, \infty), \mathbf{R}^p), \ell = 1, 2,$$

have the state-space realization

$$dx_\ell(t) = [A_{0\ell}(\eta(t))x_\ell(t) + B_{0\ell}(\eta(t))u(t)]dt \tag{6.4}$$

$$+ \sum_{k=1}^{r} [A_{k\ell}(\eta(t))x_\ell(t) + B_{k\ell}(\eta(t))u(t)]dw_k(t),$$

$$y_\ell(t) = C_\ell(\eta(t))x_\ell(t) + D_\ell(\eta(t))u(t), \ell = 1, 2,$$

then the operators $\alpha_1 T_1 + \alpha_2 T_2$ will have the state-space realization of form (6.1) with

$$A_k(i) = \begin{bmatrix} A_{k1}(i) & 0 \\ 0 & A_{k2}(i) \end{bmatrix},$$

$$B_k(i) = \begin{bmatrix} B_{k1}(i) \\ B_{k2}(i) \end{bmatrix},$$

$$C(i) = [\alpha_1 C_1(i) \quad \alpha_2 C_2(i)],$$

$$D(i) = \alpha_1 D_1(i) + \alpha_2 D_2(i), \quad \text{and}$$

$$x = \begin{bmatrix} x_1 \\ x_2 \end{bmatrix}.$$

Remark 1. For every $\tau > 0$, the system (6.1) defines a linear operator

$$T_\tau : L^2_{\eta,w}([0, \tau], \mathbf{R}^m) \rightarrow L^2_{\eta,w}([0, \tau], \mathbf{R}^p)$$

by $y = T_\tau u$ with

$$y(t) = C(\eta(t))x_u(t) + D(t)u(t), \ t \in [0, \tau], \forall u \in L^2_{\eta,w}([0, \tau], \mathbf{R}^m).$$

Based on Remark 17 of Chapter 2, one immediately deduces that T_τ is a bounded operator. One expects that the norm $\|T_\tau\|$ depends on τ. Moreover, for any $0 < \tau_1 < \tau_2$, we have

$$\|T_\tau\| \leq \|T_{\tau_2}\|.$$

If the system $(A_0, A_1, \ldots, A_r; Q)$ is stable, then

$$\|T\| = \sup_{\tau > 0} \|T_\tau\|.$$

The last assertion in the above remark is also true if the linear operator T defined by (6.2) on the space $L^2_{\eta,w}([0,\infty), \mathbf{R}^m)$ is a bounded operator with values in the space $L^2_{\eta,w}([0,\infty), \mathbf{R}^p)$.

Concerning the product and the inversion of the input–output operators we have the following proposition.

Proposition 1. (i) *If*

$$T^1_\tau : L^2_{\eta,w}\left([0,\tau], \mathbf{R}^m\right) \to L^2_{\eta,w}\left([0,\tau], \mathbf{R}^p\right),$$
$$T^2_\tau : L^2_{\eta,w}\left([0,\tau], \mathbf{R}^{m_1}\right) \to L^2_{\eta,w}\left([0,\tau], \mathbf{R}^m\right)$$

have the state-space realizations as in (6.4) with $A_{k\ell}(i) \in \mathbf{R}^{n_\ell \times n_\ell}$, $B_{k1}(i) \in \mathbf{R}^{n_1 \times m}$, $B_{k2}(i) \in \mathbf{R}^{n_2 \times m_1}$, $0 \le k \le r$, $C_1(i) \in \mathbf{R}^{p \times n_1}$, $C_2(i) \in \mathbf{R}^{m \times n_2}$, $D_1(i) \in \mathbf{R}^{p \times m}$, $D_2(i) \in \mathbf{R}^{m \times m_1}$, $i \in \mathcal{D}$, then the product

$$T^1_\tau T^2_\tau : L^2_{\eta,w}\left([0,\tau], \mathbf{R}^{m_1}\right) \to L^2_{\eta,w}\left([0,\tau], \mathbf{R}^p\right)$$

has the state-space realization of form (6.1), where

$$A_k(i) = \begin{bmatrix} A_{k1}(i) & B_{k1}(i)C_2(i) \\ 0 & A_{k2}(i) \end{bmatrix},$$

$$B_k(i) = \begin{bmatrix} B_{k1}(i)D_2(i) \\ B_{k2}(i) \end{bmatrix}, \quad 0 \le k \le r,$$

$$C(i) = [C_1(i) \quad D_1(i)C_2(i)],$$

$$D(i) = D_1(i)D_2(i), \quad i \in \mathcal{D}.$$

(ii) *Assume that in (6.1) we have $p = m$ and $\det D(i) \ne 0$, $i \in \mathcal{D}$. Then for every $\tau > 0$, the input–output operator $T_\tau : L^2_{\eta,w}([0,\tau], \mathbf{R}^m) \to L^2_{\eta,w}([0,\tau], \mathbf{R}^m)$ is invertible, and its inverse T_τ^{-1} has the state-space realization*

$$d\xi(t) = \left[\tilde{A}_0(\eta(t))\xi(t) + \tilde{B}_0(\eta(t))y(t)\right] dt$$

$$+ \sum_{k=1}^r \left[\tilde{A}_k(\eta(t))\xi(t) + \tilde{B}_k(\eta(t))y(t)\right] dw_k(t), \tag{6.5}$$

$$u(t) = \tilde{C}(\eta(t))\xi(t) + \tilde{D}(\eta(t))y(t),$$

where

$$\tilde{A}_k(i) = A_k(i) - B_k(i)D^{-1}(i)C(i),$$
$$\tilde{B}_k(i) = B_k(i)D^{-1}(i),$$
$$\tilde{C}(i) = -D^{-1}(i)C(i),$$
$$\tilde{D}(i) = D^{-1}(i), \quad i \in \mathcal{D}.$$

Moreover, if the systems $(A_0, A_1, \ldots, A_r; Q)$ and $\left(\tilde{A}_0, \tilde{A}_1, \ldots, \tilde{A}_r; Q\right)$ are stable, then the input–output operator T associated with (6.1) is invertible and its inverse T^{-1} has the realization given by (6.5).

Proof. Part (i) of the statement immediately follows by the uniqueness of the solution $x_u(.)$ of the linear system (6.1).

(ii) Denote by $\widehat{T}_\tau$ the input–output operator defined by (6.5) on $[0, \tau]$. Applying the result of part (i) one can easily check that

$$T_\tau \widehat{T}_\tau = I_{L^2_{\eta,w}([0,\tau],\mathbf{R}^m)} = \widehat{T}_\tau T_\tau,$$

where $I_{L^2_{\eta,w}([0,\tau],\mathbf{R}^m)}$ is the identity operator on $L^2_{\eta,w}([0, \tau], \mathbf{R}^m)$. The last assertion follows in the same way as above.   □

In the following we shall prove a result that will play an important role in the proof of the Bounded Real Lemma in the next section. For each continuous function $F : [0, \tau] \to \mathcal{M}^d_{m,n}$, $F(t) = (F(t, 1), \dots, F(t, d))$, consider the following Lyapunov-type equation on $\mathcal{S}^d_n$:

$$\frac{d}{dt}K(t, i) + (A_0(i) + B_0(i)F(t, i))^* K(t, i) + (A_0(i) + B_0(i)F(t, i))K(t, i)$$

$$+ \sum_{k=1}^{r}(A_k(i) + B_k(i)F(t, i))^* K(t, i)(A_k(i) + B_k(i)F(t, i))$$

$$+ \sum_{j=1}^{d} q_{ij} K(t, j) + (C(i) + D(i)F(t, i))^*(C(i) + D(i)F(t, i))$$

$$- \gamma^2 F^*(t, i)F(t, i) = 0, \ i \in \mathcal{D}. \tag{6.6}$$

For each $\gamma > 0$, denote by

$$K_\gamma(t) = \big(K_\gamma(t, 1), \dots, K_\gamma(t, d)\big)$$

the solution of equation (6.6) verifying the condition $K_\gamma(\tau, i) = 0, \ i \in \mathcal{D}$.

Lemma 2. *Assume that for a fixed $\tau > 0$ we have $\|T_\tau\| < \gamma$. Then for all ε_0 such that $0 < \varepsilon_0^2 < \gamma^2 - \|T_\tau\|^2$, we have*

$$\gamma^2 I_m - D^*(i)D(i) - \sum_{k=1}^{r} B_k^*(i)K_\gamma(t, i)B_k(i) \geq \varepsilon_0^2 I_m \tag{6.7}$$

for all $t \in [0, \tau]$, $i \in \mathcal{D}$.

Proof. Denoting

$$\Gamma_\gamma(t, i) = \gamma^2 I_m - D^*(i)D(i) - \sum_{k=1}^{r} B_k^*(i)K_\gamma(t, i)B_k(i),$$

(6.7) can be written as $\Gamma_\gamma(t, i) \geq \varepsilon_0^2 I_m \ \forall t \in [0, \tau], i \in \mathcal{D}$. The proof, then, has two stages.

Stage 1 We first prove that for each γ satisfying the condition $\gamma > \|T_\tau\|$, we have

$$\Gamma_\gamma(t, i) \geq 0, \quad \forall t \in (0, \tau), i \in \mathcal{D}. \tag{6.8}$$

If (6.8) does not hold, then it follows that there exist $t_0 \in (0, \tau)$, $i_0 \in \mathcal{D}$, $u \in \mathbf{R}^m$ with $|u_0| = 1$ such that $u_0^* \Gamma_\gamma(t_0, i_0)u_0 < 0$. Since the function $t \to u_0^* \Gamma_\gamma(t, i_0)u_0$ is continuous, it follows that there exist $\delta_0 > 0$, $\nu > 0$ such that

$$u_0^* \Gamma_\gamma(t, i_0) u_0 < -\nu < 0, \quad \forall t \in [t_0, t_0 + \delta_0], \tag{6.9}$$

with $t_0 + \delta_0 < \tau$. Let $\delta \in (0, \delta_0)$ be arbitrary but fixed and define the stochastic process

$$v_\delta(t) = \begin{cases} 0 & \text{if } t \notin [t_0, t_0 + \delta_0], \\ u_0 \chi_{\eta(t)=i_0} & \text{if } t \in [t_0, t_0 + \delta_0]. \end{cases}$$

It is obvious that $v_\delta \in L_{\eta,w}^2 ([0, \tau], \mathbf{R}^m)$. Let $x_\delta(t)$, $t \in [0, \tau]$, be the solution of the following problem with initial conditions:

$$dx(t) = \{[A_0(\eta(t)) + B_0(\eta(t))F(t, \eta(t))]x(t) + B_0(\eta(t))v_\delta(t)\} dt$$
$$+ \sum_{k=1}^{r} \{[A_k(\eta(t)) + B_k(\eta(t))F(t, \eta(t))]x(t)$$
$$+ B_k(\eta(t))v_\delta(t)\} dw_k(t), \quad t \in [0, \tau], \quad x_\delta(0) = 0. \tag{6.10}$$

Define $u_\delta(t) = v_\delta(t) + F(t, \eta(t))x_\delta(t)$, $t \in [0, \tau]$. Since

$$u_\delta(t) \in L_{\eta,w}^2 ([0, \tau], \mathbf{R}^m),$$

from (6.10) one deduces that

$$x_{u_\delta}(t) = x_\delta(t), \quad t \in [0, \tau].$$

Let $y_\delta = T_\tau u_\delta$. Therefore

$$y_\delta(t) = C(\eta(t))x_\delta(t) + D(\eta(t))u_\delta(t), \quad t \in [0, \tau].$$

By direct computation, taking into account the definition of $u_\delta(t)$, we obtain that

$$|y_\delta(t)|^2 - \gamma^2 |u_\delta(t)|^2 = x_\delta^*(t)[(C(\eta(t)) + D(\eta(t))F(t, \eta(t)))^*$$
$$\times (C(\eta(t)) + D(\eta(t))F(t, \eta(t)))$$
$$- \gamma^2 F^*(t, \eta(t))F(t, \eta(t))]x_\delta(t) + 2x_\delta^*(t)$$
$$\times [(C(\eta(t)) + D(\eta(t))F(t, \eta(t)))^* D(\eta(t))$$
$$- \gamma^2 F^*(t, \eta(t))]v_\delta(t) + v_\delta^*(t) \tag{6.11}$$
$$\times [D^*(\eta(t))D(\eta(t)) - \gamma^2 I_m]v_\delta(t).$$

Using the Itô-type formula for the function

$$v(t, x, i) = x^* K_\gamma(t, i) x$$

and for the process $x_\delta(t)$, $t \in [0, \tau]$, based on (6.6) and (6.11), one obtains that

$$E\left[\int_0^\tau \left(|y_\delta(t)|^2 - \gamma^2 |u_\delta|^2\right) dt \mid \eta(0) = i\right]$$
$$= E\left[\int_0^\tau \left\{2x_\delta^*(t)\mathcal{P}_\gamma(t, \eta(t))v_\delta(t) - v_\delta^*(t)\Gamma_\gamma(t, \eta(t))v_\delta(t)\right\} dt \mid \eta(0) = i\right]$$

for all $i \in \mathcal{D}$, where $\mathcal{P}_\gamma(t, i)$ is defined as

$$\mathcal{P}_\gamma(t, i) = K_\gamma(t, i)B_0(i) + \sum_{k=1}^r (A_k(i) + B_k(i)F(t, i))^* K_\gamma(t, i)B_k(i)$$
$$+ (C(i) + D(i)F(i))^* D(i) - \gamma^2 F^*(t, i).$$

Taking into account the definition of v_δ, we further can write:

$$E\left[\int_0^\tau \left(|y_\delta(t)|^2 - \gamma^2 |u_\delta|^2\right) dt \mid \eta(0) = i\right]$$
$$= E\left[\int_{t_0}^{t_0+\delta} \left\{2x_\delta^*(t)\mathcal{P}_\gamma(t, \eta(t))u_0 - u_0^*\Gamma_\gamma(t, \eta(t))u_0\right\} \chi_{\eta(t)=i_0} dt \mid \eta(0) = i\right]$$
$$= \sum_{j=1}^d E\left[\int_{t_0}^{t_0+\delta} \left\{2x_\delta^*(t)\mathcal{P}_\gamma(t, j)u_0 - u_0^*\Gamma_\gamma(t, j)u_0\right\} \chi_{\eta(t)=j} \chi_{\eta(t)=i_0} dt \mid \eta(0) = i\right].$$

Since $\chi_{\eta(t)=i} \chi_{\eta(t)=i_0} = 0$ for $i \neq i_0$ and $\chi_{\eta(t)=i} \chi_{\eta(t)=i_0} = \chi_{\eta(t)=i_0}$ for $i = i_0$, we obtain

$$E\left[\int_0^\tau \left(|y_\delta(t)|^2 - \gamma^2 |u_\delta|^2\right) dt \mid \eta(0) = i\right]$$
$$= E\left[\int_{t_0}^{t_0+\delta} \left\{2x_\delta^*(t)\mathcal{P}_\gamma(t, i_0)u_0 - u_0^*\Gamma_\gamma(t, i_0)u_0\right\} \chi_{\eta(t)=i_0} dt \mid \eta(0) = i\right],$$

$$(6.12)$$

$i \in \mathcal{D}$. Based on (6.9) one immediately obtains that

$$E\left[\int_0^\tau \left(|y_\delta(t)|^2 - \gamma^2 |u_\delta|^2\right) dt \mid \eta(0) = i\right]$$
$$\geq -2E\left[\int_{t_0}^{t_0+\delta} |x_\delta^*(t)\mathcal{P}_\gamma(t, i_0)u_0| \chi_{\eta(t)=i_0} dt \mid \eta(0) = i\right]$$
$$+ vE\left[\int_{t_0}^{t_0+\delta} \chi_{\eta(t)=i_0} dt \mid \eta(0) = i\right],$$

and therefore

$$E\left[\int_0^\tau \left(|y_\delta(t)|^2 - \gamma^2 |u_\delta|^2\right) dt \mid \eta(0) = i\right]$$

$$\geq -2E\left[\int_{t_0}^{t_0+\delta} \left|x_\delta^*(t)\mathcal{P}_\gamma(t, i_0)u_0\right| \chi_{\eta(t)=i_0} dt \mid \eta(0) = i\right]$$

$$+ v \int_{t_0}^{t_0+\delta} p_{i,i_0}(t)dt, \quad i \in \mathcal{D}. \tag{6.13}$$

Based on Remark 17 of Chapter 2 one deduces that there exists $c_1 > 0$ depending on τ such that

$$\sup_{0 \leq t \leq \tau} E\left[|x_\delta(t)|^2 \mid \eta(0) = i\right] \leq c_1 E\left[\int_0^\tau |v_\delta(t)|^2 dt \mid \eta(0) = i\right]$$

$$\leq c_1 \delta. \tag{6.14}$$

On the other hand, we have

$$E\left[\int_0^\tau \left|x_\delta^*(t)\mathcal{P}_\gamma(t, i_0) u_0\right| \chi_{\eta(t)=i_0} dt \mid \eta(0) = i\right]$$

$$\leq \int_0^\tau \left(E\left[|x_\delta(t)|^2 dt \mid \eta(0) = i\right]\right)^{\frac{1}{2}} \left|\mathcal{P}_\gamma(t, i_0)\right| dt.$$

Hence, using (6.14) we obtain

$$2E\left[\int_{t_0}^{t_0+\delta} \left|x_\delta^*(t)\mathcal{P}_\gamma(t, i_0) u_0\right| \chi_{\eta(t)=i_0} dt \mid \eta(0) = i\right] \leq c_2 \delta\sqrt{\delta}, \tag{6.15}$$

where $c_2 > 0$ is a constant depending on τ. Then we have

$$E\int_0^\tau \left(|y_\delta(t)|^2 - \gamma^2 |u_\delta|^2\right) dt$$

$$= \sum_{i=1}^d \pi_i E\left[\int_0^\tau \left(|y_\delta(t)|^2 - \gamma^2 |u_\delta|^2\right) dt \mid \eta(0) = i\right] \tag{6.16}$$

$$\geq \int_{t_0}^{t_0+\delta} h(t)dt - c_2 \delta\sqrt{\delta},$$

where we denoted

$$h(t) = v \sum_{i=1}^d \pi_i p_{i,i_0}(t).$$

Since $p_{i_0,i_0}(t)$ is a continuous function, it follows that there exists $\delta \in (0, \delta_0)$ such that

$$p_{i_0,i_0}(t) \geq \frac{1}{2} p_{i_0,i_0}(t_0) > 0 \qquad \forall t_0 \leq t \leq t_0 + \delta.$$

Then, for $\delta > 0$ small enough, (6.16) becomes

$$\|y_\delta\|^2 - \gamma^2 \|u_\delta\|^2 = E \int_0^\tau \left(|y_\delta(t)|^2 - \gamma^2 |u_\delta|^2 \right) dt$$

$$\geq \frac{1}{2} \delta v \pi_{i_0} p_{i_0,i_0}(t_0) - c_2 \delta \sqrt{\delta} > 0.$$

This contradicts the assumption in the statement $\|\mathcal{T}_\tau\| < \gamma$. It follows, then, that (6.8) is accomplished for $t \in (0, \tau)$. From the continuity with respect to t it results that (6.8) is accomplished for $t \in [0, \tau]$.

Stage 2 Let ε_0 be such that

$$0 < \varepsilon_0^2 < \gamma^2 - \|\mathcal{T}_\tau\|^2.$$

Then, for $\hat{\gamma} = \left(\gamma^2 - \varepsilon_0^2 \right)^{\frac{1}{2}}$, it is obvious that $\|\mathcal{T}_\tau\| < \hat{\gamma}$. According to Stage 1 we have

$$\Gamma_{\hat{\gamma}}(t, i) \geq 0, \quad t \in [0, \tau], \ i \in \mathcal{D}.$$

This leads to

$$\gamma^2 I_m - D^*(i)D(i) - \sum_{i=1}^{r} B_k^*(i) K_{\hat{\gamma}}(t, i) B_k(i) \geq \varepsilon_0 I_m. \qquad (6.17)$$

On the other hand, one can immediately check that

$$\frac{d}{dt} \left[K_{\hat{\gamma}}(t, i) - K_\gamma(t, i) \right] + \left[A_0(i) + B_0(i) F(t, i) \right]^* \left[K_{\hat{\gamma}}(t, i) - K_\gamma(t, i) \right]$$

$$+ \left[K_{\hat{\gamma}}(t, i) - K_\gamma(t, i) \right] \left[A_0(i) + B_0(i) F(t, i) \right] + \sum_{k=1}^{r} \left[A_k(i) + B_k(i) F(t, i) \right]^*$$

$$\times \left[K_{\hat{\gamma}}(t, i) - K_\gamma(t, i) \right] \left[A_k(i) + B_k(i) F(t, i) \right] + \sum_{j=1}^{d} q_{ij} \left[K_{\hat{\gamma}}(t, i) - K_\gamma(t, i) \right]$$

$$+ \varepsilon_0^2 F^*(t, i) F(t, i) = 0,$$

from which it follows that $K_{\hat{\gamma}}(t, i) - K_\gamma(t, i) \geq 0$. Therefore, from (6.17), we deduce that

$$\gamma^2 I_m - D^*(i)D(i) - \sum_{i=1}^{r} B_k^*(i) K_\gamma(t, i) B_k(i) \geq \varepsilon_0 I_m,$$

and hence the proof is complete. $\qquad \square$

Corollary 3. *If there exists $\tau > 0$ such that $\|\mathcal{T}_\tau\| < \gamma$, then $D^*(i)D(i) < \gamma^2 I_m$, $i \in \mathcal{D}$.* $\qquad \square$

Remark 2. If the system $(A_0, A_1, \ldots, A_r; Q)$ is stable and if $\|\mathcal{T}\| < \gamma$, then $\|\mathcal{T}_\tau\| < \gamma$ for all $\tau > 0$.

6.2 Stochastic version of the Bounded Real Lemma

In the present section we shall derive necessary and sufficient conditions under which
the norm of the input–output operator is less than a prescribed level of attenuation γ.
These conditions extend at the stochastic systems of form (6.1) the well-known conditions given by the Bounded Real Lemma in the deterministic framework. The results
proved in this section include as particular cases the results separately proved for
stochastic systems with multiplicative white noise and for systems with Markovian
jumps, respectively.

Consider the following system of generalized Riccati algebraic equations:

$$
A_0^*(i)X(i) + X(i)A_0(i) + \sum_{k=1}^r A_k^*(i)X(i)A_k(i)
$$

$$
+ \sum_{j=1}^d q_{ij}X(j) + \left(X(i)B_o(i) + \sum_{k=1}^r A_k^*(i)X(i)B_k(i) + C^*(i)D(i) \right)
$$

$$
\times \left(\gamma^2 I_m - D^*(i)D(i) - \sum_{k=1}^r B_k^*(i)X(i)B_k(i) \right)^{-1}
$$

$$
\times \left(B_0^*(i)X(i) + \sum_{k=1}^r B_k^*(i)X(i)A_k(i) + D^*(i)C(i) \right)
$$

$$
+ C^*(i)C(i) = 0, \quad i \in \mathcal{D}. \tag{6.18}
$$

One can notice that in the particular case when $A_k(i) = 0$, $B_k(i) = 0$, $1 \le k \le r$,
$\mathcal{D} = \{1\}$, the SGRAE (6.18) reduces to the well-known algebraic Riccati equation
used in the deterministic framework in order to determine the H^∞ norm of a linear
system. With the notations introduced in Section 3.2, the SGRAE (6.18) can be written
as the following nonlinear equation on $\mathcal{S}_n^d$:

$$
\mathcal{L}^* X - \mathcal{P}^*(X)\mathcal{R}^{-1}(X)\mathcal{P}(X) + C^*C = 0, \tag{6.19}
$$

$\mathcal{L} : \mathcal{S}_n^d \to \mathcal{S}_n^d$ being the Lyapunov-type operator defined by the system
$(A_0, A_1, \ldots, A_r; Q)$,

$$
\mathcal{P}(X) = (\mathcal{P}_1(X), \ldots, \mathcal{P}_d(X)),
$$

with

$$
\mathcal{P}_i(X) = B_0^*(i)X(i) + \sum_{k=1}^r B_k^*(i)X(i)A_k(i) + C^*(i)D(i),
$$

$$
\mathcal{R}(X) = (\mathcal{R}_1(X), \ldots, \mathcal{R}_d(X)),
$$

where

$$
\mathcal{R}_i(X) = -\gamma^2 I_m + D^*(i)D(i) + \sum_{k=1}^r B_k^*(i)X(i)B_k(i), \quad i \in \mathcal{D},
$$

and $X = (X(1), \ldots, X(d))$. We shall also use the following differential equations on $\mathcal{S}_n^d$:

$$\frac{d}{dt}X(t) + \mathcal{L}^*X(t) - \mathcal{P}^*(X)\mathcal{R}^{-1}(X)\mathcal{P}(X) + C^*C = 0, \tag{6.20}$$

$$\frac{d}{dt}K(t) = \mathcal{L}^*K(t) - \mathcal{P}^*(K(t))\mathcal{R}^{-1}(K(t))\mathcal{P}(K(t)) + C^*C = 0. \tag{6.21}$$

Remark 3. (i) Both the algebraic equation (6.19) and the differential equations (6.20) and also (6.21) are defined on the subset $\mathcal{U}_\gamma \subset \mathcal{S}_n^d$ with the elements $X = (X(1), \ldots, X(d))$ for which $det\, \mathcal{R}_i(X) \neq 0, i \in \mathcal{D}$. From Corollary 3 it follows that if $\tau > 0$ exists such that $\|T_\tau\| < \gamma$, then the null element $(0, 0, \ldots, 0) \in \mathcal{S}_n^d$ is in $\mathcal{U}_\gamma$.

(ii) A C^1-function $X : [0, \tau] \to \mathcal{U}_\gamma$ is a solution of equation (6.20) if and only if $K : [0, \tau] \to \mathcal{U}_\gamma$ defined as $K(t) = X(\tau - t)$ is a solution of (6.21).

For every $\tau > 0$, $x_0 \in \mathbf{R}^n$, $\gamma > 0$, $i \in \mathcal{D}$, consider the following cost functions:

$$\mathcal{H}_\gamma(\tau, x_0, i, .) : L^2_{\eta, w}([0, \tau]; \mathbf{R}^m) \to \mathbf{R},$$

$$\mathcal{H}_\gamma(\tau, x_0, .) : L^2_{\eta, w}([0, \tau]; \mathbf{R}^m) \to \mathbf{R},$$

defined by

$$\mathcal{H}_\gamma(\tau, x_0, i, u) = E\left[\int_0^\tau \left(|y_u(t, x_0)|^2 - \gamma^2|u(t)|^2\right) dt \mid \eta(0) = i\right]$$

and

$$\mathcal{H}_\gamma(\tau, x_0, u) = E\int_0^\tau \left(|y_u(t, x_0)|^2 - \gamma^2|u(t)|^2\right) dt,$$

where

$$y_u(t, x_0) = C(\eta(t))x_u(t, x_0) + D(\eta(t))u(t), \quad t \in [0, \tau],$$

$x_u(t, x_0)$ being the solution of the system (6.1) determined by the input $u(t)$ and the initial condition $x_u(0, x_0) = x_0$. It is obvious that

$$\mathcal{H}_\gamma(\tau, x_0, u) = \sum_{i=1}^d \pi_i \mathcal{H}_\gamma(\tau, x_0, i, u).$$

From Corollary 2 of Chapter 5 and from Remark 3(ii) one directly obtains the following Lemma.

Lemma 4. *If* $X : [0, \tau] \to \mathcal{S}_n^d$, $X(t) = (X(t, 1), \ldots, X(t, d))$ *is a solution of equation* (6.20) *and* $K(t) = X(\tau - t)$, *then*

$$\mathcal{H}_\gamma(\tau, x_0, i, u) = x_0^* X(0, i)x_0 - E\left[x^*(\tau)X(\tau, \eta(\tau))x(\tau) \mid \eta(0) = i\right]$$

$$- E\left[\int_0^\tau \left(u(t) - F^X(t, \eta(t))x(t)\right)^*\right.$$

$$\left. \times \left[\gamma^2 I_m - D^*(\eta(t))D(\eta(t)) - \sum_{k=1}^r B_k^*(\eta(t))X(t, \eta(t))B_k(\eta(t))\right]\right.$$

$$\times \left(u(t) - F^X(t, \eta(t))x(t)\right) dt \mid \eta(0) = i\bigg]$$

$$= x_0^* K(\tau, i)x_0 - E\left[x^*(\tau)K(0, \eta(t)) x(\tau) \mid \eta(0) = i\right]$$

$$-E\left[\int_0^\tau \left(u(t) - F^K(t, \eta(t))x(t)\right)^*\right.$$

$$\times\left[\gamma^2 I_m - D^*(\eta(t))D(\eta(t))\right.$$

$$\left. - \sum_{k=1}^r B_k^*(\eta(t))K(\tau - t, \eta(t)) B_k(\eta(t))\right]$$

$$\times \left(u(t) - F^K(t, \eta(t))x(t)\right) dt \mid \eta(0) = i\bigg],$$

$\forall\ x_0 \in \mathbf{R}^n,\ i \in \mathcal{D},\ u \in L_{\eta,w}^2\left([0, \tau], \mathbf{R}^m\right),\ x(t) = x_u(t, x_0),$

$$F^X(t, i) = -\left(R(i) + \sum_{k=1}^r B_k^*(i)X(t, i)B_k(i)\right)^{-1}$$

$$\times \left(B_0^*(i)X(t, i) + \sum_{k=1}^r B_k^*(i)X(t, i)A_k(i) + D^*(i)C(i)\right),$$

$$F^K(t, i) = -\left(R(i) + \sum_{k=1}^r B_k^*(i)K(\tau - t, i)B_k(i)\right)^{-1}$$

$$\times \left(B_0^*(i)K(\tau - t, i) + \sum_{k=1}^r B_k^*(i)K(\tau - t, i)A_k(i) + D^*(i)C(i)\right),$$

where $R(i) = -\gamma^2 I_m + D^*(i)D(i)$.

We prove now the following useful result.

Lemma 5. *Assume that the system* $(A_0, A_1, \ldots, A_r; Q)$ *is stable and* $\|T\| < \gamma$. *In these conditions there exists a constant* $\rho > 0$ *such that*

$$\mathcal{H}_\gamma(\tau, x_0, i, u) \le \rho|x_0|^2 \ \forall \tau > 0, \ x_0 \in \mathbf{R}^m, \ u \in L_{\eta,w}^2([0, \tau], \mathbf{R}^m).$$

Proof. Let $x_u(t, x_0)$ be the solution of the system (6.1) corresponding to the arbitrary control $u \in L_{\eta,w}^2([0, \tau], \mathbf{R}^m)$. Then one can write

$$x_u(t, x_0) = x_0(t, x_0) + x_u(t, 0),$$

where $x_0(t, x_0)$ is the solution of the system (6.1) for $u = 0$ satisfying $x_0(0, x_0) = x_0$. Therefore $x_0(t, x_0) = \Phi(t, 0)x_0$. As in the preceding subsection the process

$x_u(t) = x_u(t, 0)$ is the solution of the system (6.1) satisfying the initial condition $x_u(0, 0) = 0$. Denoting

$$y_0(t, x_0) = C(\eta(t))x_0(t, x_0) \text{ and}$$
$$y_u(t) = C(\eta(t))x_u(t) + D(t)u(t),$$

one obtains that

$$y_u(t, x_0) = y_0(t, x_0) + y_u(t). \tag{6.22}$$

Since the system $(A_0, A_1, \ldots, A_r; Q)$ is stable there exists $\rho_1 > 0$ not depending on x_0, such that

$$E\left[\int_0^\infty |y_0(t, x_0)|^2 \, dt\right] \leq \rho_1^2|x_0|^2, \quad \forall x_0 \in \mathbf{R}^n. \tag{6.23}$$

On the other hand, from the inequalities

$$\|\mathcal{T}_\tau\| \leq \|\mathcal{T}\| < \gamma$$

it follows that there exists $\nu > 0$ not depending on $u(t)$ such that

$$E\int_0^\tau \left(|y_u(t)|^2 - \gamma^2|u(t)|^2\right) dt \leq -\nu^2 E\int_0^\tau |u(t)|^2 dt \tag{6.24}$$

$\forall u \in L^2_{\eta, w}([0, \tau], \mathbf{R}^m)$. Using the decomposition (6.22) of y_u, one obtains that

$$\mathcal{H}_\gamma(\tau, x_0, u) = E\int_0^\tau |y_0(t, x_0)|^2 \, dt + 2E\int_0^\tau y_0^*(t, x_0)y_u(t)dt$$
$$+ E\int_0^\tau \left(|y_u(t)|^2 - \gamma^2|u(t)|^2\right) dt.$$

Taking into account (6.23) and (6.24), one immediately obtains

$$\mathcal{H}_\gamma(\tau, x_0, u) \leq \rho_1^2|x_0|^2 + 2\rho_1\gamma|x_0|\|u\| - \nu^2\|u\|^2 \tag{6.25}$$

$\forall u \in L^2_{\eta, w}([0, \tau], \mathbf{R}^m)$, where $\|u\| = \left(E\int_0^\tau |u(t)|^2 dt\right)^{\frac{1}{2}}$. Since the right-hand side of (6.25) is a second degree polynomial with respect with $\|u\|$, one immediately deduces that

$$\mathcal{H}_\gamma(\tau, x_0, u) \leq \rho^2|x_0|^2, \tag{6.26}$$

where $\rho = \rho_1\nu^{-1}\sqrt{\gamma^2 + \nu^2}$, and therefore the proof is complete. □

In the following we shall denote by $X_\tau(t) = (X_\tau(t, 1), \ldots, X_\tau(t, d))$ the solution of the equation (6.20) satisfying the condition $X_\tau(\tau, i) = 0$, $i \in \mathcal{D}$. Let $\mathcal{I}_\tau(\gamma) \subset [0, \tau]$ be the maximal interval on which the solution $X_\tau(.)$ is defined. From Remark 3(i) it follows that if $\|\mathcal{T}_\tau\| < \gamma$, then $\mathcal{I}_\tau(\gamma)$ is nonempty. Then from Lemma 2 one obtains the following Lemma.

Lemma 6. *If* $\sup_{\tau>0} \|T_\tau\| < \gamma$ *then*

$$\gamma^2 I_m - D^*(i)D(i) - \sum_{k=1}^{r} B_k^*(i)X_\tau(t,i)B_k(i) \geq \varepsilon_0^2 I_m, \quad t \in \mathcal{I}_\tau(\gamma), \tag{6.27}$$

$i \in \mathcal{D}$, $\tau > 0$, *where* $\varepsilon_0 > 0$ *does not depend upon* τ.

Proof. Let $\varepsilon_0 > 0$ such that $\varepsilon_0^2 < \gamma^2 - \sup_{\tau>0} \|T_\tau\|^2$. Let $\tau > 0$ and $t_1 \in \mathcal{I}_\gamma(\tau)$, $t_1 < \tau$. Obviously $[t_1, \tau] \subset \mathcal{I}_\gamma(\tau)$. Denote

$$F_\tau(t,i) = \left(\gamma^2 I_m - D^*(i)D(i) - \sum_{k=1}^{r} B_k^*(i)X_\tau(t,i)B_k(i) \right)^{-1} \tag{6.28}$$

$$\times \left(B_0^*(i)X_\tau(t,i) + \sum_{k=1}^{r} B_k^*(i)X_\tau(t,i)A_k(i) + D^*(i)C(i) \right),$$

$t \in [t_1, \tau]$, $i \in \mathcal{D}$. With Lemma 1 of Chapter 4 one immediately obtains that (6.20) verified by $X_\tau(.)$ can be written in a Lyapunov form on $\mathcal{S}_n^d$ as follows:

$$\frac{d}{dt} X_\tau(t,i) + [A_0(i) + B_0(i)F_\tau(t,i)]^* X_\tau(t,i) + X_\tau(t,i)[A_0(i) + B_0(i)F_\tau(t,i)]$$

$$+ \sum_{k=1}^{r} [A_k(i) + B_k(i)F_\tau(t,i)]^* X_\tau(t,i)[A_0(i) + B_0(i)F_\tau(t,i)]$$

$$+ \sum_{j=1}^{d} q_{ij} X_\tau(t,j) - \gamma^2 F_\tau^*(t,i)F_\tau(t,i)$$

$$+ [C(i) + D(i)F_\tau(t,i)]^* [C(i) + D(i)F_\tau(t,i)] = 0, \tag{6.29}$$

$t \in [t_1, \tau]$, $i \in \mathcal{D}$.

Let $F : [0, \tau] \to \mathcal{M}_{mn}^d$ be defined as

$$F(t) = (F(t,1), \ldots, F(t,d)), \tag{6.30}$$

$$F(t,i) = \begin{cases} F_\tau(t,i), & t \in [t_1, \tau], \\ F_\tau(t_1,i), & t \in [0, t_1], \ i \in \mathcal{D}, \end{cases}$$

and let $X(t) = (X(t,1), \ldots, X(t,d))$, with $X(\tau) = 0$ the solution of the equation (6.6) corresponding to the feedback $F(.)$ defined as in (6.30). Then, from (6.29) and (6.30), it follows that $X(t) = X_\tau(t)$, $t \in [t_1, \tau]$. Applying Lemma 2 one obtains that (6.27) is true for all $t \in [t_1, \tau]$ and the proof is complete. $\qquad\square$

In the following we shall denote by $K^0(t) = \left(K^0(t,1), \ldots, K^0(t,d) \right)$ the solution of the equation (6.21) satisfying the initial condition $K^0(0,i) = 0$, $i \in \mathcal{D}$. We also denote by $[0, t_f)$ the maximal interval on which this solution is defined. The next lemma summarizes some properties of the solution $K^0(t)$.

Lemma 7. *Assume that the system* $(A_0, A_1, \ldots, A_r; Q)$ *is stable and* $\|T\| < \gamma$. *Then the solution* $K^0(t)$ *of equation* (6.21) *has the following properties.*

(i)

$$\gamma^2 I_m - D^*(i)D(i) - \sum_{k=1}^{r} B_k^*(i)K^0(t,i)B_k(i) \geq \varepsilon_0^2 I_m$$

$\forall t \in [0, t_f)$, ε_0 *independent of* t.

(ii)

$$x_0^* K^0(\tau, i)x_0 = \mathcal{H}_\gamma (\tau, x_0, i, \tilde{u}_\tau) \geq \mathcal{H}_\gamma(\tau, x_0, i, u)$$

$\forall \tau \in (0, t_f)$, $x_0 \in \mathbf{R}^n$, $i \in \mathcal{D}$, $u \in L^2_{\eta,w}([0,\tau], \mathbf{R}^m)$, *where* $\tilde{u}_\tau(t) = F_\tau(t, \eta(t))\tilde{x}_\tau(t)$ *and*

$$F_\tau(t, i) = \left(\gamma^2 I_m - D^*(i)D(i) - \sum_{k=1}^{r} B_k^*(i)K^0(\tau - t, i)B_k(i)\right)^{-1}$$

$$\times \left(B_0^*(i)K^0(\tau - t, i) + \sum_{k=1}^{r} B_k^*(i)K^0(\tau - t, i)A_k(i) + D^*(i)C(i)\right)$$

and $\tilde{x}_\tau(t)$, $t \in [0, \tau]$, *is the solution of the equation*

$$d\tilde{x}(t) = [A_0(\eta(t)) + B_0(\eta(t))F_\tau(t, \eta(t))]\tilde{x}(t)dt$$

$$+ \sum_{k=1}^{r} [A_k(\eta(t)) + B_k(\eta(t))F_\tau(t, \eta(t))]\tilde{x}(t)dw_k(t)$$

with the initial condition $\tilde{x}_0(0) = x_0$.

(iii) *There exists* $\tilde{\rho} > 0$ *not depending on* τ *such that*

$$0 \leq K^0(\tau, i) \leq \tilde{\rho} I_n, \ \forall \tau \in [0, t_f), \ i \in \mathcal{D}.$$

(iv)

$$K^0(\tau_1, i) \leq K^0(\tau_2, i), \ \forall 0 \leq \tau_1 < \tau_2 < t_f.$$

Proof. (i) Let $\tau \in (0, t_f)$ be arbitrary but fixed and denote

$$X_\tau(t) = (X_\tau(t, 1), \ldots, X_\tau(t, d))$$

defined by

$$X_\tau(t) = K^0(\tau - t, i), \ t \in [0, \tau], \ i \in \mathcal{D}.$$

Then $X_\tau(t)$ is the solution of equation (6.20) with the final condition $X_\tau(\tau) = 0$. Based on Lemma 6 and Remark 1 one obtains

$$\gamma^2 I_m - D^*(i)D(i) - \sum_{k=1}^{r} B_k^*(i)X_\tau(t, i)B_k(i) \geq \varepsilon_0^2 I_m, \ t \in [0, \tau]. \tag{6.31}$$

Since ε_0 does not depend on τ, based on (6.31) and on the definition of X_τ the proof of part (i) is complete.

(ii) Applying Lemma 4 for $K^0(t) = X_\tau (\tau - t)$ one obtains

$$
\mathcal{H}_\gamma(\tau, x_0, i, u)
$$

$$
= x_0^* K^0(\tau, i)x_0 - E\left[\int_0^\tau (u(t) - F_\tau(t, \eta(t)x_u(t, x_0))^* \right. \tag{6.32}
$$

$$
\times \left(\gamma^2 I_m - D^*(\eta(t))D(\eta(t)) - \sum_{k=1}^r B_k^*(\eta(t))X_\tau(t, \eta(t))B_k(\eta(t)) \right)
$$

$$
\left. \times (u(t) - F_\tau(t, \eta(t)x_u(t, x_0)) \, dt \mid \eta(0) = i \right]
$$

$\forall x_0 \in \mathbf{R}^n, i \in \mathcal{D}, u \in L^2_{\eta.w}([0, \tau], \mathbf{R}^m)$. From (6.32) and (i) it immediately follows that

$$
\mathcal{H}_\gamma(\tau, x_0, i, u) \le x_0^* K^0(\tau, i)x_0, \tag{6.33}
$$

and for

$$
u(t) = F_\tau(t, \eta(t))x_u(t, x_0) = F_\tau(t, \eta(t))\tilde{x}(t)
$$

the inequality (6.33) becomes an equality.

(iii) From (6.33) one immediately deduces that

$$
0 \le \mathcal{H}_\gamma(\tau, x_0, i, 0) \le x_0^* K^0(\tau, i)x_0. \tag{6.34}
$$

On the other hand, for every $i \in \mathcal{D}$ one can write

$$
\pi_i x_0^* K^0(\tau, i)x_0 \le \sum_{j=1}^d \pi_j \mathcal{H}_\gamma(\tau, x_0, j, \tilde{u}) = \mathcal{H}_\gamma(\tau, x_0, \tilde{u}).
$$

From Lemma 5 we have

$$
\mathcal{H}_\gamma(\tau, x_0, \tilde{u}) \le \rho^2 |x_0|^2. \tag{6.35}
$$

Then from (6.34) and (6.35) it follows that (iii) is satisfied for

$$
\tilde{\rho} = \max_{i \in \mathcal{D}} \frac{\rho^2}{\pi_i}.
$$

(iv) Let $0 < \tau_1 < \tau_2 < t_f$ and consider the stochastic process u_{τ_2}, $t \in [0, \tau_2]$, as follows:

$$
u_{\tau_2}(t) = \begin{cases} \tilde{u}_{\tau_1}(t), & t \in [0, \tau_1], \\ 0, & t \in (\tau_1, \tau_2]. \end{cases}
$$

It is obvious that $u_{\tau_2} \in L^2_{\eta, w}([0, \tau_2], \mathbf{R}^m)$. Let $x_{\tau_2}(t)$, $t \in [0, \tau_2]$, be the solution of the system (6.1) determined by the input variable $u_{\tau_2}(t)$ and by the initial conditions $x_{\tau_2}(0) = x_0$. One can easily check that $x_{\tau_2}(t) = \tilde{x}_{\tau_1}(t)$ for $t \in [0, \tau_1]$ and

$$
\mathcal{H}_\gamma(\tau_1, x, i, \tilde{u}_{\tau_1}) \le \mathcal{H}_\gamma(\tau_2, x_0, i, u_{\tau_2}).
$$

Invoking again the maximality properties in (ii), one obtains

$$
x_0^* K^0(\tau_1, i) x_0 = \mathcal{H}_\gamma(\tau_1, x_0, i, \tilde{u}_{\tau_1}) \le \mathcal{H}_\gamma(\tau_2, x_0, i, \tilde{u}_{\tau_2})
$$

$$
\le x_0^* K^0(\tau_2, i)x_0 \quad \forall x_0 \in \mathbf{R}^n, \ i \in \mathcal{D},
$$

and therefore the proof is complete.   □

Remark 4. From (i) and (iii) of Lemma 7 it follows that the solution $K^0(\cdot)$ is defined on $[0, \infty)$, that is, $t_f = \infty$.

Consider the following subsets of $\mathcal{S}_n^d$:

$$\Pi = \{X = (X(1), \dots, X(d)) \in \mathcal{S}_n^d \mid \mathcal{L}^*X - \mathcal{P}^*(X)\mathcal{R}^{-1}(X)\mathcal{P}(X)$$
$$+ C^*C \leq 0,\ \mathcal{R}(X) < 0\} \tag{6.36}$$

and

$$\tilde{\Pi} = \{X = (X(1), \dots, X(d)) \in \mathcal{S}_n^d \mid \mathcal{L}^*X - \mathcal{P}^*(X)\mathcal{R}^{-1}(X)\mathcal{P}(X)$$
$$+ C^*C < 0,\ \mathcal{R}(X) < 0\}. \tag{6.37}$$

Remark 5. (i) $\tilde{\Pi} \subset \Pi$.

(ii) If the system $(A_0, A_1, \dots, A_r; Q)$ is stable, then $\Pi \subset \mathcal{S}_{n+}^d$.

(iii) Let us introduce the generalized dissipation matrix

$$\mathcal{N}(X) = (\mathcal{N}_1(X, \gamma), \dots, \mathcal{N}_d(X, \gamma))$$

associated with the system (6.1) and with the scalar γ, as follows:

$$\mathcal{N}_i(X, \gamma) = \begin{bmatrix} \mathcal{N}_{11}^i(X, \gamma) & \mathcal{N}_{12}^i(X, \gamma) \\ (\mathcal{N}_{12}^i)^*(X, \gamma) & \mathcal{N}_{22}^i(X, \gamma) \end{bmatrix},$$

where

$$\mathcal{N}_{11}^i(X, \gamma) = A_0^*(i)X(i) + X(i)A_0(i) + \sum_{k=1}^r A_k^*(i)X(i)A_k(i)$$

$$+ \sum_{j=1}^d q_{ij}X(j) + C^*(i)C(i),$$

$$\mathcal{N}_{12}^i(X, \gamma) = X(i)B_0(i) + \sum_{k=1}^r A_k^*(i)X(i)B_k(i) + C^*(i)D(i) = \mathcal{P}_i^*(X),$$

$$\mathcal{N}_{22}^i(X, \gamma) = -\gamma^2 I_m + D^*(i)D(i) + \sum_{k=1}^r B_k^*(i)X(i)B_k(i) = \mathcal{R}(X).$$

It is easy to check that

$$\Pi = \{X \in \mathcal{S}_n^d \mid \mathcal{N}(X) \leq 0,\ \mathcal{R}(X) < 0\}$$

and

$$\Pi = \{X \in \mathcal{S}_n^d \mid \mathcal{N}(X) < 0\}.$$

From the above inequalities one easily deduces that both Π and $\tilde{\Pi}$ are convex sets. The set Π includes the solutions of the equation (6.19) for which the condition $\mathcal{R}(X) < 0$ is accomplished.

Proposition 8. *Assume that the system* $(A_0, \ldots, A_r; Q)$ *is stable and* $\Pi \neq \varnothing$. *Then for all*

$$\widehat{X} = \left(\widehat{X}(1), \ldots, \widehat{X}(d)\right) \in \Pi,$$

we have

$$K^0(t) \leq \widehat{X} \quad \forall t \in [0, t_f),$$

K^0 *denoting the solution of equation* (6.21) *verifying the initial condition* $K^0(0) = 0$.

Proof. Under the above assumptions, by Remark 5(ii) it follows that there exists $X \geq 0$ with $\mathcal{R}(X) < 0$. Therefore $\gamma^2 I_m - D^*(i)D(i) > 0$, $i \in \mathcal{D}$. Thus we may conclude that the solution $K^0(t)$ is defined on an interval $[0, \tau]$, $\tau > 0$. Let $\widehat{X} = \left(\widehat{X}(1), \ldots, \widehat{X}(d)\right) \in \Pi$ arbitrary but fixed. Define

$$\widehat{M} = \left(\widehat{M}(1), \ldots, \widehat{M}(d)\right)$$

by

$$\widehat{M} = -\mathcal{L}^*\widehat{X} + \mathcal{P}^*\left(\widehat{X}\right)\mathcal{R}^{-1}\left(\widehat{X}\right)\mathcal{P}\left(\widehat{X}\right) - C^*C.$$

From the definition of $\widehat{M}$ it follows that $\widehat{X}$ verifies the algebraic equation

$$\mathcal{L}^*\widehat{X} - \mathcal{P}^*\left(\widehat{X}\right)\mathcal{R}^{-1}\left(\widehat{X}\right)\mathcal{P}\left(\widehat{X}\right) + C^*C + \widehat{M} = 0. \tag{6.38}$$

Let $\tau \in (0, t_f)$ and let $X_\tau(t) = (X_\tau(t, 1), \ldots, X_\tau(t, d))$ be defined as

$$X_\tau(t, i) = K^0(\tau - t, i), \quad t \in [0, \tau], \quad i \in \mathcal{D}.$$

Thus one deduces that $X_\tau(\cdot)$ is the solution of equation (6.20) satisfying the terminal condition $X_\tau(\tau) = 0$. Define

$$F_\tau(t) = (F_\tau(t, 1), \ldots, F_\tau(t, d)),$$
$$F_\tau(t, i) = -\mathcal{R}_i^{-1}(X_\tau(t))\mathcal{P}_i(X_\tau(t)), \quad i \in \mathcal{D}, \quad t \in [0, \tau].$$

By direct computations, similar to the proof of Lemma 1 of Chapter 4, one obtains that $\widehat{X}$ verifying (6.38) is also a solution of the equation parameterized with respect to t:

$$\mathcal{L}_{F_\tau}^*(t)X - \gamma^2 F_\tau^*(t)F_\tau(t) + (C + DF_\tau(t))^*(C + DF_\tau(t))$$
$$+ \widehat{M} - \left(F_\tau(t) - \widehat{F}\right)^*\mathcal{R}\left(\widehat{X}\right)\left(F_\tau(t) - \widehat{F}\right) = 0, \quad t \in [0, \tau], \tag{6.39}$$

where, $\mathcal{L}_{F_\tau}(t)$ denotes, as usual, the Lyapunov-type operator defined by the system $(A_0 + B_0 F_\tau, \ldots, A_r + B_r F_\tau; Q)$ and

$$\widehat{F} = \left(\widehat{F}(1), \ldots, \widehat{F}(d)\right),$$
$$\widehat{F}(i) = -\mathcal{R}_i^{-1}\left(\widehat{X}\right)\mathcal{P}_i\left(\widehat{X}\right), \quad i \in \mathcal{D}.$$

On the other side, based on (6.29), one obtains that equation (6.20) verified by $X_\tau(\cdot)$ can be rewritten as

$$\frac{d}{dt}X_\tau(t) + \mathcal{L}_{F_\tau}^*(t)X_\tau(t) - \gamma^2 F_\tau^*(t)F_\tau(t)$$
$$+ (C + DF_\tau(t))^*(C + DF_\tau(t)) = 0. \tag{6.40}$$

Let $Y(t) = \widehat{X} - X_\tau(t)$, $t \in [0, \tau]$. From (6.39) and (6.40) one obtains that

$$\frac{d}{dt}Y(t) + \mathcal{L}_{F_\tau}^*(t)Y(t) + \overline{M}(t) = 0, \tag{6.41}$$

where

$$\overline{M}(t) = -\left(F_\tau(t) - \widehat{F}\right)^* \mathcal{R}\left(\widehat{X}\right)\left(F_\tau(t) - \widehat{F}\right) + \widehat{M},$$

and it immediately follows that $\overline{M}(t) \geq 0$. Based on Remark 5(ii) it follows that $Y(\tau) = \widehat{X} \geq 0$. Based on the constant variation formula, we have

$$Y(t) = T_\tau(\tau, t)\, Y(\tau) + \int_t^\tau T_\tau^*(s, t)\, \overline{M}(s)ds, \ t \in [0, \tau], \tag{6.42}$$

where $T_\tau(t, s)$ is the linear operator of evolution on $\mathcal{S}_n^d$ defined by the differential equation

$$\frac{dY}{dt} = \mathcal{L}_{F_\tau}(t)Y(t).$$

Since $T_\tau^*(s, t)$ is a positive operator on $\mathcal{S}_n^d$ for any $s \geq t$ from (6.42) it follows that $Y(t) \geq 0$ for all $t \in [0, \tau]$, which leads to $X_\tau(t) \leq \widehat{X}, t \in [0, \tau]$, or equivalently,

$$K^0(t) \leq \widehat{X}, \ \forall t \in [0, \tau]. \tag{6.43}$$

Since τ has been arbitrarily chosen in $[0, t_f)$ it follows that (6.43) is verified for any $t \in [0, t_f)$. $\square$

Before proving the main result of this section we revisit the following known result from the theory of differential equations.

Lemma 9. *Let $F : \mathcal{X} \to \mathcal{X}$ be a continuous function defined on the Banach space $\mathcal{X}$. If $\xi : [0, \infty) \to \mathcal{X}$ is a solution of the differential equation $\dot{\xi}(t) = F(\xi(t))$ with the property $\lim_{t\to\infty} \xi(t) = \hat{\xi} \in \mathcal{X}$, then $F(\hat{\xi}) = 0$.*

Proof. Let $\varphi : \mathcal{X} \to \mathbf{R}$ be a linear and continuous functional. Then $t \to \varphi(\xi(t))$ verifies

$$\frac{d}{dt}\varphi(\xi(t)) = \varphi\left(F(\xi(t))\right)$$

and

$$\lim_{t\to\infty} \varphi(\xi(t)) = \varphi\left(\hat{\xi}\right).$$

Since

$$\varphi(\xi(t)) - \varphi(\xi(t_0)) = \int_{t_0}^t \varphi\left(F(\xi(s))\right) ds,$$

it follows that

$$\lim_{t\to\infty} \int_{t_0}^t \varphi\left(F(\xi(s))\right) ds = \varphi\left(\hat{\xi}\right) - \varphi(\xi(t_0)) \in \mathbf{R}.$$

Then the integral $\int_{t_0}^{\infty} \varphi\left(F(\xi(s))\right) ds$ is convergent. On the other hand,

$$\lim_{t \to \infty} \varphi\left(F(\xi(t))\right) = \varphi\left(F(\hat{\xi})\right).$$

From the convergence of the above integral it follows that $\varphi\left(F(\hat{\xi})\right) = 0$. Since φ is an arbitrary linear and continuous functional we deduce that $F(\hat{\xi}) = 0$ and hence the proof is complete.   □

The main result of this section is the following theorem.

Theorem 10. *(Bounded Real Lemma) The following assertions are equivalent:*
    (i) *The system $(A_0, A_1, \ldots, A_r; Q)$ is stable and $\|T\| < \gamma$.*
    (ii) *There exists $\widehat{X} = \left(\widehat{X}(1), \ldots, \widehat{X}(d)\right) \in \mathcal{S}_n^d$, $\widehat{X}(i) > 0$ satisfying the following LMI on $\mathcal{S}_{n+m}^d$:*

$$\mathcal{N}\left(\widehat{X}, \gamma\right) < 0,$$

$\mathcal{N}(X, \gamma)$ *denoting the generalized dissipation matrix associated with the system* (6.1) *and with the parameter γ.*
    (iii) *The SGRAE* (6.18) *has a stabilizing solution $\widetilde{X} = \left(\widetilde{X}(1), \ldots, \widetilde{X}(d)\right)$ satisfying $\widetilde{X}(i) \geq 0$ and*

$$\gamma^2 I_m - D^*(i)D(i) - \sum_{k=1}^{r} B_k^*(i)\widetilde{X}(i)B_k(i) > 0, \ i \in \mathcal{D}. \tag{6.44}$$

Proof. (i) $\Rightarrow$ (ii). For every $\delta > 0$ consider the linear and bounded operator

$$T_\delta : L_{\eta,w}^2\left([0, \infty), \mathbf{R}^m\right) \to L_{\eta,w}^2\left([0, \infty), \mathbf{R}^{n+p}\right)$$

defined by

$$T_\delta u = y_{u,\delta},$$

where

$$y_{u,\delta}(t) = \begin{bmatrix} C(\eta(t)) \\ \delta I_n \end{bmatrix} x_u(t) + \begin{bmatrix} D(\eta(t)) \\ 0 \end{bmatrix} u(t)$$

and where $x_u(t)$ is the solution of the system (6.1) with the initial condition $x_u(0) = 0$. Then

$$E \int_0^{\infty} |y_{u,\delta}(t)|^2 dt = E \int_0^{\infty} |y_u(t)|^2 dt + \delta^2 E \int_0^{\infty} |x_u(t)|^2 dt.$$

Applying Theorem 32(ii) of Chapter 2, one deduces that there exists $c > 0$ not depending on u such that

$$E \int_0^{\infty} |y_{u,\delta}(t)|^2 dt \leq E \int_0^{\infty} |y_u(t)|^2 dt + \delta^2 c E \int_0^{\infty} |u(t)|^2 dt$$

$$\leq \left(\|T\|^2 + \delta^2 c\right) E \int_0^{\infty} |u(t)|^2 dt$$

$\forall u \in L^2_{\eta, w}([0, \infty), \mathbf{R}^m)$. Hence we obtained that $\|T_\delta\|^2 \le \|T\|^2 + \delta^2 c$. Therefore, there exists $\delta_0 > 0$ such that

$$\sup_{0 \le \delta < \delta_0} \|T_\delta\| < \gamma. \tag{6.45}$$

For $0 < \delta < \delta_0$ let us denote by $K^0_\delta(t)$ the solution of the differential equation

$$\frac{d}{dt} K(t) = \mathcal{L}^* K(t) - \mathcal{P}^*(K(t)) \mathcal{R}^{-1}(K(t)) \mathcal{P}(K(t)) \tag{6.46}$$

$$+ C^* C + \delta^2 J_d$$

satisfying the initial condition $K^0_\delta(0) = 0$. Since the system $(A_0, A_1, \ldots, A_r; Q)$ is stable and $\|T_\delta\| < \gamma$ it follows that one can apply Lemma 7 and Remark 4 to the solution $K^0_\delta(t)$, $\delta \in (0, \delta_0]$. Therefore, there exists $\tilde{\rho} > 0$ such that

$$0 \le K^0_\delta(t, i) \le \tilde{\rho} I_n, \ t \ge 0, \ i \in \mathcal{D}, \tag{6.47}$$

$$K^0_\delta(\tau_1, i) \le K^0_\delta(\tau_2, i), \ \forall 0 \le \tau_1 < \tau_2, \tag{6.48}$$

$$\gamma^2 I_m - D^*(i) D(i) - \sum_{k=1}^r B^*_k(i) K^0_\delta(t, i) B_k(i) \ge \varepsilon_0^2 I_m, \tag{6.49}$$

where $\varepsilon_0 > 0$.

From (6.47) and (6.48) it also follows that

$$\tilde{K}_\delta = \left(\tilde{K}_\delta(1), \ldots, \tilde{K}_\delta(d) \right),$$

with

$$\tilde{K}_\delta(i) = \lim_{t \to \infty} K^0_\delta(t, i), \ i \in \mathcal{D}, \tag{6.50}$$

is well defined.

From Lemma 9 it follows that $\tilde{K}_\delta$ is a stationary solution of the differential equation (6.46) and hence it verifies

$$\mathcal{L}^* \tilde{K}_\delta - \mathcal{P}^* \left(\tilde{K}_\delta \right) \mathcal{R}^{-1} \left(\tilde{K}_\delta \right) \mathcal{P} \left(\tilde{K}_\delta \right) + C^* C + \delta^2 J_d = 0. \tag{6.51}$$

Using (6.49) one also obtains that $\tilde{K}_\delta$ defined by (6.50) verifies

$$\mathcal{R} \left(\tilde{K}_\delta \right) \le -\varepsilon_0^2 I_m, \ i \in \mathcal{D}. \tag{6.52}$$

Since $(A_0, A_1, \ldots, A_r; Q)$ is stable, one easily obtains the following representation:

$$\tilde{K}_\delta = \int_0^\infty e^{\mathcal{L}^* s} \left[C^* C + \delta^2 J_d - \mathcal{P}^* \left(\tilde{K}_\delta \right) \mathcal{R}^{-1} \left(\tilde{K}_\delta \right) \mathcal{P} \left(\tilde{K}_\delta \right) \right] ds.$$

Taking into account the positivity of the operator $e^{\mathcal{L}^* s}$ and the inequality (6.52) it follows that

$$\tilde{K}_\delta \ge \delta^2 \int_0^\infty e^{\mathcal{L}^* s} J_d ds. \tag{6.53}$$

From Remark 2 of Chapter 2 it follows that there exists $\nu > 0$ such that $e^{\mathcal{L}^* s} J_d \ge e^{-\nu s} J_d$. Therefore (6.53) reduces to $\tilde{K}_\delta \ge \frac{\delta^2}{\nu} J_d > 0$. Finally, notice that for all

$\delta \in (0, \delta_0)$, $\tilde{K}_\delta$ verifies

$$\mathcal{L}^* \tilde{K}_\delta - \mathcal{P}^* \left(\tilde{K}_\delta \right) \mathcal{R}^{-1} \left(\tilde{K}_\delta \right) \mathcal{P} \left(\tilde{K}_\delta \right) + C^* C < 0.$$

This shows, together with (6.52), that $\tilde{K}_\delta$ verifies $\mathcal{N}(\tilde{K}_\delta) < 0$ and therefore (ii) is true.

(ii) $\Rightarrow$ (iii) From Remark 5(iii) it follows that $\hat{X} \in \tilde{\Pi}$. Therefore $\mathcal{L}^* \hat{X} < 0$ and $\hat{X} > 0$. By using Theorem 15(iv) of Chapter 2 one concludes that $(A_0, \ldots, A_r; Q)$ is stable. Hence $(\mathbf{A}, \mathbf{B}; Q)$ is stabilizable. Now, by virtue of Theorem 10 of Chapter 4, where

$$M(i) = -C^*(i)C(i),$$
$$L(i) = -C^*(i)D(i),$$
$$R(i) = \gamma^2 I_m - D^*(i)D(i), \ \ i \in \mathcal{D},$$

we may conclude that the SGRAE (6.18) has a stabilizing solution $\tilde{X} = (\tilde{X}(1), \ldots, \tilde{X}(d))$ verifying $\mathcal{R}(\tilde{X}) < 0$. It remains only to show that $\tilde{X} \geq 0$. Indeed, since the system $(A_0, \ldots, A_r; Q)$ is stable and $\mathcal{R}(\tilde{X}) < 0$, from (6.19) for $\tilde{X}$ and Proposition 14 of Chapter 2, it follows directly that $\tilde{X} \geq 0$.

(iii) $\Rightarrow$ (i) Assume that the SGRAE (6.18) has a stabilizing solution $\tilde{X} \geq 0$ verifying (6.44). To prove that the system $(A_0, \ldots, A_r; Q)$ is stable we write the SGRAE (6.18) verified by $\tilde{X}$ in the equivalent form:

$$\mathcal{L}^* \tilde{X} + \overline{C}^* \overline{C} = 0, \tag{6.54}$$

where

$$\overline{C} = (\overline{C}(1), \ldots, \overline{C}(d)), \ \ \overline{C}(i) = \begin{bmatrix} \overline{C}_1(i) \\ \overline{C}_2(i) \end{bmatrix},$$

with

$$\overline{C}_1(i) = \left(\gamma^2 I_m - D^*(i)D(i) - \sum_{k=1}^{r} B_k^*(i) \tilde{X}(i) B_k(i) \right)^{-\frac{1}{2}}$$

$$\times \left(B_0^*(i) \tilde{X}(i) + \sum_{k=1}^{r} B_k^*(i) \tilde{X}(i) A_k(i) + D^*(i)C(i) \right),$$

$$\overline{C}_2(i) = C(i), \ \ i \in \mathcal{D}.$$

Further, take

$$H_k = (H_k(1), \ldots, H_k(d)), \ \ k = 0, 1, \ldots, r,$$

where

$$H_k(i) = \left[B_k(i) \left(-\mathcal{R}_i \left(\tilde{X} \right) \right)^{-\frac{1}{2}} \quad 0 \right], \ \ i \in \mathcal{D}.$$

With the above notations, one obtains that

$$\left(A_0 + H_0 \overline{C}, \ldots, A_r + H_r \overline{C}; Q \right) = \left(A_0 + B_0 \tilde{F}, \ldots, A_r + B_r \tilde{F}; Q \right)$$

is stable. If $x(t)$, $t \geq 0$, is an arbitrary solution of the equation

$$dx(t) = A_0(\eta(t))x(t)dt + \sum_{k=1}^{r} A_k(\eta(t))x(t)dw_k(t),$$

then we can write

$$dx(t) = \left[\left(A_0(\eta(t)) + H_0(\eta(t))\overline{C}(\eta(t))\right) x(t) + f_0(t)\right]dt \tag{6.55}$$

$$+ \sum_{k=1}^{r} \left[\left(A_k(\eta(t)) + H_k(\eta(t))\overline{C}(\eta(t))\right) x(t) + f_k(t)\right]dw_k(t).$$

Based on similar reasoning as in the proof of Lemma 15 in Chapter 4 one deduces that the null solution of the equation (6.55) is ESMS. It remains to prove that $\|T\| < \gamma$. Applying the Itô-type formula for the function $x^* \widetilde{X}(i) x$ and to the system (6.1) one obtains that

$$E \int_0^\infty \left\{ |y_u(t)|^2 - \gamma^2 |u(t)|^2 dt \right\} \tag{6.56}$$

$$= -E \int_0^\infty \left| \left(-\mathcal{R}_{\eta(t)}\left(\widetilde{X}\right)\right)^{\frac{1}{2}} \left(u(t) - \widetilde{F}(\eta(t))x_u(t)\right) \right|^2 dt$$

for any $u \in L^2_{\eta,w}([0, \infty), \mathbf{R}^m)$, $x_u(t)$, $t \geq 0$, denoting the solution of (6.1) with the input $u(t)$, $t \geq 0$, and with zero initial conditions. The equality (6.56) can be rewritten as follows:

$$\|Tu\|^2 - \gamma^2 \|u\|^2 = -\|g_u\|^2, \tag{6.57}$$

where

$$g_u(t) = \left(\gamma^2 I_m - D^*(i)D(i) - \sum_{k=1}^{r} B_k^*(i)\widetilde{X}(i)B_k(i)\right)^{\frac{1}{2}} \tag{6.58}$$

$$\times \left(u(t) - \widetilde{F}(\eta(t))x_u(t)\right).$$

From (6.57) it follows that

$$\|T\| \leq \gamma. \tag{6.59}$$

It remains to prove that the equality cannot take place in (6.59). Indeed, if $\|T\| = \gamma$ there exist a sequence of stochastic processes $u_l, l \geq 0$, $\{u_l\} \subset L^2_{\eta,w}([0, \infty), \mathbf{R}^n)$ with

$$\|u_l\| = 1, \ \forall l \geq 0, \tag{6.60}$$

and

$$\lim_{l \to \infty} \|Tu_l\| = \gamma. \tag{6.61}$$

Let $x_l(t)$, $t \geq 0$, be the solution of the system (6.1) determined by the input $u_l(t)$ and having the initial conditions $x_l(0) = 0$, $l \geq 0$. We also denote by $g_l(t)$ the process defined by (6.58) in which u has been replaced by u_l. Using (6.60), the equality (6.57) becomes

$$\|\mathcal{T}u_l\|^2 - \gamma^2 = -\|g_l\|^2.$$

Then, taking into account (6.61), one obtains

$$\lim_{l\to\infty} \|g_l\| = 0. \tag{6.62}$$

Further, from (6.58) and (6.44) it results that

$$\lim_{l\to\infty} \|\tilde{g}_l\| = 0, \tag{6.63}$$

where we denoted $\tilde{g}_l(t) = u_l(t) - \tilde{F}(\eta(t))x_l(t),\ t \geq 0$. The differential equation verified by x_l can be rewritten as

$$dx_l(t) = \left\{\left[A_0(\eta(t)) + B_0(\eta(t))\tilde{F}(\eta(t))\right]x_l(t) + B_0(\eta(t))\tilde{g}_l(t)\right\}dt$$
$$+ \sum_{k=1}^{r} \left\{\left[A_k(\eta(t)) + B_k(\eta(t))\tilde{F}(\eta(t))\right]x_l(t) + B_k(\eta(t))\tilde{g}_l(t)\right\}dw_k(t).$$

Since the system $\left(A_0 + B_0\tilde{F}, \ldots, A_r + B_r\tilde{F}; Q\right)$ is stable, combining the result in Theorem 32 in Chapter 2 and (6.63), we obtain that $\lim_{l\to\infty} \|x_l\| = 0$, and then, again using (6.63), it immediately follows that $\lim_{l\to\infty} \|u_l\| = 0$, which contradicts (6.60), and thus the proof is complete.   □

Remark 6. (i) From the above theorem it follows that

$$\|\mathcal{T}\| = \inf\{\gamma > 0, \text{ for which it exists } X = (X(1), \ldots, X(d)) \in \mathcal{S}_n^d,.$$
$$X > 0 \text{ such that } \mathcal{N}_i(X) < 0\}$$
$$= \inf\{\gamma > 0, \text{ SGRAE } (6.18) \text{ has the stabilizing solution.}$$
$$\tilde{X} = (\tilde{X}(1), \ldots, \tilde{X}(d)) \text{ verifying } \tilde{X}(i) \geq 0,\ \mathcal{R}_i(\tilde{X}) < 0,\ i \in \mathcal{D}\}.$$

(ii) Let us notice that in contrast with the H^2 norms associated with a stochastic linear system that can be directly computed by the results in Theorems 18 and 21 of Chapter 5, the norm of the input–output operator associated with a stochastic linear system cannot be directly computed. This norm can be estimated using a γ-procedure as in the deterministic case.

(iii) From the numerical point of view, the equivalence (i)$\Leftrightarrow$(ii) is more effective for computing $\|\mathcal{T}\|$ since for every γ it reduces to testing the feasibility of an LMI system. The equivalence (i) $\Leftrightarrow$ (iii) of Theorem 10 is useful for developing mixed H^2/H^∞ procedures for robust stabilization.

(iv) In the particular case when there exists $r_1 \geq 1$, such that $A_k(i) = 0,\ r_1 \leq k \leq r$, and $B_k(i) = 0,\ 0 \leq k \leq r_1 - 1,\ C^*(i)D(i) = 0,\ i \in \mathcal{D}$, SGRAE (6.18) reduces to the following Lyapunov-type equation:

$$A_0^*(i)X(i) + X(i)A_0(i) + \sum_{k=1}^{r_1-1} A_k^*(i)X(i)A_k(i)$$

$$+ \sum_{j=1}^{d} q_{ij}X(j) + C^*(i)C(i) = 0, \ i \in \mathcal{D}. \tag{6.64}$$

By convention, if $r_1 = 1$, the first sum in (6.64) is missing. If the system $(A_0, \ldots, A_{r_1-1}; Q)$ is stable, then the equation (6.64) has a unique solution $\tilde{X} = (\tilde{X}(1), \ldots, \tilde{X}(d)) \geq 0$. Moreover, if Theorem 10(i) is fulfilled, then the solution of the equation (6.64) verifies the condition

$$D^*(i)D(i) + \sum_{k=r_1}^{r} B_k^*(i)\tilde{X}(i)B_k(i) < \gamma^2 I_m, \ i \in \mathcal{D}.$$

Remark 7. $L_{\eta,w}^2([0, \infty), \mathbf{R}^m)$ can also be organized as a real Hilbert space, taking the inner product

$$(u, v) = \sum_{i=1}^{d} E\left[\int_0^\infty u^*(t)v(t)dt \mid \eta(0) = i\right].$$

The corresponding induced norm will be denoted by $||| \cdot |||$.

Proposition 11. *Suppose that $(A_0, \ldots, A_r; Q)$ is stable. Then $\|\mathcal{T}\| = ||| \mathcal{T} |||$.*

Proof. It is easy to see that all preceding results and remarks hold if the norm $\| \cdot \|$ is replaced by $||| \cdot |||$. In this case the performance index $H_\gamma(\tau, x_0, u)$ is replaced by $\sum_{i=1}^{d} H_\gamma(\tau, x_0, i, u)$. Therefore, taking into account Remark 6(i), we have

$$||| \mathcal{T} ||| = \inf \{\gamma > 0, \text{ SGRAE (6.18) has a stabilizing solution}$$
$$\tilde{X} \geq 0 \text{ with } R_i(\tilde{X}) < 0, \ i \in \mathcal{D}\}.$$

Hence $||| \mathcal{T} ||| = \|\mathcal{T}\|$, and thus the proof is complete. $\square$

From Theorem 10 and Remark 6(i) one immediately obtains the following corollary.

Corollary 12. *Consider the system*

$$dx(t) = A_0(\eta(t))x(t)dt + \sum_{k=1}^{r_1-1} A_k(\eta(t))x(t)dw_k(t)$$

$$+ \sum_{k=r_1}^{r} B_k(\eta(t))u(t)dw_k(t), \tag{6.65}$$

$$y(t) = C(\eta(t))x(t) + D(\eta(t))u(t)$$

with $C^*(i)D(i) = 0$, $i \in \mathcal{D}$. Assume that the system $(A_0, \ldots, A_{r_1-1}; Q)$ is stable and denote by

$$\mathcal{T} : L_{\eta,w}^2([0, \infty); \mathbf{R}^m) \to L_{\eta,w}^2([0, \infty); \mathbf{R}^p)$$

the input–output operator associated with the system (6.65). *Then*

$$\|\mathcal{T}\| = \max_{i \in \mathcal{D}} \sqrt{\lambda_{\max}(i)},$$

where $\lambda_{\max}(i)$ *is the largest eigenvalue of the matrix*

$$D^*(i)D(i) + \sum_{k=r_1}^{r} B_k^*(i)\widetilde{X}(i)B_k(i), \ i \in \mathcal{D},$$

$\widetilde{X} = (\widetilde{X}(1), \dots, \widetilde{X}(d))$ *being the unique solution of* (6.64).   □

Proposition 13. *Let* $\mathbf{D} : L_{\eta,w}^2 ([0, \infty); \mathbf{R}^m) \to L_{\eta,w}^2 ([0, \infty); \mathbf{R}^p)$ *be the linear bounded operator defined by*

$$(\mathbf{D}u)(t) = D(\eta(t))u(t), \ u \in L_{\eta,w}^2 ([0, \infty); \mathbf{R}^m).$$

Then

$$\|\mathbf{D}\| = |D| = \max \{|D(i)|, \ i \in \mathcal{D}\}.$$

Proof. Since $D^*(i)D(i) \leq |D|^2 I_m$, we have

$$\|\mathbf{D}u\|^2 = E \int_0^\infty u^*(t)D^*(\eta(t))D(\eta(t))u(t)dt$$

$$\leq |D|^2 E \int_0^\infty |u(t)|^2 dt = |D|^2 \|u\|^2$$

for every $u \in L_{\eta,w}^2 ([0, \infty); \mathbf{R}^m)$. Hence $\|\mathbf{D}\| \leq |D|$.

Further, let $i \in \mathcal{D}$, $u \in \mathbf{R}^m$ arbitrary but fixed. Take

$$\hat{u}(t) = \begin{cases} u\chi_{\eta(t)=i} & \text{if } t \in [0, 1], \\ 0 & \text{if } t > 1. \end{cases}$$

Obviously $\hat{u} \in L_\eta^2 ([0, \infty); \mathbf{R}^m)$ and therefore $\hat{u} \in L_{\eta,w}^2 ([0, \infty); \mathbf{R}^m)$. The inequality

$$\|\mathbf{D}\hat{u}\|^2 \leq \|\mathbf{D}\|^2 \|\hat{u}\|^2$$

becomes

$$\int_0^1 E\left(|D(\eta(t))u|^2 E\chi_{\eta(t)=i}\right) dt \leq \|\mathbf{D}\|^2 \int_0^1 |u|^2 E\chi_{\eta(t)=i}dt.$$

Therefore

$$\int_0^1 |D(i)u|^2 E\chi_{\eta(t)=i}dt \leq \|\mathbf{D}\|^2 |u|^2 \int_0^1 E\chi_{\eta(t)=i}dt.$$

But

$$\int_0^1 E\chi_{\eta(t)=i}dt = \int_0^1 \sum_{j=1}^d \pi_j E\left[\chi_{\eta(t)=i} \mid \eta(0) = j\right]dt$$

$$= \sum_{j=1}^d \int_0^1 \pi_j p_{ji}(t)dt \geq \int_0^1 \pi_i p_{ii}(t)dt > 0.$$

Thus we may conclude that

$$|D(i)u| \le \|D\| \, |u| \, ,$$

which leads to $|D| \le \|D\|$ and thus the proof is complete. $\qquad\qquad\square$

Remark 8. (i) Evidently, if $u \in L^2_{\eta,w}([0,\infty); \mathbf{R}^m)$, then $\mathbf{D}u \in L^2_\eta([0,\infty); \mathbf{R}^p)$. The proof of Proposition 13 shows that $\|\widehat{\mathbf{D}}\| = |D| = \|\mathbf{D}\|$, where $\widehat{\mathbf{D}}$ is the restriction of the operator $\mathbf{D}$ to the subspace $L^2_\eta([0,\infty); \mathbf{R}^m) \subset L^2_{\eta,w}([0,\infty); \mathbf{R}^m)$.

   (ii) The conclusion of Proposition 13 can be obtained directly from Corollary 12. Indeed, if we take $C(i) = 0$, $i \in \mathcal{D}$, it follows that $\widetilde{X}(i) = 0$, $i \in \mathcal{D}$, and therefore $\|\mathbf{D}\|^2 = \max_{i\in\mathcal{D}} |D(i)|^2$.

The following result allows us to increase the number of relations of equivalence in Theorem 10, and it is useful in some applications.

Proposition 14. *Let $\mathcal{N}(X) = (\mathcal{N}_1(X), \dots, \mathcal{N}_d(X))$ be the generalized dissipation matrix associated with the matrices $A_k(i)$, $B_k(i)$, $C(i)$, $D(i)$, and with the scalar $\gamma > 0$. Then the following assertions are equivalent:*

   (i) *There exists $X = (X(1), \dots, X(d)) \in \mathcal{S}^d_n$, $X > 0$, such that $\mathcal{N}_i(X) < 0$ $\forall i \in \mathcal{D}$.*

   (ii) *There exists $Y = (Y(1), \dots, Y(d)) \in \mathcal{S}^d_n$, $Y > 0$, such that*

$$\begin{bmatrix}
\mathcal{W}_{0,0}(Y,i) & \mathcal{W}_{0,1}(Y,i) & \cdots & \mathcal{W}_{0,r}(Y,i) & \mathcal{W}_{0,r+1}(Y,i) & \mathcal{W}_{0,r+2}(Y,i) \\
\mathcal{W}^*_{0,1}(Y,i) & \mathcal{W}_{1,1}(Y,i) & \cdots & \mathcal{W}_{1,r}(Y,i) & \mathcal{W}_{1,r+1}(Y,i) & \mathcal{W}_{1,r+2}(Y,i) \\
\vdots & \vdots & \ddots & \vdots & \vdots & \vdots \\
\mathcal{W}^*_{0,r}(Y,i) & \mathcal{W}^*_{1,r}(Y,i) & \cdots & \mathcal{W}_{r,r}(Y,i) & \mathcal{W}_{r,r+1}(Y,i) & \mathcal{W}_{r,r+2}(Y,i) \\
\mathcal{W}^*_{0,r+1}(Y,i) & \mathcal{W}^*_{1,r+1}(Y,i) & \cdots & \mathcal{W}^*_{r,r+1}(Y,i) & \mathcal{W}_{r+1,r+1}(Y,i) & \mathcal{W}_{r+1,r+2}(Y,i) \\
\mathcal{W}^*_{0,r+2}(Y,i) & \mathcal{W}^*_{1,r+2}(Y,i) & \cdots & \mathcal{W}^*_{r,r+2}(Y,i) & \mathcal{W}^*_{r+1,r+2}(Y,i) & \mathcal{W}_{r+2,r+2}(Y,i)
\end{bmatrix} < 0,$$

$$\tag{6.66}$$

$i \in \mathcal{D}$, *where*

$$\mathcal{W}_{0,0}(Y,i) = \left(A_0(i) + \frac{1}{2}q_{ii}I_n\right)Y(i) + Y(i)\left(A_0(i) + \frac{1}{2}q_{ii}I_n\right)^* + B_0(i)B_0^*(i),$$

$$\mathcal{W}_{0,k}(Y,i) = Y(i)A_k^*(i) + B_0(i)B_k^*(i), \ \ k = 1, \dots, r,$$

$$\mathcal{W}_{0,r+1}(Y,i) = Y(i)C^*(i) + B_0(i)D^*(i),$$

$$\mathcal{W}_{0,r+2}(Y,i) = \left(\sqrt{q_{i1}}Y(i), \dots, \sqrt{q_{i,i-1}}Y(i) \ \sqrt{q_{i,i+1}}Y(i), \dots, \ \sqrt{q_{id}}Y(i)\right),$$

$$\mathcal{W}_{l,k}(Y,i) = B_l(i)B_k^*(i), \ 1 \le l, k \le r, \ l \ne k,$$

$$\mathcal{W}_{l,l}(Y,i) = B_l(i)B_l^*(i) - Y(i), \ 1 \le l \le r,$$

$$\mathcal{W}_{l,r+1}(Y,i) = B_l(i)D^*(i), \ 1 \le l \le r,$$

$$\mathcal{W}_{r+1,r+1}(Y,i) = D(i)D^*(i) - \gamma^2 I_p,$$

$$\mathcal{W}_{l,r+2}(Y,i) = 0, \ 1 \le l \le r+1,$$

$$\mathcal{W}_{r+2,r+2}(Y,i) = diag\left(-Y(1), \dots, -Y(i-1) \ -Y(i+1), \dots, -Y(d)\right).$$

Proof. It is easy to see that the existence of $X = (X(1), \ldots, X(d)) > 0$ such that $\mathcal{N}_i(X) < 0$ is equivalent to the existence of $X = (X(1), \ldots, X(d)) > 0$ such that

$$\begin{bmatrix} \mathcal{V}_{11}(X, i) & \mathcal{V}_{12}(X, i) & \mathcal{V}_{13}(X, i) \\ \mathcal{V}_{12}^*(X, i) & \mathcal{V}_{22}(X, i) & \mathcal{V}_{23}(X, i) \\ \mathcal{V}_{13}^*(X, i) & \mathcal{V}_{23}^*(X, i) & \mathcal{V}_{33}(X, i) \end{bmatrix} < 0, \qquad (6.67)$$

where $\mathcal{V}_{11}(X, i)$ is an $(n + m) \times (n + m)$ matrix given by

$$\mathcal{V}_{11}(X, i) = \begin{bmatrix} A_0^*(i)X(i) + X(i)A_0(i) + \sum_{j=1}^d q_{ij}X(j) & X(i)B_0(i) \\ B_0^*(i)X(i) & -\gamma^2 I_m \end{bmatrix},$$

$\mathcal{V}_{12}(X, i)$ is an $(n + m) \times (r \cdot n)$ matrix

$$\mathcal{V}_{12}(X, i) = \begin{bmatrix} A_1^*(i)X(i) & \ldots & A_r^*(i)X(i) \\ B_1^*(i)X(i) & \ldots & B_r^*(i)X(i) \end{bmatrix},$$

$\mathcal{V}_{13}(X, i)$ is an $(n + m) \times p$ matrix defined by

$$\mathcal{V}_{13}(X, i) = \begin{bmatrix} C^*(i) \\ D^*(i) \end{bmatrix},$$

$\mathcal{V}_{22}(X, i)$ is an $(n \cdot r) \times (n \cdot r)$ matrix

$$\mathcal{V}_{22}(X, i) = diag\,(-X(i), \ldots, -X(i)),$$

$\mathcal{V}_{23}(X, i)$ is an $(n \cdot r) \times p$ matrix given by

$$\mathcal{V}_{23}(X, i) = 0,$$

and

$$\mathcal{V}_{33}(X, i) = -I_p.$$

Let us introduce

$$\Psi(i) = diag\,\left(X^{-1}(i) \quad I_m \quad -\mathcal{V}_{22}^{-1}(X, i) \quad I_p \right).$$

It is obvious that $\Psi(i) = \Psi^*(i) > 0$. Through pre- and postmultiplication of (6.67) by $\Psi(i)$, one obtains that there exists $X = (X(1), \ldots, X(d)) > 0$, such that

$$\begin{bmatrix} \tilde{\mathcal{V}}_{11}(X, i) & B_0(i) & \tilde{\mathcal{V}}_{13}(X, i) & X^{-1}(i)C^*(i) \\ B_0^*(i) & -\gamma^2 I_m & \tilde{\mathcal{V}}_{23}(X, i) & D^*(i) \\ \tilde{\mathcal{V}}_{13}^*(X, i) & \tilde{\mathcal{V}}_{23}^*(X, i) & \tilde{\mathcal{V}}_{33}(X, i) & \tilde{\mathcal{V}}_{34}(X, i) \\ C(i)X^{-1}(i) & D(i) & \tilde{\mathcal{V}}_{34}^*(X, i) & -I_p \end{bmatrix} < 0, \qquad (6.68)$$

where

$$\tilde{\mathcal{V}}_{11}(X, i) = A_0(i)X^{-1}(i) + X^{-1}(i)A_0^*(i) + \sum_{j=1}^d q_{ij}X^{-1}(i)X(j)X^{-1}(i),$$

$$\tilde{\mathcal{V}}_{13}(X, i) = \left[X^{-1}(i)A_1^*(i) \quad \ldots \quad X^{-1}(i)A_r^*(i) \right],$$

$$\tilde{\mathcal{V}}_{23}(X, i) = \left[B_1^*(i) \quad \ldots \quad B_r^*(i) \right],$$

$\widetilde{V}_{33}(X, i)$ is an $rn \times rn$ matrix defined by

$$\widetilde{V}_{33}(X, i) = diag\left(-X^{-1}(i), \ldots, -X^{-1}(i)\right),$$

and $\widetilde{V}_{34}(X, i)$ is an $rn \times p$ matrix, $\widetilde{V}_{34}(X, i) = 0$. Denoting $Z(i) = X^{-1}(i)$ one immediately obtains that (6.68) is equivalent to the existence of $Z = (Z(1), \ldots, Z(d)) > 0$ satisfying

$$\begin{bmatrix} \widehat{V}_{11}(Z, i) & \widehat{V}_{12}(Z, i) & Z(i)C^*(i) & \widehat{V}_{14}(Z, i) & B_0(i) \\ \widehat{V}_{12}^*(Z, i) & \widehat{V}_{22}(Z, i) & \widehat{V}_{23}(Z, i) & \widehat{V}_{24}(Z, i) & \widehat{V}_{25}(Z, i) \\ C(i)Z(i) & \widehat{V}_{23}(Z, i) & -I_p & \widehat{V}_{34}(Z, i) & D(i) \\ \widehat{V}_{14}^*(Z, i) & \widehat{V}_{24}^*(Z, i) & \widehat{V}_{34}^*(Z, i) & \widehat{V}_{44}(Z, i) & \widehat{V}_{45}(Z, i) \\ B_0^*(i) & \widehat{V}_{25}^*(Z, i) & D^*(i) & \widehat{V}_{45}^*(Z, i) & -\gamma^2 I_m \end{bmatrix} < 0, \qquad (6.69)$$

where

$$\widehat{V}_{11}(Z, i) = \left(A_0(i) + \frac{1}{2}q_{ii}I_n\right)Z(i) + Z(i)\left(A_0(i) + \frac{1}{2}q_{ii}I_n\right)^*,$$

$$\widehat{V}_{12}(Z, i) = \left[Z(i)A_1^*(i) \ \ldots \ Z(i)A_r^*(i)\right],$$

$$\widehat{V}_{14}(Z, i) = \left[\sqrt{q_{i1}}Z(i) \ \ldots \ \sqrt{q_{i,i-1}}Z(i) \ \sqrt{q_{i,i+1}}Z(i) \ \ldots \ \sqrt{q_{id}}Z(i)\right]$$

is an $n \times (d - 1)n$ matrix,

$$\widehat{V}_{22}(Z, i) = diag\left(-Z(i) \ \ldots \ -Z(i)\right)$$

has the dimensions $rn \times rn$, $\widehat{V}_{23}(Z, i) = 0$ is an $nr \times p$ matrix, $\widehat{V}_{24}(Z, i) = 0$ is an $nr \times (d - 1)n$ matrix,

$$\widehat{V}_{25}(Z, i) = \begin{bmatrix} B_1(i) \\ \vdots \\ B_r(i) \end{bmatrix}$$

is an $nr \times m$ matrix, $\widehat{V}_{34}(Z, i) = 0$ has the dimensions $p \times (d - 1)n$,

$$\widehat{V}_{44}(Z, i) = diag\left(-Z(1) \ \ldots \ -Z(i - 1) \ -Z(i + 1) \ \ldots \ -Z(d)\right)$$

is a $(d - 1)n \times (d - 1)n$ matrix, and $\widehat{V}_{45}(Z, i) = 0$ has the dimensions $(r - 1)n \times m$.

Taking the Schur complement of the block $-\gamma^2 I_m$ of (6.69) it follows that this condition is accomplished if and only if there exists $Z = (Z(1), \ldots, Z(d)) > 0$ such that

$$\begin{bmatrix} \widehat{W}_{11}(Z, i) & \widehat{W}_{12}(Z, i) & Z(i)C^*(i) + \gamma^{-2}B_0(i)D^*(i) & \widehat{W}_{14}(Z, i) \\ \widehat{W}_{12}^*(Z, i) & \widehat{W}_{22}(Z, i) & \widehat{W}_{23}(Z, i) & \widehat{W}_{24}(Z, i) \\ C(i)Z(i) + \gamma^{-2}D(i)B_0^*(i) & \widehat{W}_{23}^*(Z, i) & -I_p + \gamma^{-2}D(i)D^*(i) & \widehat{W}_{34}(Z, i) \\ \widehat{W}_{14}^*(Z, i) & \widehat{W}_{24}^*(Z, i) & \widehat{W}_{34}^*(Z, i) & \widehat{W}_{44}(Z, i) \end{bmatrix} < 0,$$

$$(6.70)$$

where

$$\widehat{W}_{11}(Z, i) = \left(A_0(i) + \frac{1}{2}q_{ii} I_n\right) Z(i) + Z(i) \left(A_0(i) + \frac{1}{2}q_{ii} I_n\right)^* + \gamma^{-2} B_0(i) B_0^*(i),$$

$$\widehat{W}_{12}(Z, i) = \left[Z(i)A_1^*(i) + \gamma^{-2} B_1(i) B_1^*(i) \ldots Z(i)A_r^*(i) + \gamma^{-2} B_r(i) B_r^*(i)\right],$$

$$\widehat{W}_{14}(Z, i) = \widehat{V}_{14}(Z, i), \quad \widehat{W}_{22}(Z, i) = \widehat{V}_{22}(Z, i),$$

$$\widehat{W}_{23}(Z, i) = \widehat{V}_{23}(Z, i), \quad \widehat{W}_{24}(Z, i) = \widehat{V}_{24}(Z, i),$$

$$\widehat{W}_{34}(Z, i) = \widehat{V}_{34}(Z, i), \quad \widehat{W}_{44}(Z, i) = \widehat{V}_{44}(Z, i).$$

Consider the $(2n(r+d)+p) \times (2n(r+d)+p)$ matrix

$$\Gamma = diag\left(\gamma I_n, \gamma I_{rn}, \gamma I_p, \gamma I_{n(d-1)}\right).$$

By pre- and postmultiplication of (6.70) with Γ and denoting $Y(i) = \gamma^2 Z(i)$, $i \in \mathcal{D}$, one obtains (6.66) and therefore the proof is complete. $\square$

At the end of this section we consider the particular cases when the system (6.1) is subjected either only to Markov perturbations or to white noise multiplicative perturbations.

Assume that in (6.1) we have $A_k(i) = 0$, $B_k(i) = 0$, $k = 1, \ldots, r$, $i \in \mathcal{D}$. Then (6.1) becomes

$$\dot{x}(t) = A_0(\eta(t))x(t) + B_0(\eta(t))u(t), \tag{6.71}$$
$$y(t) = C(\eta(t))x(t) + D(\eta(t))u(t).$$

The generalized dissipation matrix is in this case

$$\mathcal{N}(X) = (\mathcal{N}_1(X), \ldots, \mathcal{N}_d(X))$$

with

$$\mathcal{N}_i(X) = \begin{bmatrix} A_0^*(i)X(i) + X(i)A_0(i) \\ + \sum_{j=1}^d q_{ij}X(j) + C^*(i)C(i) & X(i)B_0(i) + C^*(i)D(i) \\ B_0^*(i)X(i) + D^*(i)C(i) & -\gamma^2 I_m + D^*(i)D(i) \end{bmatrix} \tag{6.72}$$

for any $X = (X(1), \ldots, X(d)) \in \mathcal{S}_n^d$, $i \in \mathcal{D}$. The SGRAE (6.17) becomes in this case

$$A_0^*(i)X(i) + X(i)A_0(i) + \sum_{j=1}^d q_{ij}X(j) + \left[X(i)B_0(i) + C^*(i)D(i)\right]$$

$$\times \left[\gamma^2 I_m - D^*(i)D(i)\right]^{-1} \left[B_0^*(i)X(i) + D^*(i)C(i)\right] + C^*(i)C(i) = 0, \tag{6.73}$$

$i \in \mathcal{D}$. Combining Theorem 10 and Proposition 14 one directly obtains the Bounded Real Lemma in the case of systems subjected to Markov perturbations.

Corollary 15. *For the system (6.71) and for a $\gamma > 0$ the following assertions are equivalent:*

(i) *The pair $(A_0; Q)$ is stable and the input–output operator T defined by the system (6.71) satisfies*

$$\|T\| < \gamma.$$

(ii) *There exists $X = (X(1), \ldots, X(d)) > 0$ such that $\mathcal{N}_i(X, \gamma) < 0 \ \forall i \in \mathcal{D}$.*

(iii) *$\gamma^2 I_m - D^*(i)D(i) > 0$ and the SGRAE (6.73) has a stabilizing solution $\tilde{X} = (\tilde{X}(1), \ldots, \tilde{X}(d)) \geq 0$.*

(iv) *There exists $Y = (Y(1), \ldots, Y(d)) > 0$ satisfying the following system of LMIs:*

$$\begin{bmatrix} \mathcal{W}_{0,0}(Y,i) & \mathcal{W}_{0,r+1}(Y,i) & \mathcal{W}_{0,r+2}(Y,i) \\ \mathcal{W}^*_{0,r+1}(Y,i) & \mathcal{W}_{r+1,r+1}(Y,i) & \mathcal{W}_{r+1,r+2}(Y,i) \\ \mathcal{W}^*_{0,r+2}(Y,i) & \mathcal{W}^*_{r+1,r+2}(Y,i) & \mathcal{W}_{r+2,r+2}(Y,i) \end{bmatrix} < 0, \ i \in \mathcal{D},$$

where $\mathcal{W}_{ij}(Y,i)$ are the same as in (6.66). □

In the following we assume that $\mathcal{D} = \{1\}$, $q_{11} = 0$, and $r \geq 1$. In this case the system (6.1) becomes

$$dx(t) = [A_0 x(t) + B_0 u(t)] dt + \sum_{k=1}^r [A_k x(t) + B_k u(t)] dw_k(t)$$

$$y(t) = Cx(t) + Du(t). \tag{6.74}$$

Then the generalized dissipation matrix is

$$\mathcal{N}(X)$$
$$= \begin{bmatrix} A_0^* X + X A_0 + \sum_{k=1}^r A_k^* X A_k + C^* C & X B_0 + \sum_{k=1}^r A_k^* X B_k + C^* D \\ B_0^* X + \sum_{k=1}^r B_k^* X A_k + D^* C & -\gamma^2 I_m + D^* D + \sum_{k=1}^r B_k^* X B_k \end{bmatrix}$$

for any $X \in \mathcal{S}_n$. The SGRAE (6.18) becomes in this case

$$A_0^* X + X A_0 + \sum_{k=1}^r A_k^* X A_k + \left[X B_0 + \sum_{k=1}^r A_k^* X B_k + C^* D \right]$$

$$\times \left[\gamma^2 I_m - D^* D - \sum_{k=1}^r B_k^* X B_k \right]^{-1} \left[B_0^* X + \sum_{k=1}^r B_k^* X A_k + D^* C \right]$$

$$+ C^* C = 0. \tag{6.75}$$

Again applying Theorem 10 and Proposition 14, one directly obtains the Bounded Real Lemma for systems subjected only to multiplicative white noise perturbations.

Corollary 16. *For the system (6.74) and for a $\gamma > 0$, the following are equivalent:*

(i) *The system $(A_0, \ldots, A_r)$ is stable and the input–output operator T associated with the system (6.74) satisfies the condition $\|T\| < \gamma$.*

(ii) *There exists a matrix* $\widehat{X} > 0$ *satisfying* $\mathcal{N}\left(\widehat{X}\right) < 0$.

(iii) *The SGRAE* (6.75) *has a positive semidefinite stabilizing solution* $\widetilde{X}$ *satisfying*
$\gamma^2 I_m - D^* D - \sum_{k=1}^{r} B_k^* X B_k > 0$.

(iv) *There exists* $Y > 0$, $Y \in \mathcal{S}_n^d$, *verifying the following LMI:*

$$
\begin{bmatrix}
A_0 Y + Y A_0^* + B_0 B_0^* & Y A_1^* + B_0 B_1^* & \cdots & Y A_r + B_0 B_r^* & Y C_0^* + B_0 D^* \\
A_1 Y + B_1 B_0^* & -Y + B_1 B_1^* & \cdots & B_1 B_r^* & B_1 D^* \\
\vdots & \vdots & \ddots & \vdots & \vdots \\
A_r Y + B_r B_0^* & B_r B_1^* & \cdots & -Y + B_r B_r^* & B_r D^* \\
C Y + D^* B_0 & D B_1^* & \cdots & D B_r^* & -\gamma^2 I_p + D D^*
\end{bmatrix} < 0.
$$

Remark 9. It is easy to see that in the case $\mathcal{D} = \{1\}$, $A_k = 0$, $B_k = 0$, $k = 1, \ldots, r$, the results stated in Corollaries 15 and 16 reduce to the well-known version of the Bounded Real Lemma of the deterministic case.

6.3 Robust stability with respect to linear structured uncertainty

At the beginning of this section we shall prove the stochastic version of the so-called *Small Gain Theorem* (SGT). As is known from the deterministic framework, this is a powerful tool in analyzing the robust stabilization with respect to different classes of linear perturbations.

6.3.1 Small gain theorem

We first prove the following result.

Theorem 17. *Assume the following.*

(a) *The system* $(A_0, \ldots, A_r; Q)$ *is stable.*

(b) *The system* (6.1) *has the same number of inputs and outputs.*

(c) *The input–output operator* T *defined by the system* (6.1) *satisfies the condition* $\|T\| < 1$.

Then we have the following.

(i) *The matrices* $I_m \pm D(i)$, $i \in \mathcal{D}$ *are invertible.*

(ii) *The system* $\left(\overline{A}_0, \ldots, \overline{A}_1; Q\right)$ *is stable, where*

$$
\overline{A}_k(i) = A_k(i) \pm B_k(i)(I_m \mp D(i))^{-1} C(i), \quad k = 0, 1, \ldots, r.
$$

Proof. (i) Using Corollary 3 and Remark 2 for the case $\gamma = 1$ one obtains that $I_m - D^*(i)D(i) > 0$, $i \in \mathcal{D}$. It follows that all eigenvalues of the matrices $D(i)$, $i \in \mathcal{D}$, are inside the unit circle, and therefore $\det(I_m \pm D(i)) \neq 0$, which shows that $I_m \pm D(i)$, $i \in \mathcal{D}$, are invertible.

(ii) From the implication (i) $\Rightarrow$ (ii) of Theorem 10 for $\gamma = 1$ we deduce that there exists $\widehat{X} = \left(\widehat{X}(1), \ldots, \widehat{X}(d)\right) > 0$ satisfying

$$
\mathcal{N}_i\left(\widehat{X}, 1\right) < 0, \quad i \in \mathcal{D}. \tag{6.76}
$$

Using the Schur complement of the block $(2,2)$ one obtains that (6.76) is equivalent to the condition

$$\mathcal{L}^*\widehat{X} - \mathcal{P}^*\left(\widehat{X}\right)\mathcal{R}^{-1}\left(\widehat{X}\right)\mathcal{P}\left(\widehat{X}\right) + C^*C + \widehat{M} = 0, \quad \mathcal{R}\left(\widehat{X}\right) < 0 \qquad (6.77)$$

for a certain $\widehat{M} > 0$, $\widehat{M} = \left(\widehat{M}(1), \ldots, \widehat{M}(d)\right) \in \mathcal{S}_n^d$. By direct computations similar to those in Lemma 1 of Chapter 4 one obtains that (6.77) can be rewritten as

$$\mathcal{L}_G^*\widehat{X} - G^*G + (C + DG)^*(C + DG)$$
$$- \left(G - \widehat{F}\right)^*\mathcal{R}\left(\widehat{X}\right)\left(G - \widehat{F}\right) + \widehat{M} = 0, \qquad (6.78)$$

where

$$G = (G(1), \ldots, G(d)), \quad G(i) = \pm\left(I_m \mp D(i)\right)^{-1}C(i),$$
$$\widehat{F} = \left(\widehat{F}(1), \ldots, \widehat{F}(d)\right), \quad \widehat{F}(i) = -\mathcal{R}_i^{-1}\left(\widehat{X}\right)\mathcal{P}_i\left(\widehat{X}\right), \quad i \in \mathcal{D}.$$

Then one obtains

$$(C(i) + D(i)G(i))^*(C(i) + D(i)G(i)) - G^*(i)G(i)$$
$$= C^*(i)\left[I_m \pm \left(I_m \mp D^*(i)\right)^{-1}D^*(i)\right]\left[I_m \pm D(i)\left(I_m \mp D(i)\right)^{-1}\right]C(i)$$
$$- G^*(i)G(i)$$
$$= C^*(i)\left(I_m \mp D^*(i)\right)^{-1}\left(I_m \mp D(i)\right)^{-1}C(i) - G^*(i)G(i)$$
$$= G^*(i)G(i) - G^*(i)G(i) = 0.$$

Thus it follows that (6.78) reduces to

$$\mathcal{L}_G^*\widehat{X} - \left(G - \widehat{F}\right)^*\mathcal{R}\left(\widehat{X}\right)\left(G - \widehat{F}\right) + \widehat{M} = 0.$$

Since $\widehat{M} - (G - \widehat{F})^*\mathcal{R}(\widehat{X})(G - \widehat{F}) > 0$ and $\widehat{X} > 0$, using Theorem 20 of Chapter 2 one obtains that the system $(A_0 + B_0G, \ldots, A_r + B_rG; Q)$ is stable. But $A_k(i) + B_k(i)G(i) = \overline{A}_k(i)$ and thus the proof is complete. $\qquad\square$

Theorem 18. *(The first small gain theorem) Assume that the assumptions in Theorem 17 hold. Then the operators*

$$I \mp T : L_{\eta,w}^2\left\{[0, \infty), \mathbf{R}^m\right\} \to L_{\eta,w}^2\left\{[0, \infty), \mathbf{R}^m\right\}$$

are invertible and the operators

$$(I \mp T)^{-1} : L_{\eta,w}^2\left\{[0, \infty), \mathbf{R}^m\right\} \to L_{\eta,w}^2\left\{[0, \infty), \mathbf{R}^m\right\}$$

have the following state-space realization:

$$dx(t) = \left[\overline{A}_0(\eta(t))x(t) + \overline{B}_0(\eta(t))y(t)\right]dt \qquad (6.79)$$

$$+ \sum_{k=1}^r\left[\overline{A}_k(\eta(t))x(t) + \overline{B}_k(\eta(t))y(t)\right]dw_k(t),$$

$$u(t) = \overline{C}(\eta(t))x(t) + \overline{D}(\eta(t))y(t),$$

$\overline{A}(i)$ being defined as in Theorem 17, $\overline{B}_k(i) = B_k(i)(I_m \mp D(i))^{-1}$, $\overline{C}_k(i) = \pm(I_m \mp D(i))^{-1}C(i)$, $\overline{D}(i) = (I_m \mp D(i))^{-1}$, $0 \leq k \leq r$, $i \in \mathcal{D}$.

The proof immediately follows using Theorem 17 and part (ii) of Proposition 1.

□

Remark 10. If $\|T\| < 1$ then the invertibility of the operators $I \mp T$ can also be obtained by a well-known result from the theory of linear and bounded operators on a Banach space. Theorem 18 additionally shows that the operators $(I \mp T)^{-1}$ have realizations in the state space.

Consider the following systems:

$$dx_1(t) = [A_{01}(\eta(t))x_1(t) + B_{01}(\eta(t))u_1(t)]dt$$

$$+ \sum_{k=1}^{r}[A_{k1}(\eta(t))x_1(t) + B_{k1}(\eta(t))u_1(t)]dw_k(t), \quad (6.80)$$

$$y_1(t) = C_1(\eta(t))x_1(t),$$

$$dx_2(t) = [A_{02}(\eta(t))x_2(t) + B_{02}(\eta(t))u_2(t)]dt$$

$$+ \sum_{k=1}^{r}[A_{k2}(\eta(t))x_2(t) + B_{k2}(\eta(t))u_2(t)]dw_k(t) \quad (6.81)$$

$$y_2(t) = C_2(\eta(t))x_2(t) + D_2(\eta(t))u_2(t),$$

with the states $x_l \in \mathbf{R}^{n_l}$, $l = 1, 2$; the output variables $y_1 \in \mathbf{R}^p$, $y_2 \in \mathbf{R}^m$; and the inputs $u_1 \in \mathbf{R}^m$, $u_2 \in \mathbf{R}^p$. When coupling (6.80) and (6.81) by taking $u_2 = y_1$ and $u_1 = y_2$ one obtains the following resulting system:

$$d\xi(t) = A_{0cl}(\eta(t))\xi(t)dt + \sum_{k=1}^{r}A_{kcl}(\eta(t))\xi(t)dw_k(t), \quad (6.82)$$

where

$$A_{kcl}(i) = \begin{bmatrix} A_{k1}(i) + B_{k1}(i)D_2(i)C_1(i) & B_{k1}(i)C_2(i) \\ B_{k2}(i)C_1(i) & A_{k2}(i) \end{bmatrix}, \quad k = 0, 1, \ldots, r.$$

Then another consequence of Theorem 17 is as follows.

Theorem 19. *(The second small gain theorem) Assume that the following assumptions hold:*

(i) *The systems* $(A_{0l}, \ldots, A_{rl}; Q)$, $l = 1, 2$ *are stable.*

(ii) $\|T_1\| < \gamma$, $\|T_2\| < \gamma^{-1}$ *for a certain* $\gamma > 0$, *where*

$$T_1 : L^2_{\eta,w}\{[0, \infty), \mathbf{R}^m\} \to L^2_{\eta,w}\{[0, \infty), \mathbf{R}^p\},$$

$$T_2 : L^2_{\eta,w}\{[0, \infty), \mathbf{R}^p\} \to L^2_{\eta,w}\{[0, \infty), \mathbf{R}^m\}$$

are the input–output operators defined by the systems (6.80) and (6.81), respectively. In these conditions the zero solution of the system (6.82) is ESMS.

Proof. From Proposition 1 one deduces that a state-space realization of the operator $T_1 T_2$ is

$$dx(t) = [A_0(\eta(t))x(t) + B_0(\eta(t))u(t)] \, dt$$

$$+ \sum_{k=1}^{r} [A_k(\eta(t))x(t) + B_k(\eta(t))u(t)] \, dw_k(t), \qquad (6.83)$$

$$y(t) = C(\eta(t))x(t),$$

where $A_k(\cdot)$, $B_k(\cdot)$ are defined as in Proposition 1 and

$$C(i) = [C_1(i) \, 0], \quad x = \begin{bmatrix} x_1 \\ x_2 \end{bmatrix}.$$

It is easy to see that

$$A_{kcl}(i) = A_k(i) + B_k(i)C(i) = \overline{A}_k(i), \quad k = 0, \dots, r, \ i \in \mathcal{D},$$

$\overline{A}_k(i)$ being the ones in Theorem 17 with $D(i) = 0$. The conclusion in the statement follows, applying Theorem 17 to the system (6.83). We show now that the assumptions in this theorem are fulfilled. Thus, from assumption (a) in the statement and from the triangular structure of the matrices $A_k(i)$, using Theorem 32 of Chapter 2 one deduces that the zero solution of the system (6.83) for $u(t) = 0$ is ESMS. From assumption (b) we have $\|T_1 T_2\| \leq \|T_1\| \|T_2\| < 1$, and hence the proof is complete. $\qquad \square$

Remark 11. Without important changes, the result in Theorem 19 also remains valid in the case when the output equation of (6.80) has the form

$$y_1(t) = C_1(\eta(t))x_1(t) + D_1(\eta(t))u_1(t).$$

From Theorem 19(ii), it immediately results that $I_m - D_1(i)D_2(i)$ is invertible for all $i \in \mathcal{D}$. The coefficients of the closed-loop system will be changed accordingly. We shall not detail them since they will be not used in the following developments.

An interesting case is the one when in the system (6.80) we have $n_1 > 0$, and in (6.81) $n_2 = 0$. In this situation the resulting system obtained by coupling (6.80) with (6.81) reduces to

$$dx_1(t) = [A_{01}(\eta(t)) + B_{01}(\eta(t))D_2(\eta(t))C_1(\eta(t))] x_1(t)dt \qquad (6.84)$$

$$+ \sum_{k=1}^{r} [A_{k1}(\eta(t)) + B_{k1}(\eta(t))D_2(\eta(t))C_1(\eta(t))] x_1(t)dw_k(t).$$

The input–output operator T_2 associated with the system (6.81) becomes

$$(T_2 u_2)(t) = D_2(\eta(t))u(t), \quad t \geq 0 \ \forall u \in L^2_{\eta,w}([0, \infty), \mathbf{R}^p).$$

From Proposition 13 it follows that $\|T_2\| = |D| = \max\{|D(i)|, \ i \in \mathcal{D}\}$. Consider the system

$$dx(t) = [A_0(\eta(t))x(t) + B_0(\eta(t))u(t)]\,dt$$

$$+ \sum_{k=1}^{r} [A_k(\eta(t))x(t) + B_k(\eta(t))u(t)]\,dw_k(t), \qquad (6.85)$$

$$y(t) = C(\eta(t))x(t).$$

Then we have the following corollary.

Corollary 20. *Assume as follows.*
   (i) *The system $(A_0, \ldots, A_r; Q)$ is stable.*
   (ii) *$\|T\| < \gamma$ and $|D| < \gamma^{-1}$, where*

$$\mathcal{T} : L_{\eta,w}^2([0, \infty), \mathbf{R}^m) \to L_{\eta,w}^2([0, \infty), \mathbf{R}^p)$$

denotes the input–output operator associated with the system (6.85) and $D = (D(1), \ldots, D(d)) \in \mathcal{M}_{mp}^d$.
   Then the zero solution of the system

$$dx(t) = [A_0(\eta(t)) + B_0(\eta(t))D(\eta(t))C(\eta(t))]\,x(t)dt$$

$$+ \sum_{k=1}^{r} [A_k(\eta(t)) + B_k(\eta(t))D(\eta(t))C(\eta(t))]\,x(t)dw_k(t)$$

is ESMS.

6.3.2 Robust stability with respect to linear parametric uncertainty

It is a known fact that the exponential stability of a solution of a linear deterministic system is not essentially influenced when the coefficients of the equation describing the system are subjected to "small perturbations." Taking into account the equivalence between the ESMS of a zero solution of a stochastic differential equation and the exponential stability of the zero solution of a Lyapunov-type linear differential equation, one expects the ESMS not to be affected by the small perturbations of the coefficients in the given equation. When analyzing the robustness of the solution of a system of stochastic differential equations we refer to the preservation of the stability property when the system is subjected to coefficient variations that are not necessarily small. Such variations or uncertainties are due to the inaccurate knowledge of the system coefficients or to some simplifications of the mathematical model. One must take into account that a controller designed for the simplified model will be used for the real system subjected to perturbations.

In the present section the robust stability with respect to a class of linear uncertainty will be investigated. Consider the linear system described by

$$dx(t) = [A_0(\eta(t)) + B_0(\eta(t))\Delta(\eta(t))C(\eta(t))]\,x(t)dt \qquad (6.86)$$

$$+ \sum_{k=1}^{r} [A_k(\eta(t) + B_k(\eta(t))\Delta(\eta(t))C(\eta(t))]\,x(t)dw_k(t),$$

where $A_k(i) \in \mathbf{R}^{n \times n}$, $0 \leq k \leq r$, $B_k(i) \in \mathbf{R}^{n \times m}$, $0 \leq k \leq r$, $C(i) \in \mathbf{R}^{p \times n}$, $i \in \mathcal{D}$, are assumed known and $\Delta(i) \in \mathbf{R}^{n \times p}$ are unknown matrices. Thus the system (6.86) is the perturbed system of the nominal one:

$$dx(t) = A_0(\eta(t))x(t)dt + \sum_{k=1}^{r} A_k(\eta(t)x(t)dw_k(t), \qquad (6.87)$$

and the matrices $B_k(i)$, $C(i)$ determine the structure of the uncertainty. If the zero solution of the nominal system (6.87) is ESMS we shall analyze if the zero solution of the perturbed system (6.86) remains ESMS for $\Delta(i) \neq 0$. This is a primary formulation of the robust stability with respect to structured linear uncertainty for a stochastic system. For a more precise formulation we shall introduce a norm in the set of uncertainties. If $\Delta = (\Delta(1), \ldots, \Delta(d)) \in \mathcal{M}_{mp}^d$, one defines

$$|\Delta| = \max\{|\Delta(i)|, \, i \in \mathcal{D}\} = \max_{i \in \mathcal{D}} \sqrt{\lambda_{\max}(i)},$$

where $\lambda_{\max}(i)$ is the largest eigenvalue of the matrix $\Delta^*(i)\Delta(i)$.

As a measure of the stability robustness we introduce the *stability radius* with respect to linear structured uncertainty.

Definition 1. *The stability radius of the pair* $(A_0, \ldots, A_r; Q)$ *with respect to the structure of linear uncertainty described by* $(B_0, \ldots, B_r; C)$ *is the number*

$$\rho_L(\mathbf{A}, Q \mid \mathbf{B}, C) = \inf\{\rho > 0 \mid \exists \; \Delta = (\Delta(1), \ldots, \Delta(d)) \in \mathcal{M}_{mp}^d$$
$$\text{with } |\Delta| \leq \rho \text{ for which the zero solution of the corresponding}$$
$$\text{system of type (6.86) is not ESMS}\}.$$

The result stated in Corollary 20 allows us to obtain a lower bound of the stability radius defined above. To this end, let us introduce the fictitious system:

$$dx(t) = [A_0(\eta(t))x(t) + B_0(\eta(t))u(t)]\,dt$$
$$+ \sum_{k=1}^{r}[A_k(\eta(t)\,x(t) + B_k(\eta(t))u(t)]\,dw_k(t), \qquad (6.88)$$
$$y(t) = C(\eta(t)x(t))$$

with the known matrices of the perturbed system (6.86).

Corollary 21. *Assume that the zero solution of the nominal system (6.87) is ESMS. Let*

$$\mathcal{T} : L_{\eta,w}^2([0, \infty), \mathbf{R}^m) \to L_{\eta,w}^2([0, \infty), \mathbf{R}^p)$$

be the input–output operator associated with the fictious system (6.88). Then

$$\rho_L(\mathbf{A}, Q \mid \mathbf{B}, C) \geq \|\mathcal{T}\|^{-1}. \qquad (6.89)$$

Proof. Let $\rho < \|T\|^{-1}$ be an arbitrarily fixed number. We show that for any $\Delta \in \mathcal{M}_{mp}^d$ with $|\Delta| < \rho$ the zero solution of the perturbed system (6.86) is ESMS. Let Δ with $|\Delta| < \rho < \|T\|^{-1}$. Denoting $\gamma = \rho^{-1}$, we have $\|T\| < \gamma$ and $|\Delta| < \gamma^{-1}$. Applying the result of Corollary 20 one deduces that the zero solution of the system (6.86) is ESMS for the considered perturbation Δ. Therefore $\rho_L (\mathbf{A}, \mathbf{Q} \mid \mathbf{B}, \mathbf{C}) \geq \rho$. Since ρ is arbitrary it follows that (6.89) holds and thus the proof is complete. $\square$

At the end of this subsection we shall show that certain structures of the linear uncertainty frequently used in the literature can be embedded in the general form of the system (6.86).

Consider first the perturbed system

$$dx(t) = \left[A_0(\eta(t)) + \widehat{B}_0(\eta(t))\Delta_0(\eta(t))C(\eta(t))\right] x(t)dt \qquad (6.90)$$

$$+ \sum_{k=1}^{r} \left[A_k \left(\eta(t) + \widehat{B}_k(\eta(t))\Delta_k(\eta(t))C(\eta(t))\right] x(t)dw_k(t),$$

where $A_k(i) \in \mathbf{R}^{n \times n}$, $\widehat{B}_k(i) \in \mathbf{R}^{n \times m_k}$, $0 \leq k \leq r$, $i \in \mathcal{D}$, are known and $\Delta_k(i) \in \mathbf{R}^{m_k \times p}$, $0 \leq k \leq r$, $i \in \mathcal{D}$, are assumed unknown. In order to show that the system (6.90) is in fact a particular case of the system (6.86), we define $B_k(i) \in \mathbf{R}^{n \times m}$, $m = \sum_{k=0}^{r} m_k$ as follows:

$$B_0(i) = \left[\widehat{B}_0(i) \quad 0 \quad \cdots \quad 0\right],$$

$$B_k(i) = \left[0 \quad 0 \quad \cdots \widehat{B}_k(i) \quad \cdots 0\right], \qquad (6.91)$$

$$1 \leq k \leq r, \ i \in \mathcal{D}, \ \Delta(i) = \begin{bmatrix} \Delta_0(i) \\ \vdots \\ \Delta_r(i) \end{bmatrix}.$$

With these notations the system (6.90) can be rewritten in the equivalent form (6.86). Further we have

$$|\Delta(i)|^2 = \lambda_{\max}\left[\Delta^*(i)\Delta(i)\right] = \lambda_{\max}\left[\sum_{k=0}^{r} \Delta_k^*(i)\Delta_k(i)\right].$$

Another interesting structure of perturbations is the situation when

$$dx(t) = \left[A_0(\eta(t)) + \widehat{B}_0(\eta(t))\Delta_0(\eta(t))\widehat{C}_0(\eta(t))\right] x(t)dt \qquad (6.92)$$

$$+ \sum_{k=1}^{r} \left[A_k(\eta(t)) + \widehat{B}_k(\eta(t))\Delta_k(\eta(t))\widehat{C}_k(\eta(t))\right] x(t)dw_k(t),$$

where $A_k(i) \in \mathbf{R}^{n \times n}$, $\widehat{B}_k(i) \in \mathbf{R}^{n \times m_k}$, $\widehat{C}_k(i) \in \mathbf{R}^{p_k \times n}$, $0 \leq k \leq r$, $i \in \mathcal{D}$, are assumed known and $\Delta_k(i) \in \mathbf{R}^{m_k \times p_k}$, $0 \leq k \leq r$, $i \in \mathcal{D}$, are unknown matrices describing the modeling uncertainty. Define $B_k(i) \in \mathbf{R}^{n \times m}$, $m = \sum_{k=0}^{r} m_k$ as in (6.91):

$$C(i) \in \mathbf{R}^{p \times n}, \quad p = \sum_{k=0}^{r} p_k, \quad C(i) = \begin{bmatrix} \widehat{C}_0(i) \\ \vdots \\ \widehat{C}_r(i) \end{bmatrix},$$

and $\Delta(i) = diag(\Delta_0, \ldots, \Delta_r(i))$.

With these notations the system (6.92) can be written in (6.86) form. Obviously we have

$$|\Delta(i)|^2 = \lambda_{\max} \left[\Delta^*(i)\Delta(i) \right] = \max_{0 \le k \le r} \lambda_{\max} \left[\Delta_k^*(i)\Delta_k(i) \right]$$
$$= \max_{0 \le k \le r} |\Delta_k(i)|^2.$$

6.3.3 Robust stability with respect to a class of nonlinear uncertainty

In this section we shall consider the case when a stochastic linear system is subjected to a class of nonlinear uncertainty. We shall also define the stability radius and provide an estimation of its lower bound.

Consider the system

$$dx(t) = [A_0(\eta(t))x(t) + B_0(\eta(t))\Delta(t, y(t), \eta(t))] \, dt \qquad (6.93)$$
$$+ \sum_{k=1}^{r} [A_k(\eta(t)x(t) + B_k(\eta(t))\Delta(t, y(t), \eta(t))] \, dw_k(t),$$

$$y(t) = C(\eta(t))x(t),$$

where $A_k(i) \in \mathbf{R}^{n \times n}$, $B_k(i) \in \mathbf{R}^{n \times m}$, $0 \le k \le r$, $C(i) \in \mathbf{R}^{p \times n}$ are assumed known and $\Delta : \mathbf{R}_+ \times \mathbf{R}^p \times \mathcal{D} \to \mathbf{R}^m$ are functions with the following properties:

(i) For any $i \in \mathcal{D}$, $(t, y) \to \Delta(t, y, i)$ is a continuous function on $\mathbf{R}_+ \times \mathbf{R}^p$ and $\Delta(t, 0, i) = 0$ for all $t \ge 0$.

(ii) For every $\tau > 0$ there exists $\nu(\tau) > 0$ such that

$$|\Delta(t, y_1, i) - \Delta(t, y_2, i)| \le \nu(\tau) |y_1 - y_2|$$

for all $t \in [0, \tau]$, $y_1, y_2 \in \mathbf{R}^p, i \in \mathcal{D}$.

(iii) There exists $\delta > 0$ such that $|\Delta(t, y, i)| \le \delta |y| \ \forall (t, y, i) \in \mathbf{R}_+ \times \mathbf{R}^p \times \mathcal{D}$.

In this section we shall denote by $\mathbf{\Delta}$ the set of all functions $\Delta : \mathbf{R}_+ \times \mathbf{R}^p \times \mathcal{D} \to \mathbf{R}^m$ satisfying the above conditions. Let us notice that both constants $\nu(\tau)$ and δ in (ii) and in (iii) depend on the function $\Delta(\cdot, \cdot) \in \mathbf{\Delta}$.

For every Δ in $\mathbf{\Delta}$ denote

$$\|\Delta\| = \sup \left\{ \frac{|\Delta(t, y, i)|}{|y|}; \ t \ge 0, \ y \ne 0, \ i \in \mathcal{D} \right\}. \qquad (6.94)$$

Let $\mathcal{X}_{t_0}$ be the set of all random n-dimensional $\mathcal{H}_{t_0}$-measurable vectors ξ which additionally satisfy $E|\xi|^2 < \infty$. It is obvious that $\mathbf{R}^n \subset \mathcal{X}_{t_0} \ \forall t_0 \ge 0$. For every

$t_0 \geq 0$, $\xi \in \mathcal{X}_{t_0}$, and $\Delta \in \mathbf{\Delta}$, denote by $x_\Delta(t, t_0, \xi)$ the solution of the perturbed system (6.93) satisfying the initial condition $x_\Delta(t_0, t_0, \xi) = \xi$. Applying Theorem 36 of Chapter 1, one deduces that $x_\Delta(\cdot, t_0, \xi) \in L^2_{\eta, w}([t_0, T], \mathbf{R}^n)$ for every $T > t_0$. Moreover, if $E|\xi|^{2b} < \infty$, $b \geq 1$, then

$$\sup_{t_0 \leq t \leq T} \left\{ E\left[|x_\Delta(t, t_0, \xi)|^{2b} \mid \eta(t_0) = i\right] \right\} \leq K\left(1 + E\left[|\xi|^{2b} \mid \eta(t_0) = i\right]\right),$$

where K depends on T and on $T - t_0$.

Definition 2. *The zero solution of the perturbed system* (6.93) *is called* exponentially stable in mean square (ESMS) *if there exist* $\alpha > 0$ *and* $\beta \geq 1$ *such that*

$$E\left[|x_\Delta(t, t_0, x_0)|^2 \mid \eta(t_0) = i\right] \leq \beta e^{-\alpha(t - t_0)}|x_0|^2$$

for any $t \geq t_0 \geq 0$, $x_0 \in \mathbf{R}^n$, $i \in \mathcal{D}$.

The constant α, β of the above definition may depend on the perturbation $\Delta \in \mathbf{\Delta}$, but they do not depend on t, t_0, x_0.

In order to characterize the robustness of the nominal system (6.87) with respect to the nonlinear perturbations $\Delta \in \mathbf{\Delta}$, we introduce the following definition.

Definition 3. *The robustness radius* with respect to nonlinear stochastic uncertainty *which structure is determined by* $\mathbf{B} = (B_0, \dots, B_r)$ *and* C, *is given by*

$$\rho_{NL}(\mathbf{A}, Q \mid \mathbf{B}, C) = \inf \{\rho > 0 \mid \exists \; \Delta \in \mathbf{\Delta} \text{ with } \|\Delta\| \leq \rho$$
$$\text{for which the zero solution of the}$$
$$\text{system (6.93) is not ESMS}\}.$$

Remark 12. Since the class of uncertainty $\mathbf{\Delta}$ also includes the functions $\Delta(t, y, i) = \Delta(i)y$ modeling the linear uncertainty considered in the previous section, it is easy to check that
$$\rho_{NL}(\mathbf{A}, Q \mid \mathbf{B}, C) \leq \rho_L(\mathbf{A}, Q \mid \mathbf{B}, C).$$

In order to prove the main result of this section, two additional results are required.

Lemma 22. *Let* $\varphi : \mathbf{R}^n \times \Omega \to \mathbf{R}_+$ *be measurable with respect to* $\mathcal{B}(\mathbf{R}^n) \otimes \mathcal{R}_t$ *and* $g : \Omega \to \mathbf{R}^n$ *be measurable with respect to* $\mathcal{H}_t$, $t \geq 0$, *being fixed, where* $\mathcal{R}_t$ *and* $\mathcal{H}_t$ *are as defined in Chapter 1. Let*

$$h(x, i) = E[\varphi(x, \cdot) \mid \eta(t) = i] \; \forall x \in \mathbf{R}^n, \; i \in \mathcal{D}, \; \text{and} \; \hat{\varphi}(\omega) = \varphi(g(\omega), \omega).$$

If $\hat{\varphi}(\cdot)$ *and* $\varphi(x, \cdot)$ *are integrable, then*

$$h(g(\omega), \eta(t, \omega)) = E[\hat{\varphi} \mid \mathcal{H}_t](\omega) \quad a.s. \tag{6.95}$$

Proof. We first prove (6.95) for the case when $\varphi(x, \omega) = \varphi_1(X)\varphi_2(\omega)$, with $\varphi_1(X) \geq 0$ measurable with respect to $\mathcal{B}(\mathbf{R}^n)$ and bounded and $\varphi_2(\cdot) \geq 0$, $\mathcal{R}_t$-measurable and bounded. From Theorem 34 of Chapter 1 one obtains

$$E[\varphi_2 \mid \mathcal{H}_t] = E[\varphi_2 \mid \eta(t)] \quad a.s.$$

Therefore

$$E\left[\hat{\varphi} \mid \mathcal{H}_t\right](\omega) = E\left[\varphi_1(g)\varphi_2 \mid \mathcal{H}_t\right](\omega)$$
$$= \varphi_1(g(\omega)) E\left[\varphi_2 \mid \mathcal{H}_t\right](\omega)$$
$$= \varphi_1(g(\omega)) E\left[\varphi_2 \mid \eta(t)\right](\omega).$$

On the other hand

$$h(x, \eta(t, \omega)) = E\left[\varphi_1(x)\varphi_2 \mid \eta(t)\right](\omega)$$
$$= \varphi_1(x) E\left[\varphi_2 \mid \eta(t)\right](\omega),$$

and then

$$h(g(\omega), \eta(t, \omega)) = \varphi_1(g(\omega)) E\left[\varphi_2 \mid \eta(t)\right](\omega) \quad \text{a.s.},$$

which shows that (6.95) is true for the special considered case.

Further, let

$$\mathcal{M} = \left\{A \in \mathcal{B}(\mathbf{R}^n) \otimes \mathcal{R}_t \mid \chi_A \text{ verifies (6.95)}\right\},$$
$$\mathcal{C} = \left\{U \times S \mid U \in \mathcal{B}(\mathbf{R}^n), S \in \mathcal{N}_t\right\}.$$

Since $\chi_{U \times S}(x, \omega) = \chi_U(x)\chi_S(\omega)$ it follows that $\mathcal{C} \subset \mathcal{M}$. One can easily verify that $\mathcal{C}$ is a π-system and $\mathcal{M}$ satisfies the conditions (i), (ii), (iii) of Theorem 1 of Chapter 1. Thus it results that $\mathcal{M}$ contains $\sigma[\mathcal{C}]$, $\sigma[\mathcal{C}]$ denoting the smallest σ-algebra containing $\mathcal{C}$, namely $\sigma[\mathcal{C}] = \mathcal{B}(\mathbf{R}^n) \otimes \mathcal{R}_t$. It results that (6.95) is verified by any $A \in \mathcal{B}(\mathbf{R}^n) \otimes \mathcal{R}_t$. Further, let $0 \le \varphi_k \le \varphi_{k+1} \le \varphi$, $\varphi_k(x, \omega)$ being a measurable function with respect to $\mathcal{B}(\mathbf{R}^n) \otimes \mathcal{R}_t$, $\varphi_k(x, \omega) \to \varphi(x, \omega) \, \forall x, \, \omega$. Since (6.95) is true for φ_k, from Legesgue's Theorem (see Theorem 11 of Chapter 1) one obtains that this relation is also true for a function φ verifying the assumptions in the statement, and therefore the proof is complete. □

Now consider the nonlinear system of stochastic nonlinear differential equations:

$$dx(t) = F_0(t, x(t), \eta(t))dt + \sum_{k=1}^{r} F_k(t, x(t), \eta(t))dw_k(t), \tag{6.96}$$

where the functions $F_k : \mathbf{R}_+ \times \mathbf{R}^n \times \mathcal{D} \to \mathbf{R}^n$ have the following properties:

(i) $(t, x) \to F_k(t, x, i) : \mathbf{R}_+ \times \mathbf{R}^n \to \mathbf{R}^n$ are continuous functions and $F_k(t, 0, i) = 0$, $t \ge 0, i \in \mathcal{D}$, $0 \le k \le r$.

(ii) For any $\tau > 0$ there exists $\nu(\tau) > 0$ such that

$$|F_k(t, x_1, i) - F_k(t, x_2, i)| \le \nu(\tau)|x_1 - x_2|, \, i \in \mathcal{D}, \, 0 \le k \le r,$$

$\forall x_1, x_2 \in \mathbf{R}^n, \, t \in [0, \tau]$.

(iii) There exists $\delta > 0$ such that

$$|F_k(t, x, i)| \le \delta|x|, \, \forall t \ge 0, x \in \mathbf{R}^n, \, i \in \mathcal{D}, \, 0 \le k \le r.$$

It is obvious that for any $\Delta \in \Delta$ the perturbed system (6.93) satisfies the conditions (i), (ii), and (iii). Applying Theorem 36 of Chapter 1 it follows that for any $t_0 \ge 0$ and $\xi \in \mathcal{X}_{t_0}$ the system (6.96) has a unique solution $x(t, t_0, \xi), \, t \ge 0$, such that $x(t_0, t_0, \xi) = \xi_0$.

Definition 4. *The zero solution of the system* (6.96) *is ESMS if there exist* $\alpha > 0$, $\beta > 0$ *such that*

$$E\left[|x(t, t_0, \xi)|^2 \mid \eta(t_0) = i\right] \leq \beta e^{-\alpha(t-t_0)}|\xi|^2,$$

$\forall t \geq t_0 \geq 0$, $\xi \in \mathbf{R}^n$, $i \in \mathcal{D}$.

The next result extends to the nonlinear case some results proved in Chapter 2 for the linear case.

Theorem 23. *The following assertions are equivalent:*
(i) *The zero solution of the system* (6.96) *is ESMS.*
(ii) *There exists* $c > 0$ *such that*

$$\int_t^\infty E\left[|x(s, t, \xi)|^2 \mid \eta(t) = i\right] ds \leq c|\xi|^2 \tag{6.97}$$

$\forall t \geq 0$, $\xi \in \mathbf{R}^n$, *the constant c being independent of t and ξ.*
(iii) *There exist* $\alpha > 0$ *and* $\beta \geq 1$ *such that*

$$E\left[|x(t, t_0, \xi)|^2 \mid \eta(t_0) = i\right] \leq \beta e^{-\alpha(t-t_0)} E\left[|\xi|^2 \mid \eta(t_0) = i\right],$$

$\forall t \geq t_0 \geq 0$, $\xi \in \mathcal{X}_{t_0}$, $i \in \mathcal{D}$.

Proof. (i) $\Rightarrow$ (ii) and (iii) $\Rightarrow$ (i) are obvious. We prove that (ii) $\Rightarrow$ (iii). Define

$$v(t, x, i) = \int_t^\infty h(s, t, x, i)ds,$$

where

$$h(s, t, x, i) = E\left[|x(s, t, x, \cdot)|^2 \mid \eta(t) = i\right]$$

with $s \geq t \geq 0$, $x \in \mathbf{R}^n$, $i \in \mathcal{D}$. By virtue of Theorem 38 of Chapter 1 we can apply Lemma 22 for the function $\varphi(x, \omega) = |x(s, t, x, \omega)|^2 \,\forall (x, \omega) \in \mathbf{R}^n \times \Omega$, where $s \geq t$ are fixed and for the function $g(\omega) = x(t, t_0, \xi, \omega)$ with $t \geq t_0$, $\xi \in \mathcal{X}_{t_0}$ fixed. Therefore one obtains that

$$h(s, t, x(t, t_0, \xi, \omega), \eta(t, \omega)) = E\left[|x(s, t, x(t, t_0, \xi, \omega), \omega)|^2 \mid \mathcal{H}_t\right]$$
$$= E\left[|x(s, t_0, \xi, \omega)|^2 \mid \mathcal{H}_t\right]. \tag{6.98}$$

In the following we shall omit to write the argument ω explicitly. Define

$$v_i(t) = E\left[v(t, x(t, t_0, \xi), \eta(t)) \mid \eta(t_0) = i\right].$$

From (6.97) one deduces

$$v_i(t) \leq cE\left[|x(t, t_0, \xi)|^2 \mid \eta(t_0) = i\right]. \tag{6.99}$$

Further, from (6.98) one obtains

$$v_i(t) = E\left[\int_t^\infty h(s, t, x(t, t_0, \xi), \eta(t)) ds \mid \eta(t_0) = i\right]$$
$$= E\left[\int_t^\infty E\left[|x(s, t_0, \xi)|^2 \mid \mathcal{H}_t\right] ds \mid \eta(t_0) = i\right],$$

from which, using the properties of conditional mean values, it immediately follows that

$$v_i(t) = \int_t^\infty E\left[|x(s, t_0, \xi)|^2 \mid \eta(t_0) = i\right] ds, \tag{6.100}$$

$\forall t \geq t_0 \geq 0$, $\xi \in \mathcal{X}_{t_0}$, $i \in \mathcal{D}$. From (6.100) it follows that the function $t \mapsto v_i(t)$ is absolutely continuous on $[t_0, \infty)$, and therefore it is derivable a.e. on $[t_0, \infty)$, and then from (6.100) one obtains that

$$\frac{d}{dt} v_i(t) = -E\left[|x(t, t_0, \xi)|^2 \mid \eta(t_0) = i\right].$$

Based on (6.99) it results that

$$\frac{d}{dt} v_i(t) \leq -\frac{1}{c} v_i(t) \quad \text{a.e.} \quad t \geq t_0. \tag{6.101}$$

Applying Theorem 35 of Chapter 1 to the function $|x|^2$ and to the system (6.96), one obtains

$$E\left[|x(t, t_0, \xi)|^2 \mid \eta(t_0) = i\right] - E\left[|\xi|^2 \mid \eta(t_0) = i\right]$$

$$= E\left[\int_{t_0}^t \left\{ 2x^*(s, t_0, \xi) F_0(s, x(s, t_0, \xi), \eta(s)) \right.\right.$$

$$+ \sum_{k=1}^r |F_k(s, x(s, t_0, \xi), \eta(s))|^2$$

$$\left.\left. + \sum_{j=1}^d q_{\eta(s),j} |x(s, t_0, \xi)|^2 \right\} ds \mid \eta(t_0) = i\right]. \tag{6.102}$$

Taking into account (jjj) one obtains that

$$\left| 2x^* F_0(t, x, i) + \sum_{k=1}^r |F_k(t, x, i)|^2 \right| \leq \delta_0 |x|^2, \tag{6.103}$$

where $\delta_0 = \delta(2 + r\delta)$. Hence

$$2x^* F_0(t, x, i) + \sum_{k=1}^r |F_k(t, x, i)|^2 \geq -\delta_0 |x|^2.$$

Denoting

$$g_i(t) = E\left[|x(t, t_0, \xi)|^2 \mid \eta(t_0) = i\right],$$

from (6.102), $g_i(\cdot)$ is an absolute continuous function on $[t_0, \infty)$ and

$$\frac{d}{dt} g_i(t) = E\left[2x^*(t, t_0, \xi) F_0(t, x(t, t_0, \xi), \eta(t)) + \sum_{j=1}^d q_{\eta(t),j} |x(t, t_0, \xi)|^2 \right.$$

$$\left. + \sum_{k=1}^r |F_k(t, x(t, t_0, \xi), \eta(t))|^2 \mid \eta(t_0) = i\right].$$

Using (6.103) one obtains that there exists $\delta_1 > 0$ such that

$$\frac{d}{dt} g_i(t) \geq -\delta_1 g_i(t) \quad \text{a.e.,} \ t \geq t_0,$$

which is equivalent to

$$\frac{d}{dt} \left[g_i(t) e^{\delta_1 t} \right] \geq 0,$$

which leads to

$$E \left[|x(t, t_0, \xi)|^2 \mid \eta(t_0) = i \right] \geq e^{-\delta_1(t-t_0)} E \left[|\xi|^2 \mid \eta(t_0) = i \right],$$

$t \geq t_0 \geq 0$, $\xi \in \mathcal{X}_{t_0}$, $i \in \mathcal{D}$. From the last inequality one immediately obtains

$$h(s, t, x, i) \geq e^{-\delta_1(s-t)} |x|^2$$

for all $s \geq t \geq 0$, $x \in \mathbf{R}^n$, $i \in \mathcal{D}$. Therefore, $v(t, x, i) \geq \delta_1^{-1} |x|^2$, $t \geq 0$, $x \in \mathbf{R}^n$, $i \in \mathcal{D}$,

$$v_i(t) \geq \delta_1^{-1} E \left[|x(t, t_0, \xi)|^2 \mid \eta(t_0) = i \right].$$

From the above inequality and from (6.99) and (6.101) one obtains directly

$$E \left[|x(t, t_0, \xi)|^2 \mid \eta(t_0) = i \right] ds \leq \beta e^{-\alpha(t-t_0)} E \left[|\xi|^2 \mid \eta(t_0) = i \right]$$

with $\beta = \delta_1 c$ and $\alpha = 1/c$, and thus the proof is complete. □

Before proving the main result of this section, let us notice that using the known constant matrices $A_k(i)$, $B_k(i)$, and $C(i)$ of the realization of the perturbed system (6.93), one can associate the following auxiliary system:

$$dx(t) = [A_0(\eta(t))x(t) + B_0(\eta(t))u(t)] \, dt$$

$$+ \sum_{k=1}^{r} [A_k(\eta(t))x(t) + B_k(\eta(t))u(t)] \, dw_k(t), \qquad (6.104)$$

$$y(t) = C(\eta(t))x(t).$$

Then we have the following theorem.

Theorem 24. *Assume that the system* $(A_0, \dots, A_r; Q)$ *is stable. Then*

$$\rho_{NL} \{A; Q \mid B, C\} \geq \|T\|^{-1},$$

where

$$T : L^2_{\eta, w}([0, \infty), \mathbf{R}^m) \to L^2_{\eta, w}([0, \infty), \mathbf{R}^p)$$

is the input–output operator associated with the auxiliary system (6.104) defined by the matrices $A_k(i)$, $B_k(i)$, *and* $C(i)$, $0 \leq k \leq r$, $i \in \mathcal{D}$.

Proof. We show that for every $\rho < \|T\|^{-1}$ and for all $\Delta \in \mathbf{\Delta}$ with $\|\Delta\| < \rho$, the zero solution of the perturbed system (6.93) is ESMS. Denoting $\gamma = \rho^{-1}$ it follows that $\|T\| < \gamma$ and $\|\Delta\| < \gamma^{-1}$, or

$$\sup\left\{\frac{|\Delta(t, y, i)|}{|y|}; \ t \geq 0, \ y \neq 0, \ i \in \mathcal{D}\right\} < \gamma^{-1}. \tag{6.105}$$

Using the implication (i) $\Rightarrow$ (iii) of Theorem 10 one deduces that the equation

$$A_0^*(i)X(i) + X(i)A_0(i) + \sum_{k=1}^{r} A_k^*(i)X(i)A_k(i) + \sum_{j=1}^{d} q_{ij}X(j)$$

$$+ \left[X(i)B_0(i) + \sum_{k=1}^{r} A_k^*(i)X(i)B_k(i)\right]\left[\gamma^2 I_m - \sum_{k=1}^{r} B_k^*(i)X(i)B_k(i)\right]^{-1}$$

$$\times \left[B_0^*(i)X(i) + \sum_{k=1}^{r} B_k^*(i)X(i)A_k(i)\right] + C^*(i)C(i) = 0 \tag{6.106}$$

has a stabilizing solution $\tilde{X} = (\tilde{X}(1), \ldots, \tilde{X}(d)) \geq 0$ such that

$$\gamma^2 I_m - \sum_{k=1}^{r} B_k^*(i)X(i)B_k(i) > 0 \tag{6.107}$$

for any $i \in \mathcal{D}$. Applying the Itô-type formula for the function $x^*\tilde{X}(i)x$ and for the process $x(t) = x_\Delta(t, t_0, x_0)$, one obtains, using (6.106), that

$$E\left[\int_{t_0}^{\tau} \{|y(t)|^2 - \gamma^2 |\Delta(t, y(t), \eta(t))|^2\} \, dt \mid \eta(t_0) = i\right]$$

$$= x_0^*\tilde{X}(i)x_0 - E\left[x^*(\tau)\tilde{X}(\eta(\tau))x(\tau) \mid \eta(t_0) = i\right]$$

$$- E\left[\int_{t_0}^{\tau} \left(\Delta(t, y(t), \eta(t)) - \tilde{F}(\eta(t))x(t)\right)^* \right. \tag{6.108}$$

$$\times \left(\gamma^2 I_m - \sum_{k=1}^{r} B_k^*(\eta(t))\tilde{X}(\eta(t))B_k(\eta(t))\right)$$

$$\left. \times \left(\Delta(t, y(t), \eta(t)) - \tilde{F}(\eta(t))x(t)\right) dt \mid \eta(t_0) = i\right],$$

where $y(t) = C(\eta(t))x(t)$, $t \geq t_0$, and $\tilde{F}(i)$ denotes the stabilizing feedback associated with the solution $\tilde{X}(i)$, $i \in \mathcal{D}$. Taking into account (6.107), it follows that:

$$E\left[\int_{t_0}^{\tau} \{|y(t)|^2 - \gamma^2 |\Delta(t, y(t), \eta(t))|^2\} \, dt \mid \eta(t_0) = i\right] \leq x_0^*\tilde{X}(i)x_0,$$

for any $\tau \geq t_0 \geq 0$, $x_0 \in \mathbf{R}^n$, $i \in \mathcal{D}$, which leads to

$$E\left[\int_{t_0}^{\infty} \{|y(t)|^2 - \gamma^2 |\Delta(t, y(t), \eta(t))|^2\} \, dt \mid \eta(t_0) = i\right] \tag{6.109}$$

$$\leq \tilde{\delta}|x_0|^2, \quad \forall t_0 \geq 0, \ x \in \mathbf{R}^n, \ i \in \mathcal{D}.$$

But

$$|\Delta(t, y(t), \eta(t))| \le \|\Delta\| |y(t)|,$$

$\forall t \ge 0$, $i \in \mathcal{D}$, $y \in \mathbf{R}^p$. On the other hand, (6.105) gives $1 - \gamma^2 \|\Delta\|^2 > 0$, and then we deduce from (6.109) that

$$E\left[\int_{t_0}^{\infty} |y(t)|^2 dt \mid \eta(t_0) = i\right] \le \frac{\tilde{\delta}}{(1 - \gamma^2)\|\Delta\|^2}|x_0|^2, \tag{6.110}$$

$\forall t_0 \ge 0$, $x_0 \in \mathbf{R}^n$, $i \in \mathcal{D}$. Finally, applying Theorem 32 of Chapter 2 and using (6.110) one obtains

$$E\left[\int_{t_0}^{\infty} |x_\Delta(t, t_0, x_0)|^2 dt \mid \eta(t_0) = i\right] \le c|x_0|^2,$$

$\forall t_0 \ge 0$, $x_0 \in \mathbf{R}^n$, $i \in \mathcal{D}$, $c > 0$, being independent of t_0, x_0, i. Applying Theorem 23 we obtain that the zero solution of the perturbed system (6.93) is ESMS. Therefore $\rho_{NL}\{A; Q \mid B, C\} \ge \rho$. Since $\rho < \|\mathcal{T}\|^{-1}$ is arbitrary, it follows that $\rho_{NL}\{A; Q \mid B, C\} \ge \|\mathcal{T}\|^{-1}$ and thus the proof is complete. $\qquad\square$

At the end of this section we show that in a particular case of the system (6.93) we can obtain the exact value of the stability radius $\rho_{NL}\{A; Q \mid B, C\}$. To be more precise, consider the perturbed system

$$dx(t) = A_0 x(t)dt + \sum_{k=1}^{r_1-1} A_k x(t)dw_k(t) + \sum_{k=r_1}^{r} B_k \Delta(t, y(t))dw_k(t), \tag{6.111}$$

$$y(t) = Cx(t).$$

The system (6.111) is a perturbation of the nominal system

$$dx(t) = A_0 x(t)dt + \sum_{k=1}^{r_1-1} A_k x(t)dw_k(t) \tag{6.112}$$

and it represents a particular case of the system (6.93), namely $\mathcal{D} = \{1\}$, $A_k = 0$, $r_1 \le k \le r$, $B_k = 0$, $1 \le k \le r_1 - 1$, $q_{11} = 0$. In this particular case, instead of $\rho_{NL}\{A; Q \mid B, C\}$, we shall denote the stability radius by $\rho_{NL}\{A \mid B, C\}$. Then the stability radius is given by the following result.

Theorem 25. *Assume that the zero solution of the nominal system (6.112) is ESMS. Then*

$$\rho_{NL}\{A \mid B, C\} = \tilde{\lambda}^{-\frac{1}{2}}, \tag{6.113}$$

where $\tilde{\lambda}$ denotes the maximal eigenvalue of the matrix $\sum_{k=r_1}^{r} B_k^* \tilde{X} B_k$, $\tilde{X} \ge 0$, denoting the unique solution of the linear Lyapunov-type equation

$$A_0^* X + X A_0 + \sum_{k=1}^{r_1-1} A_k^* X A_k + C^* C = 0. \tag{6.114}$$

Proof. From Corollary 12 with $\mathcal{D} = \{1\}$ one obtains that $\tilde{\lambda}^{\frac{1}{2}} = \|\mathcal{T}\|$, where

$$\mathcal{T} : L^2_w([0, \infty), \mathbf{R}^m) \rightarrow L^2_w([0, \infty), \mathbf{R}^p)$$

is the input–output operator associated with the auxiliary system:

$$dx(t) = A_0 x(t)dt + \sum_{k=1}^{r_1-1} A_k x(t)dw_k(t) + \sum_{k=r_1}^{r} B_k u(t)dw_k(t), \qquad (6.115)$$

$$y(t) = Cx(t).$$

From Theorem 24 it follows that

$$\rho_{NL} \{\mathbf{A} \mid \mathbf{B}, C\} \geq \tilde{\lambda}^{-\frac{1}{2}}. \qquad (6.116)$$

It order to prove (6.113) it is sufficient to show that for any $\varepsilon > 0$ there exists $\Delta_\varepsilon \in \Delta$ with $\|\Delta_\varepsilon\| < \tilde{\lambda}^{-\frac{1}{2}} + \varepsilon$ for which the zero solution of (6.111) is not ESMS. Let $\lambda_\varepsilon \in (\tilde{\lambda}^{-\frac{1}{2}}, \tilde{\lambda}^{-\frac{1}{2}} + \varepsilon)$. Since $\lambda_\varepsilon^{-2} < \tilde{\lambda}$ there exists $u_\varepsilon \in \mathbf{R}^m$ with $|u_\varepsilon| = 1$ and

$$u_\varepsilon^* \left(I_m - \lambda_\varepsilon^2 \sum_{k=r_1}^{r} B_k^* \tilde{X} B_k \right) u_\varepsilon < 0. \qquad (6.117)$$

Let

$$\Delta_\varepsilon(y) = \lambda_\varepsilon u_\varepsilon |y|. \qquad (6.118)$$

Then it is obvious that $\Delta_\varepsilon \in \Delta$ and $\|\Delta_\varepsilon\| = \lambda_\varepsilon$. We show that the zero solution of the system

$$dx(t) = A_0 x(t)dt + \sum_{k=1}^{r_1-1} A_k x(t)dw_k(t) + \sum_{k=r_1}^{r} B_k \Delta_\varepsilon(t, y(t))dw_k(t), \qquad (6.119)$$

$$y(t) = Cx(t)$$

is not ESMS. If the zero solution of (6.119) is ESMS, then there exists $\delta > 0$ such that

$$E \int_{t_0}^{\infty} |Cx(t, t_0, x_0)|^2 \, dt \leq \delta |x_0|^2, \quad \forall t_0 \geq 0, \quad x_0 \in \mathbf{R}^n. \qquad (6.120)$$

On the other hand, applying the Itô-type formula to the function $x^* \tilde{X} x$ and to the system (6.119) and using (6.114), one obtains that

$$E \int_0^\tau \left(|y(t)|^2 - \sum_{k=r_1}^{r} \Delta_\varepsilon^*(y(t)) B_k^* \tilde{X} B_k \Delta_\varepsilon(y(t)) \right) dt \qquad (6.121)$$

$$= x_0^* \tilde{X} x_0 - E \left[x^*(\tau) \tilde{X} x(\tau) \right]$$

$\forall \tau > 0$, $x(t) = x(t, x_0)$ being the solution of (6.119) verifying $x(0, x_0) = x_0$ and $y(t) = Cx(t, x_0)$. If the zero solution of (6.119) is ESMS, then (6.121) gives

$$E \int_0^\infty \left(|y(t)|^2 - \sum_{k=r_1}^r \Delta_\varepsilon^*(y(t)) B_k^* \widetilde{X} B_k \Delta_\varepsilon(y(t)) \right) dt = x_0^* \widetilde{X} x_0 \quad (6.122)$$

$\forall x_0 \in \mathbf{R}^n$. Taking into account (6.118) one obtains that (6.122) becomes

$$u_\varepsilon^* \left(I_m - \lambda_\varepsilon^2 \sum_{k=r_1}^r B_k^* \widetilde{X} B_k \right) u_\varepsilon E \int_0^\infty |y(t)|^2 dt = x_0^* \widetilde{X} x_0, \quad \forall x_0 \in \mathbf{R}^n,$$

which contradicts (6.117), taking $x_0 \neq 0$ such that $x_0^* \widetilde{X} x_0 > 0$ (since $\widetilde{X} \geq 0$, (6.117) implies that there exists $x_0 \in R^n$ such that $x_0^* \widetilde{X} x_0 > 0$). Thus the proof is complete. $\square$

Notes and references

The theoretical developments presented in this chapter are new. They provide a unified approach of the stochastic version of the Bounded Real Lemma and stability radius for systems subjected both to multiplicative white noise and to Markovian jumping. The stochastic version of the Bounded Real Lemma for systems with multiplicative white noise has been studied in [64], [99], [93], [9], and for stochastic systems subjected to Markov perturbations we cite [92]. For the case of stochastic systems subjected to both multiplicative white noise and Markovian jumping, a stochastic version of the Bounded Real Lemma was proved in [33]. The stochastic counterpart of the Small Gain Theorem for systems with multiplicative white noise is given in [40] and [41] for systems subjected to Markov perturbations. As concerns the stability radius for systems with multiplicative white noise, we cite [44], [65], [93], [90], [91], and for systems with Markovian jumping, see [92]. Some estimations for the stability radius in the case of stochastic systems with state multiplicative white noise and Markov jump perturbations are given in [33]. A different approach to estimating the stability radius for systems subjected both to multiplicative white noise and to Markovian jumping can be found in [46].

7

Robust Stabilization of Linear Stochastic Systems

In the present chapter we consider the robust stabilization problem of systems subjected to both multiplicative white noise and Markovian jumps with respect to some classes of parametric uncertainty. As is already known, a wide variety of aspects of the robust stabilization problem can be embedded in a general disturbance attenuation problem which extends the well-known H^∞ control problem in the case of deterministic invariant linear systems. Special attention will be paid in this chapter to the attenuation problem of exogenous perturbations with a specified level of attenuation. At the same time, some particular robust stabilization problems, the solutions of which are derived using the results in the preceding chapter, will be presented. The solution of the general attenuation problem will be given in terms of some linear matrix inequalities, which provide necessary and sufficient solvability conditions.

7.1 Formulation of the disturbance attenuation problem

As shown in the preceding chapter, a measure of the robustness radius of stabilization with respect to a wide class of static or dynamic uncertainty can be characterized using the norm of the input–output operator associated with the nominal system. Based on this fact it follows that in order to achieve a certain level of robustness of stability, one can design a stabilizing controller such that the norm of the input–output operator associated with the resulting system is less than the inverse of the imposed robustness radius.

The design problem of a stabilizing controller such that the norm of the input–output operator is less than a given level of attenuation is usually called in the literature the *disturbance attenuation problem*. In this section the formulation of this problem will be given for the case of the stochastic linear systems considered in the present book.

Consider the following stochastic linear system:

$$dx(t) = [A_0(\eta(t))x(t) + G_0(\eta(t))v(t) + B_0(\eta(t))u(t)]\,dt$$

$$+ \sum_{k=1}^{r}[A_k(\eta(t))x(t) + G_k(\eta(t))v(t) + B_k(\eta(t))u(t)]\,dw_k(t),$$

$$z(t) = C_z(\eta(t))x(t) + D_{zv}(\eta(t))v(t) + D_{zu}(\eta(t))u(t), \qquad (7.1)$$

$$y(t) = C_0(\eta(t))x(t) + D_0(\eta(t))v(t),$$

with two inputs, namely $v(t) \in \mathbf{R}^{m_1}$, $u(t) \in \mathbf{R}^{m_2}$, and two outputs, $z(t) \in \mathbf{R}^{p_1}$, $y(t) \in \mathbf{R}^{p_2}$. The input variable $v(t)$ denotes exogenous signals, $u(t)$ includes the control variables, $z(t)$ is the regulated output, and $y(t)$ denotes the measured output. As usual, the state vector $x(t) \in \mathbf{R}^n$. The coefficients $A_k(i)$, $G_k(i)$, $B_k(i)$, $0 \le k \le r$, $C_z(i)$, $D_{zv}(i)$, $D_{zu}(i)$, $C_0(i)$, $D_0(i)$, $i \in \mathcal{D}$, are known matrices with real coefficients with appropriate dimensions. The stochastic processes $\{\eta(t)\}_{t\ge0}$, $\{w(t)\}_{t\ge0}$, $w(t) = (w_1(t), \ldots, w_r(t))^*$ are defined as in the preceding chapters. The class of admissible controllers is described by the following equations:

$$dx_c(t) = [A_c(\eta(t))x_c(t) + B_c(\eta(t))y(t)]\,dt \qquad (7.2)$$

$$u(t) = C_c(\eta(t))x_c(t) + D_c(\eta(t))y(t),$$

where $x_c \in \mathbf{R}^{n_c}$. In fact, the controller (7.2) is characterized by the set of parameters $\{n_c, A_c(i), B_c(i), C_c(i), D_c(i), i \in \mathcal{D}\}$, where $n_c \ge 0$ is an integer number denoting the order of the controller and $A_c(i) \in \mathbf{R}^{n_c \times n_c}$, $B_c(i) \in \mathbf{R}^{n_c \times p_2}$, $C_c(i) \in \mathbf{R}^{m_2 \times n_c}$, $D_c(i) \in \mathbf{R}^{m_2 \times p_2}$, $i \in \mathcal{D}$. When coupling the controller (7.2) at the system (7.1) one obtains the following resulting system:

$$dx_{cl}(t) = [A_{0cl}(\eta(t))x_{cl}(t) + G_{0cl}(\eta(t))v(t)]\,dt \qquad (7.3)$$

$$+ \sum_{k=1}^{r}[A_{kcl}(\eta(t))x_{cl}(t) + G_{kcl}(\eta(t))v(t)]\,dw_k(t),$$

$$z(t) = C_{cl}(\eta(t))x_{cl}(t) + D_{cl}(\eta(t))v(t),$$

where

$$A_{0cl}(i) = \begin{bmatrix} A_0(i) + B_0(i)D_c(i)C_0(i) & B_0(i)C_c(i) \\ B_c(i)C_0(i) & A_c(i) \end{bmatrix},$$

$$A_{kcl}(i) = \begin{bmatrix} A_k(i) + B_k(i)D_c(i)C_0(i) & B_k(i)C_c(i) \\ 0 & 0 \end{bmatrix}, \; 1 \le k \le r,$$

$$G_{0cl}(i) = \begin{bmatrix} G_0(i) + B_0(i)D_c(i)D_0(i) \\ B_c(i)D_0(i) \end{bmatrix}, \qquad (7.4)$$

$$G_{kcl} = \begin{bmatrix} G_k(i) + B_k(i)D_c(i)D_0(i) \\ 0 \end{bmatrix}, \; 1 \le k \le r,$$

$$C_{cl}(i) = [C_z(i) + D_{zu}(i)D_c(i)C_0(i) \quad D_{zu}(i)C_c(i)],$$

$$D_{cl}(i) = D_{zv}(i) + D_{zu}(i)D_c(i)D_o(i), \; i \in \mathcal{D}.$$

Definition 1. *A controller in the class* (7.2) *is a stabilizing controller of the system* (7.1) *if the zero solution of the system*

$$d\xi(t) = A_{0cl}(\eta(t))\xi(t)dt + \sum_{k=1}^{r} A_{kcl}(\eta(t))\xi(t)dw_k(t)$$

is ESMS.

For every stabilizing controller, define by

$$\mathcal{T}_{cl} : L_{\eta,w}^2 \left([0, \infty); \mathbf{R}^{m_1}\right) \to L_{\eta,w}^2 \left([0, \infty); \mathbf{R}^{p_1}\right)$$

the input–output operator defined by the closed-loop system (7.3), namely:

$$(\mathcal{T}_{cl}v)(t) = C_{cl}(\eta(t))x_{cl}(t, v) + D_{cl}(\eta(t))v(t), \ t \geq 0,$$

$\forall v \in L_{\eta,w}^2 \left([0, \infty); \mathbf{R}^{m_1}\right)$, where $x_{cl}(t, v)$ denotes the solution of the system (7.3) with the initial condition $x_{cl}(0, v) = 0$. As shown in Section 4.1 the input–output operator $\mathcal{T}_{cl}$ is a linear and bounded operator. We are now in the position to formulate the disturbance attenuation problem (DAP) for the system (7.1) with an imposed level of attenuation $\gamma > 0$.

Problem formulation. Given $\gamma > 0$, find necessary and sufficient conditions for the existence of a stabilizing controller for (7.1) such that $\|\mathcal{T}_{cl}\| < \gamma$. If such conditions are fulfilled, give a procedure to determine a controller with the required properties.

Remark 1. Based on the definition of $\|\mathcal{T}_{cl}\|$ it follows that the γ-attenuation problem formulated above is equivalent to

$$\sup_{\substack{v \in L_{\eta,w}^2 \left([0,\infty);\mathbf{R}^{m_1}\right) \\ v \neq 0}} \frac{\|z\|}{\|v\|} < \gamma.$$

7.2 Robust stabilization of linear stochastic systems. The case of full state access

7.2.1 The solution of the disturbance attenuation problem in the case of complete state measurement

Consider the linear stochastic system described by

$$dx(t) = [A_0(\eta(t))x(t) + G_0(\eta(t))v(t) + B_0(\eta(t))u(t)] dt \tag{7.5}$$

$$+ \sum_{k=1}^{r} [A_k(\eta(t))x(t) + G_k(\eta(t))v(t) + B_k(\eta(t))u(t)] dw_k(t),$$

$$z(t) = C_z(\eta(t))x(t) + D_{zv}(\eta(t))v(t) + D_{zu}(\eta(t))u(t),$$

where $x(t) \in \mathbf{R}^n$, $v(t) \in \mathbf{R}^{m_1}$, $u(t) \in \mathbf{R}^{m_2}$, and $z(t) \in \mathbf{R}^{p_1}$ have the same meaning as in the system (7.1). Assume that the whole state vector is available for measurement.

In fact, the system (7.5) is a particular case of (7.1) with $p_2 = n$, $C_0(i) = I_n$, $D_0(i) = 0$, $i \in \mathcal{D}$. The class of admissible controllers is given by (7.2). We shall first solve the disturbance attenuation problem in the case when zero-order controllers are used, namely $n_c = 0$. In this case (7.2) reduces to

$$u(t) = D_c(\eta(t))x(t)$$

or, with standard notation, $u(t) = F(\eta(t))x(t)$, where $F(i) \in \mathbf{R}^{m_2 \times n}$, $i \in \mathcal{D}$. The closed-loop system obtained with this controller is

$$
\begin{aligned}
dx(t) = &\{[A_0(\eta(t)) + B_0(\eta(t))F(\eta(t))]x(t) \\
&+ G_0(\eta(t))v(t)\} dt \\
&+ \sum_{k=1}^{r} \{[A_k(\eta(t)) + B_k(\eta(t))F(\eta(t))]x(t) \\
&\qquad + G_k(\eta(t))v(t)\} dw_k(t), \\
z(t) = &[C_z(\eta(t)) + D_{zu}(\eta(t))F(\eta(t))]x(t) + D_{zv}(\eta(t))v(t).
\end{aligned}
\tag{7.6}
$$

If $F = (F(1), \ldots, F(d))$ is a stabilizing state feedback for the system (7.5) we denote by

$$\mathcal{T}_F : L^2_{\eta,w}\left([0, \infty); \mathbf{R}^{m_1}\right) \rightarrow L^2_{\eta,w}\left([0, \infty); \mathbf{R}^{p_1}\right)$$

the input–output operator associated with (7.6). Therefore the control $u(t) = F(\eta(t))x(t)$ solves the disturbance attenuation problem with the level of attenuation γ if $\|\mathcal{T}_F\| < \gamma$. The following result provides necessary and sufficient conditions for the existence of such state feedback control.

Theorem 1. *For a given $\gamma > 0$ the following are equivalent:*
  (i) *There exists a control $u(t) = F(\eta(t))x(t)$ that stabilizes the system (7.5) and* $\|\mathcal{T}_F\| < \gamma$.
  (ii) *There exist $Y = (Y(1), \ldots, Y(d)) \in \mathcal{S}_n^d$ and $\Gamma = (\Gamma(1), \ldots, \Gamma(d)) \in \mathcal{M}_{m_2,n}^d$, $Y > 0$ satisfying the following system of LMIs:*

$$
\begin{bmatrix}
\mathcal{W}_{0,0}(Y,i) & \mathcal{W}_{0,1}(Y,i) & \cdots & \mathcal{W}_{0,r}(Y,i) & \mathcal{W}_{0,r+1}(Y,i) & \mathcal{W}_{0,r+2}(Y,i) \\
\mathcal{W}_{0,1}^*(Y,i) & \mathcal{W}_{1,1}(Y,i) & \cdots & \mathcal{W}_{1,r}(Y,i) & \mathcal{W}_{1,r+1}(Y,i) & \mathcal{W}_{1,r+2}(Y,i) \\
\vdots & \vdots & \ddots & \vdots & \vdots & \vdots \\
\mathcal{W}_{0,r}^*(Y,i) & \mathcal{W}_{1,r}^*(Y,i) & \cdots & \mathcal{W}_{r,r}(Y,i) & \mathcal{W}_{r,r+1}(Y,i) & \mathcal{W}_{r,r+2}(Y,i) \\
\mathcal{W}_{0,r+1}^*(Y,i) & \mathcal{W}_{1,r+1}^*(Y,i) & \cdots & \mathcal{W}_{r,r+1}^*(Y,i) & \mathcal{W}_{r+1,r+1}(Y,i) & \mathcal{W}_{r+1,r+2}(Y,i) \\
\mathcal{W}_{0,r+2}^*(Y,i) & \mathcal{W}_{1,r+2}^*(Y,i) & \cdots & \mathcal{W}_{r,r+2}^*(Y,i) & \mathcal{W}_{r+1,r+2}^*(Y,i) & \mathcal{W}_{r+2,r+2}(Y,i)
\end{bmatrix} < 0,
\tag{7.7}
$$

$i \in \mathcal{D}$, *where*

$$
\begin{aligned}
\mathcal{W}_{0,0}(Y,i) = &\, A_0(i)Y(i) + Y(i)A_0^*(i) + q_{ii}Y(i) + B_0(i)\Gamma(i) \\
&+ \Gamma^*(i)B_0^*(i) + G_0(i)G_0^*(i), \\
\mathcal{W}_{0,k}(Y,i) = &\, Y(i)A_k^*(i) + \Gamma^*(i)B_k^*(i) + G_0(i)G_k^*(i), \quad 1 \leq k \leq r,
\end{aligned}
$$

$$\mathcal{W}_{0,r+1}(i) = Y(i)C_z^*(i) + \Gamma^*(i)D_{zu}^*(i) + G_0(i)D_{zv}^*(i),$$

$$\mathcal{W}_{0,r+2} = \left[\sqrt{q_{i1}}Y(i)\ldots\sqrt{q_{i,i-1}}Y(i)\ \sqrt{q_{i,i+1}}Y(i)\ldots\sqrt{q_{id}}Y(i)\right]$$

$$\mathcal{W}_{l,k} = G_l(i)G_k^*(i),\ 1 \leq l, k \leq r,\ l \neq k,$$

$$\mathcal{W}_{l,l} = G_l(i)G_l^*(i) - Y(i),\ 1 \leq l \leq r,$$

$$\mathcal{W}_{l,r+1}(i) = G_l(i)D_{zv}^*(i),\ 1 \leq l \leq r,$$

$$\mathcal{W}_{l,r+2}(i) = 0,\ 1 \leq l \leq r+1,$$

$$\mathcal{W}_{r+1,r+1}(i) = D_{zv}(i)D_{zv}^*(i) - \gamma^2 I_{p_1},$$

$$\mathcal{W}_{r+2,r+2}(i) = diag\,(-Y(1)\ldots - Y(i-1) - Y(i+1)\ldots - Y(d)).$$

Moreover, if $(Y, \Gamma) \in \mathcal{S}_n^d \times \mathcal{M}_{m_2,n}^d$ *is a solution of (7.7) with* $Y > 0$, *then the control* $u(t) = F(\eta(t))x(t)$ *with* $F(i) = \Gamma(i)Y^{-1}(i)$ *solves the* γ*-attenuation problem for the system (7.5).*

Proof. The proof immediately follows applying Theorem 10 together with Proposition 14 of Chapter 6 to the system (7.6). □

In the following we display the particular cases when the system (7.5) is subjected only to Markovian jumping or to multiplicative white noise, respectively. Consider the linear stochastic system described by

$$\dot{x}(t) = A_0(\eta(t))x(t) + G_0(\eta(t))v(t) + B_0(\eta(t))u(t), \qquad (7.8)$$

$$z(t) = C_z(\eta(t))x(t) + D_{zv}(\eta(t))v(t) + D_{zu}(\eta(t))u(t)$$

obtained from (7.5), with $A_k(i) = 0$, $G_k(i) = 0$, $B_k(i) = 0$, $1 \leq k \leq r$, and $i \in \mathcal{D}$. For the control $u(t) = F(\eta(t))x(t)$ one obtains the resulting system:

$$\dot{x}(t) = [A_0(\eta(t)) + B_0(\eta(t))F(\eta(t))]\,x(t) + G_0(\eta(t))v(t), \qquad (7.9)$$

$$z(t) = [C_z(\eta(t)) + D_{zu}(\eta(t))F(\eta(t))]\,x(t) + D_{zv}(\eta(t))v(t).$$

Applying Corollary 5 of Chapter 6 for the system (7.9) we get the following corollary.

Corollary 2. *For a given* $\gamma > 0$ *the following are equivalent:*

(i) *There exists a control* $u(t) = F(\eta(t))x(t)$ *stabilizing the system (7.8) such that the input–output operator* T_F *associated with (7.9) verifies* $\|T_F\| < \gamma$.

(ii) *There exist* $Y = (Y(1),\ldots,Y(d)) \in \mathcal{S}_n^d$, $Y(i) > 0$, *and* $\Gamma = (\Gamma(1),\ldots,\Gamma(d)) \in \mathcal{M}_{m_2,n}^d$, *verifying the following system of LMIs:*

$$\begin{bmatrix} \mathcal{W}_{0,0}(Y,i) & \mathcal{W}_{0,r+1}(Y,i) & \mathcal{W}_{0,r+2}(Y,i) \\ \mathcal{W}_{0,r+1}^*(Y,i) & \mathcal{W}_{r+1,r+1}(Y,i) & \mathcal{W}_{r+1,r+2}(Y,i) \\ \mathcal{W}_{0,r+2}^*(Y,i) & \mathcal{W}_{r+1,r+2}^*(Y,i) & \mathcal{W}_{r+2,r+2}(Y,i) \end{bmatrix} < 0, \qquad (7.10)$$

where $\mathcal{W}_{ij}(Y,i)$ *are the same as in (7.7). Moreover, if the pair* $(Y, \Gamma) \in \mathcal{S}_n^d \times \mathcal{M}_{m_2,n}^d$ *is a solution of (7.10) with* $Y(i) > 0$, *then the control* $u(t) = F(\eta(t))x(t)$ *with* $F(i) = \Gamma(i)Y^{-1}(i)$ *solves the* γ*-attenuation problem for the system (7.8).* □

In the case when $\mathcal{D} = \{1\}$ and $q_{11} = 0$ the system (7.5) becomes

$$dx(t) = \left[A_0 x(t) + G_0 v(t) + B_0 u(t)\right] dt$$
$$+ \sum_{k=1}^{r} \left[A_k x(t) + G_k v(t) + B_k u(t)\right] dw_k(t), \tag{7.11}$$
$$z(t) = C_z x(t) + D_{zv} v(t) + D_{zu} u(t).$$

Assuming that the whole state is available for measurement and taking $u(t) = F x(t)$, one obtains the closed-loop system

$$dx(t) = \left[(A_0 + B_0 F) x(t) + G_0 v(t)\right] dt$$
$$+ \sum_{k=1}^{r} \left[(A_k + B_k F) x(t) + G_k v(t)\right] dw_k(t), \tag{7.12}$$
$$z(t) = (C_z + D_{zu} F) x(t) + D_{zv} v(t).$$

Using Corollary 16 of Chapter 6 for the system (7.12) one obtains the following corollary.

Corollary 3. *For a given* $\gamma > 0$ *the following are equivalent:*

(i) *There exists* F *stabilizing the system* (7.11) *such that* $\|\mathcal{T}_F\| < \gamma$, *where* $\mathcal{T}_T$ *denotes the input–output operator associated with* (7.12).

(ii) *There exists* $Y \in \mathcal{S}_n$, $Y > 0$, $\Gamma \in \mathbf{R}^{m_2 \times n}$ *solving the following LMI:*

$$\begin{bmatrix} \mathcal{W}_{0,0}(Y) & \mathcal{W}_{0,1}(Y) & \cdots & \mathcal{W}_{0,r}(Y) & \mathcal{W}_{0,r+1}(Y) \\ \mathcal{W}_{0,1}^*(Y) & \mathcal{W}_{1,1}(Y) & \cdots & \mathcal{W}_{1,r}(Y) & \mathcal{W}_{1,r+1}(Y) \\ \vdots & \vdots & \ddots & \vdots & \vdots \\ \mathcal{W}_{0,r}^*(Y) & \mathcal{W}_{1,r}^*(Y) & \cdots & \mathcal{W}_{r,r}(Y) & \mathcal{W}_{r,r+1}(Y) \\ \mathcal{W}_{0,r+1}^*(Y) & \mathcal{W}_{1,r+1}^*(Y) & \cdots & \mathcal{W}_{r,r+1}^*(Y) & \mathcal{W}_{r+1,r+1}(Y) \end{bmatrix} < 0, \tag{7.13}$$

where

$$\mathcal{W}_{0,0}(Y) = A_0 Y + Y A_0^* + B_0 \Gamma + \Gamma^* B_0^* + G_0 G_0^*,$$
$$\mathcal{W}_{0,k}(Y) = Y A_k^* + \Gamma^* B_k^* + G_0 G_k^*, \ 1 \le k \le r,$$
$$\mathcal{W}_{0,r+1}(Y) = Y C_z^* + \Gamma^* D_{zu}^* + G_0 D_{zv}^*,$$
$$\mathcal{W}_{l,k}(Y) = G_l G_k^*, \ 1 \le l, k \le, \ l \ne k,$$
$$\mathcal{W}_{l,l}(Y) = G_l G_l^* - Y, \ 1 \le l \le r,$$
$$\mathcal{W}_{l,r+1}(Y) = G_l D_{zv}^*, \ 1 \le l \le r,$$
$$\mathcal{W}_{r+1,r+1}(Y) = D_{zv} D_{zv}^* - \gamma^2 I_{p_1}.$$

Moreover, if the pair $(Y, \Gamma) \in \mathcal{S}_n \times \mathbf{R}^{m_2 \times n}$, $Y > 0$, *is a solution of* (7.13), *then the control* $u(t) = \Gamma Y^{-1} x(t)$ *solves the* γ-*attenuation problem for the system* (7.11). $\square$

Now consider a controller in the set (7.2) defined by

$$(n_c, A_c(i), B_c(i), C_c(i), D_c(i); i \in \mathcal{D})$$

with $n_c > 0$, $A_c(i) \in \mathbf{R}^{n_c \times n_c}$, $B_c(i) \in \mathbf{R}^{n_c \times n}$, $C_c(i) \in \mathbf{R}^{m_2 \times n_c}$, $D_c(i) \in \mathbf{R}^{m_2 \times n}$, $i \in \mathcal{D}$. When coupling the controller to the system (7.5), one obtains a resulting system of form (7.3) with the matrix coefficients given by

$$A_{0cl}(i) = \begin{bmatrix} A_0(i) + B_0(i)D_c(i) & B_0(i)C_c(i) \\ B_c(i) & A_c(i) \end{bmatrix},$$

$$A_{kcl}(i) = \begin{bmatrix} A_k(i) + B_k(i)D_c(i) & B_k(i)C_c(i) \\ 0 & 0 \end{bmatrix}, \quad 1 \le k \le r,$$

$$G_{kcl}(i) = \begin{bmatrix} G_k(i) \\ 0 \end{bmatrix}, \quad 0 \le k \le r, \tag{7.14}$$

$$C_{cl}(i) = \begin{bmatrix} C_z(i) + D_{zu}(i)D_c(i) & D_{zu}(i)C_c(i) \end{bmatrix},$$

$$D_{cl}(i) = D_{zv}(i), \quad i \in \mathcal{D}.$$

The next result shows that if the γ-attenuation problem can be solved with a dynamic controller (i.e., $n_c > 0$), then the same problem also has a solution expressed as a state feedback (i.e., $n_c = 0$).

Theorem 4. *For a $\gamma > 0$ the following are equivalent:*

(i) There exists a dynamic controller (7.2) with $n_c > 0$ solving the DAP with the level of attenuation γ.

(ii) There exists a zero-order controller solving the DAP with the same level of attenuation γ.

Proof. (i) $\Rightarrow$ (ii). Assume that there exists a dynamic controller of order $n_c > 0$ solving the γ-attenuation problem for the system (7.5). Therefore this controller stabilizes the system (7.5) and the input–output operator $\mathcal{T}_{cl}$ associated with the closed-loop system verifies the condition $\|\mathcal{T}_{cl}\| < \gamma$. Applying Theorem 10 and Proposition 14 of Chapter 6 for the system (7.3) with the coefficients (7.14), we deduce that there exists $\widetilde{Y} = \left(\widetilde{Y}(1), \dots, \widetilde{Y}(d) \right) \in \mathcal{S}_{n+n_c}^d$, $\widetilde{Y}(i) > 0$, $i \in \mathcal{D}$, satisfying the following system of LMI:

$$\begin{bmatrix} \mathcal{W}_{0,0}(\widetilde{Y},i) & \mathcal{W}_{0,1}(\widetilde{Y},i) & \cdots & \mathcal{W}_{0,r}(\widetilde{Y},i) & \mathcal{W}_{0,r+1}(\widetilde{Y},i) & \mathcal{W}_{0,r+2}(\widetilde{Y},i) \\ \mathcal{W}_{0,1}^*(\widetilde{Y},i) & \mathcal{W}_{1,1}(\widetilde{Y},i) & \cdots & \mathcal{W}_{1,r}(\widetilde{Y},i) & \mathcal{W}_{1,r+1}(\widetilde{Y},i) & \mathcal{W}_{1,r+2}(\widetilde{Y},i) \\ \vdots & \vdots & \ddots & \vdots & \vdots & \vdots \\ \mathcal{W}_{0,r}^*(\widetilde{Y},i) & \mathcal{W}_{1,r}^*(\widetilde{Y},i) & \cdots & \mathcal{W}_{r,r}(\widetilde{Y},i) & \mathcal{W}_{r,r+1}(\widetilde{Y},i) & \mathcal{W}_{r,r+2}(\widetilde{Y},i) \\ \mathcal{W}_{0,r+1}^*(\widetilde{Y},i) & \mathcal{W}_{1,r+1}^*(\widetilde{Y},i) & \cdots & \mathcal{W}_{r,r+1}^*(\widetilde{Y},i) & \mathcal{W}_{r+1,r+1}(\widetilde{Y},i) & \mathcal{W}_{r+1,r+2}(\widetilde{Y},i) \\ \mathcal{W}_{0,r+2}^*(\widetilde{Y},i) & \mathcal{W}_{1,r+2}^*(\widetilde{Y},i) & \cdots & \mathcal{W}_{r,r+2}^*(\widetilde{Y},i) & \mathcal{W}_{r+1,r+2}^*(\widetilde{Y},i) & \mathcal{W}_{r+2,r+2}(\widetilde{Y},i) \end{bmatrix} < 0,$$

$$\tag{7.15}$$

where

$$\mathcal{W}_{0,0}\left(\tilde{Y},i\right) = A_{0cl}(i)\tilde{Y}(i) + \tilde{Y}(i)A^*_{0cl}(i) + q_{ii}\tilde{Y}(i) + G_{0cl}(i)G^*_{0cl}(i),$$

$$\mathcal{W}_{0,k}\left(\tilde{Y},i\right) = \tilde{Y}(i)A^*_{kcl}(i) + G_{0cl}(i)G^*_{kcl}(i),\ 1 \le k \le r,$$

$$\mathcal{W}_{0,r+1}\left(\tilde{Y},i\right) = \tilde{Y}(i)C^*_{cl}(i) + G_{0cl}(i)D^*_{cl}(i),$$

$$\mathcal{W}_{0,r+2}\left(\tilde{Y},i\right) = \left[\sqrt{q_{i1}}\tilde{Y}(i)\dots\sqrt{q_{i,i-1}}\tilde{Y}(i)\ \sqrt{q_{i,i+1}}\tilde{Y}(i)\dots\sqrt{q_{id}}\tilde{Y}(i)\right],$$

$$\mathcal{W}_{l,k}\left(\tilde{Y},i\right) = G_{lcl}(i)G^*_{kcl}(i),\ 1 \le l \ne k \le r,$$

$$\mathcal{W}_{l,l}\left(\tilde{Y},i\right) = G_{lcl}(i)G^*_{lcl}(i) - \tilde{Y}(i),\ 1 \le l \le r,$$

$$\mathcal{W}_{l,r+1}\left(\tilde{Y},i\right) = G_{lcl}(i)D^*_{cl}(i),\ 1 \le l \le r,$$

$$\mathcal{W}_{r+1,r+1}\left(\tilde{Y},i\right) = D_{cl}(i)D^*_{cl}(i) - \gamma^2 I_{p_1},$$

$$\mathcal{W}_{l,r+2}\left(\tilde{Y},i\right) = 0,\ 1 \le l \le r+1,$$

$$\mathcal{W}_{r+2,r+2}\left(\tilde{Y},i\right) = diag\left(-\tilde{Y}(1)\ \dots\ -\tilde{Y}(i-1) - \tilde{Y}(i+1)\ \dots\ -\tilde{Y}(d)\right).$$

Let

$$\tilde{Y}(i) = \begin{bmatrix} \tilde{Y}_{11}(i) & \tilde{Y}_{12}(i) \\ \tilde{Y}^*_{12}(i) & \tilde{Y}_{22}(i) \end{bmatrix},\ i \in \mathcal{D},$$

be the partition of $\tilde{Y}(i)$ conformably with the partition of the matrix coefficients in (7.14), that is, $\tilde{Y}_{11}(i) \in \mathcal{S}_n$, $\tilde{Y}_{22}(i) \in \mathcal{S}_{n_c}$. Define $\Psi \in \mathbf{R}^{\hat{n}\times\tilde{n}}$, $\hat{n} = n(r+d)+p_1$, $\tilde{n} = (n+n_c)(r+d)+p_1$:

$$\Psi^* = diag\left(\underbrace{\Psi_0, \dots, \Psi_0}_{r+1\ \text{times}},\ I_{p_1},\ \underbrace{\Psi_0, \dots, \Psi_0}_{d-1\ \text{times}}\right),$$

where $\Psi_0 = \begin{bmatrix} I_n & 0_{n\times n_c} \end{bmatrix}$. By pre- and postmultiplication of (7.15) by Ψ^* and Ψ, respectively, one obtains the following system of LMIs:

$$\begin{bmatrix} \mathcal{V}_{0,0}\left(\tilde{Y},i\right) & \mathcal{V}_{0,1}\left(\tilde{Y},i\right) & \cdots & \mathcal{V}_{0,r}\left(\tilde{Y},i\right) & \mathcal{V}_{0,r+1}\left(\tilde{Y},i\right) & \mathcal{V}_{0,r+2}\left(\tilde{Y},i\right) \\ \mathcal{V}^*_{0,1}\left(\tilde{Y},i\right) & \mathcal{V}_{1,1}\left(\tilde{Y},i\right) & \cdots & \mathcal{V}_{1,r}\left(\tilde{Y},i\right) & \mathcal{V}_{1,r+1}\left(\tilde{Y},i\right) & \mathcal{V}_{1,r+2}\left(\tilde{Y},i\right) \\ \vdots & \vdots & \ddots & \vdots & \vdots & \vdots \\ \mathcal{V}^*_{0,r}\left(\tilde{Y},i\right) & \mathcal{V}^*_{1,r}\left(\tilde{Y},i\right) & \cdots & \mathcal{V}_{r,r}\left(\tilde{Y},i\right) & \mathcal{V}_{r,r+1}\left(\tilde{Y},i\right) & \mathcal{V}_{r,r+2}\left(\tilde{Y},i\right) \\ \mathcal{V}^*_{0,r+1}\left(\tilde{Y},i\right) & \mathcal{V}^*_{1,r+1}\left(\tilde{Y},i\right) & \cdots & \mathcal{V}^*_{r,r+1}\left(\tilde{Y},i\right) & \mathcal{V}_{r+1,r+1}\left(\tilde{Y},i\right) & \mathcal{V}_{r+1,r+2}\left(\tilde{Y},i\right) \\ \mathcal{V}^*_{0,r+2}\left(\tilde{Y},i\right) & \mathcal{V}^*_{1,r+2}\left(\tilde{Y},i\right) & \cdots & \mathcal{V}^*_{r,r+2}\left(\tilde{Y},i\right) & \mathcal{V}^*_{r+1,r+2}\left(\tilde{Y},i\right) & \mathcal{V}_{r+2,r+2}\left(\tilde{Y},i\right) \end{bmatrix} < 0,$$

$$(7.16)$$

where

$$\mathcal{V}_{0,0}\left(\tilde{Y},i\right) = A_0(i)\tilde{Y}_{11}(i) + \tilde{Y}_{11}(i)A^*_0(i) + q_{ii}\tilde{Y}_{11}(i)$$
$$+ B_0(i)\left(D_c(i)\tilde{Y}_{11}(i) + C_c(i)\tilde{Y}^*_{12}(i)\right)$$
$$+ \left(D_c(i)\tilde{Y}_{11}(i) + C_c(i)Y^*_{12}(i)\right)^* B^*_0(i)$$
$$+ G_0(i)G^*_0(i),$$

$$\mathcal{V}_{0,k}\left(\tilde{Y},i\right) = \tilde{Y}_{11}(i)A_k^*(i) + \left(D_c(i)\tilde{Y}_{11}(i) + C_c(i)\tilde{Y}_{12}^*(i)\right)^* B_k^*(i)$$
$$+ G_0(i)G_k^*(i),\ 1 \le k \le r,$$
$$\mathcal{V}_{0,r+1}\left(\tilde{Y},i\right) = \tilde{Y}_{11}(i)C_z^*(i) + \left(D_c(i)\tilde{Y}_{11}(i) + C_c(i)\tilde{Y}_{12}^*(i)\right)^* D_{zu}^*(i)$$
$$+ G_0(i)D_{zv}^*(i),$$
$$\mathcal{V}_{0,r+2}\left(\tilde{Y},i\right) = \left[\sqrt{q_{i1}}\tilde{Y}_{11}(i)\dots\sqrt{q_{i,i-1}}\tilde{Y}_{11}(i)\ \sqrt{q_{i,i+1}}\tilde{Y}_{11}(i)\dots\sqrt{q_{id}}\tilde{Y}_{11}(i)\right],$$
$$\mathcal{V}_{l,k}\left(\tilde{Y},i\right) = G_l(i)G_k^*(i),\ 1 \le l \ne k \le r,$$
$$\mathcal{V}_{l,l}\left(\tilde{Y},i\right) = G_l(i)G_l^*(i) - \tilde{Y}_{11}(i),\ 1 \le l \le r,$$
$$\mathcal{V}_{l,r+1}\left(\tilde{Y},i\right) = G_l(i)D_{zv}^*(i),\ 1 \le l \le r,$$
$$\mathcal{V}_{r+1,r+1}\left(\tilde{Y},i\right) = D_{zv}(i)D_{zv}^*(i) - \gamma^2 I_{p_1},$$
$$\mathcal{V}_{l,r+2}\left(\tilde{Y},i\right) = 0,\ 1 \le l \le r+1,$$
$$\mathcal{V}_{r+2,r+2}\left(\tilde{Y},i\right) = diag\left(-\tilde{Y}_{11}(1)\dots-\tilde{Y}_{11}(i-1)-\tilde{Y}_{11}(i+1)\dots-\tilde{Y}_{11}(d)\right).$$

One can see that the LMI system (7.16) coincides with the LMI system (7.7) in Theorem 1, with Y replaced by $\tilde{Y}_{11}$ and $\Gamma(i)$ replaced by $D_c(i)\tilde{Y}_{11}(i) + C_c(i)\tilde{Y}_{12}^*(i)$, $i \in \mathcal{D}$. Applying Theorem 1 it follows that there exists a control $u(t) = F(\eta(t))x(t)$ solving the γ-attenuation problem for the system (7.5). More precisely,

$$F(i) = \left[D_c(i) + C_c(i)\tilde{Y}_{12}^*(i)\right]\tilde{Y}_{11}^{-1}(i),\ i \in \mathcal{D}.$$

Hence the first part of the proof is complete.

(ii) $\Rightarrow$ (i) Assume that there exists a stabilizing control state feedback $u(t) = F(\eta(t))x(t)$ solving the DAP with the level of attenuation γ for (7.1). Let $n_c > 0$ be a fixed integer and let $A_c(i) \in \mathbf{R}^{n_c \times n_c}$ be such that the zero solution of the system

$$\dot{x}_c(t) = A_c(\eta(t))x_c(t)$$

is ESMS. Then consider the controller $\left(n_c, A_c(i), 0_{n_c \times n}, 0_{m \times n_c}, F(i); i \in \mathcal{D}\right)$. It is easy to check that this controller is stabilizing and the input–output operator associated with the closed-loop system coincides with the input–output operator given by the state feedback control. Thus the proof is complete. $\qquad\square$

Remark 2. The smallest γ can be obtained by solving a semidefinite programming problem. Indeed, considering γ^2 as a new positive variable, the LMI (7.7) can be seen as a linear constraint in the minimization of γ^2.

7.2.2 Solution of some robust stabilization problems

Consider the system described by

$$dx(t) = \left\{\left[A_0(\eta(t)) + \hat{G}_0(\eta(t))\Delta_1(\eta(t))\hat{C}(\eta(t))\right]x(t)\right.$$
$$+ \left[B_0(\eta(t)) + \hat{B}_0(\eta(t))\Delta_2(\eta(t))\hat{D}(\eta(t))\right]u(t)\right\}dt$$
$$+ \sum_{k=1}^{r}\left\{\left[A_k(\eta(t)) + \hat{G}_k(\eta(t))\Delta_1(\eta(t))\hat{C}(\eta(t))\right]x(t)\right.$$
$$+ \left[B_k(\eta(t)) + \hat{B}_k(\eta(t))\Delta_2(\eta(t))\hat{D}(\eta(t))\right]u(t)\right\}dw_k(t),$$

(7.17)

where $x(t) \in \mathbf{R}^n$ is the state, $u(t) \in \mathbf{R}^m$ is the control variable, $A_k(i) \in \mathbf{R}^{n \times n}$, $B_k(i) \in \mathbf{R}^{n \times m}$, $\widehat{G}_k(i) \in \mathbf{R}^{n \times \hat{m}}$, $\widehat{B}_k(i) \in \mathbf{R}^{n \times \hat{m}}$, $\widehat{C}(i) \in \mathbf{R}^{\hat{p} \times n}$, $\widehat{D}(i) \in \mathbf{R}^{\hat{p} \times m}$, $0 \le k \le r$, $i \in \mathcal{D}$ are assumed known. The matrices $\Delta_1(i) \in \mathbf{R}^{\hat{m} \times \hat{p}}$, $\Delta_2(i) \in \mathbf{R}^{\hat{m} \times \hat{p}}$ are unknown and they describe the magnitude of the system (7.17). It is assumed that the whole state vector is accessible for measurement. The robust stabilization problem considered here can be stated as follows: For a given $\rho > 0$ determine a control $u(t) = F(\eta(t))x(t)$ stabilizing (7.17) for any $\Delta_1 = (\Delta_1(1), \ldots, \Delta_1(d))$ and $\Delta_2 = (\Delta_2(1), \ldots, \Delta_2(d))$ such that

$$\max \left(|\Delta_1|, |\Delta_2| \right) < \rho,$$

where

$$|\Delta_k| = \max_{i \in D} \lambda_{\max}^{\frac{1}{2}} \left(\Delta_k^*(i) \Delta_k(i) \right).$$

The closed-loop system obtained with $u(t) = F(\eta(t))x(t)$ is given by

$$dx(t) = \left[A_0(\eta(t)) + B_0(\eta(t))F(\eta(t)) + \widehat{G}_0(\eta(t))\Delta_1(\eta(t))\widehat{C}(\eta(t)) \right.$$
$$\left. + \widehat{B}_0(\eta(t))\Delta_2(\eta(t))\widehat{D}(\eta(t))F(\eta(t)) \right] x(t)dt \qquad (7.18)$$
$$+ \sum_{k=1}^{r} \left[A_k(\eta(t)) + B_k(\eta(t))F(\eta(t)) + \widehat{G}_k(\eta(t))\Delta_1(\eta(t))\widehat{C}(\eta(t)) \right.$$
$$\left. + \widehat{B}_k(\eta(t))\Delta_2(\eta(t))\widehat{D}(\eta(t))F(\eta(t)) \right] x(t)dw_k(t).$$

Denoting by

$$G_k(i) = \left[\widehat{G}_k(i) \quad \widehat{B}_k(i) \right],$$
$$C(i) = \left[\begin{array}{c} \widehat{C}(i) \\ 0 \end{array} \right],$$
$$D(i) = \left[\begin{array}{c} 0 \\ \widehat{D}(i) \end{array} \right],$$
$$\Delta(i) = \left[\begin{array}{cc} \Delta_1(i) & 0 \\ 0 & \Delta_2(i) \end{array} \right],$$

the system (7.18) can be rewritten as

$$dx(t) = \left\{ A_0(\eta(t)) + B_0(\eta(t))F(\eta(t)) + G_0(\eta(t))\Delta(\eta(t)) \right.$$
$$\times \left[C(\eta(t)) + D(\eta(t))F(\eta(t)) \right] \right\} x(t)dt \qquad (7.19)$$
$$+ \sum_{k=1}^{r} \left\{ A_k(\eta(t)) + B_k(\eta(t))F(\eta(t)) + G_k(\eta(t))\Delta(\eta(t)) \right.$$
$$\times \left[C(\eta(t)) + D(\eta(t))F(\eta(t)) \right] \right\} x(t)dw_k(t).$$

Assume that $F(i)$ is such that the zero solution of the system

$$dx(t) = \left[A_0(\eta(t)) + B_0(\eta(t))F(\eta(t)) \right] x(t)dt$$
$$+ \sum_{k=1}^{r} \left[A_k(\eta(t)) + B_k(\eta(t))F(\eta(t)) \right] x(t)dw_k(t)$$

is ESMS. Then, applying Corollary 21 of Chapter 6, it follows that the zero solution of (7.19) is ESMS for all Δ with $|\Delta| < \rho$ if the input–output operator $\mathcal{T}_F$ associated with the system

$$dx(t) = [(A_0(\eta(t)) + B_0(\eta(t))F(\eta(t)))\,x(t) + G_0(\eta(t))v(t)]\,dt$$

$$+ \sum_{k=1}^{r} [(A_k(\eta(t)) + B_k(\eta(t))F(\eta(t)))\,x(t) + G_k(\eta(t))v(t)]\,dw_k(t),$$

$$z(t) = (C(\eta(t)) + D(\eta(t))F(\eta(t)))\,x(t)$$

satisfies the condition $\|\mathcal{T}_F\| < 1/\rho$. Further, notice that

$$|\Delta| = \max_{i \in D} \lambda_{\max}^{\frac{1}{2}}\left(\Delta^*(i)\Delta(i)\right) = \max\left(|\Delta_1|, |\Delta_2|\right).$$

Therefore, F is a robust stabilizing state feedback with the robustness radius ρ if it is a solution of the DAP with level of attenuation $\gamma = 1/\rho$ for the following system:

$$dx(t) = [A_0(\eta(t))x(t) + G_0(\eta(t))v(t) + B_0(\eta(t))u(t)]\,dt$$

$$+ \sum_{k=1}^{r} [A_k(\eta(t))x(t) + G_k(\eta(t))v(t) + B_k(\eta(t))u(t)]\,dw_k(t),$$

$$y(t) = x(t),$$

$$z(t) = C(\eta(t))x(t) + D(\eta(t))u(t),$$

with $G_k(i)$, $C(i)$, $D(i)$, $i \in D$ defined above.

Applying Theorem 1 we obtain the following theorem.

Theorem 5. *Suppose that there exist* $Y = (Y(1), \dots, Y(d)) \in \mathcal{S}_n^d$, $Y(i) > 0$, $\Gamma = (\Gamma(1), \dots, \Gamma(d)) \in \mathcal{M}_{m,n}^d$ *verifying the following system of LMIs:*

$$\begin{bmatrix} \mathcal{W}_{0,0}(Y,i) & \mathcal{W}_{0,1}(Y,i) & \cdots & \mathcal{W}_{0,r}(Y,i) & \mathcal{W}_{0,r+1}(Y,i) & \mathcal{W}_{0,r+2}(Y,i) \\ \mathcal{W}_{0,1}^*(Y,i) & \mathcal{W}_{1,1}(Y,i) & \cdots & \mathcal{W}_{1,r}(Y,i) & \mathcal{W}_{1,r+1}(Y,i) & \mathcal{W}_{1,r+2}(Y,i) \\ \vdots & \vdots & \ddots & \vdots & \vdots & \vdots \\ \mathcal{W}_{0,r}^*(Y,i) & \mathcal{W}_{1,r}^*(Y,i) & \cdots & \mathcal{W}_{r,r}(Y,i) & \mathcal{W}_{r,r+1}(Y,i) & \mathcal{W}_{r,r+2}(Y,i) \\ \mathcal{W}_{0,r+1}^*(Y,i) & \mathcal{W}_{1,r+1}^*(Y,i) & \cdots & \mathcal{W}_{r,r+1}^*(Y,i) & \mathcal{W}_{r+1,r+1}(Y,i) & \mathcal{W}_{r+1,r+2}(Y,i) \\ \mathcal{W}_{0,r+2}^*(Y,i) & \mathcal{W}_{1,r+2}^*(Y,i) & \cdots & \mathcal{W}_{r,r+2}^*(Y,i) & \mathcal{W}_{r+1,r+2}^*(Y,i) & \mathcal{W}_{r+2,r+2}(Y,i) \end{bmatrix} < 0,$$

$$(7.20)$$

where

$$\mathcal{W}_{0,0}(Y,i) = A_0(i)Y(i) + Y(i)A_0^*(i) + q_{ii}Y(i) + B_0(i)\Gamma(i) + \Gamma^*(i)B_0^*(i)$$
$$+ \widehat{G}_0(i)\widehat{G}_0^*(i) + \widehat{B}_0(i)\widehat{B}_0^*(i),$$

$$\mathcal{W}_{0,k}(Y,i) = Y(i)A_k^*(i) + \Gamma^*(i)B_k^*(i) + \widehat{G}_0(i)\widehat{G}_k^*(i) + \widehat{B}_0(i)\widehat{B}_k^*(i), \quad 1 \le k \le r,$$

$$\mathcal{W}_{0,r+1}(Y,i) = \left[Y(i)\widehat{C}^*(i) \quad \Gamma^*(i)\widehat{D}^*(i)\right],$$

$$\mathcal{W}_{0,r+2}(Y,i) = \left[\sqrt{q_{i1}}Y(i)\dots\sqrt{q_{i,i-1}}Y(i)\ \sqrt{q_{i,i+1}}Y(i)\dots\sqrt{q_{id}}Y(i)\right],$$

$$\mathcal{W}_{l,k}(Y,i) = \widehat{G}_l(i)\widehat{G}_k^*(i) + \widehat{B}_l(i)\widehat{B}_k^*(i), \quad 1 \le k \neq l \le r,$$

$$\mathcal{W}_{l,l}(Y, i) = \widehat{G}_l(i)\widehat{G}_l^*(i) + \widehat{B}_l(i)\widehat{B}_l^*(i) - Y(i), \ 1 \le l \le r,$$
$$\mathcal{W}_{l,r+1}(Y, i) = 0, \ 1 \le l \le r,$$
$$\mathcal{W}_{r+1,r+1}(Y, i) = -\rho^{-2}I_{\hat{p}+\tilde{p}},$$
$$\mathcal{W}_{r+2,r+2}(Y, i) = diag(-Y(1)\dots - Y(i-1) - Y(i+1)\dots - Y(d)).$$

Then the state feedback gain $F(i) = \Gamma(i)Y^{-1}(i)$, $i \in \mathcal{D}$, *is a solution of the robust stabilization problem.* □

Now consider the system described by

$$dx(t) = [A_0(\eta(t))x(t) + G_0(\eta(t))\Delta (\varphi(t), \eta(t)) + B_0(\eta(t))u(t)]\, dt \qquad (7.21)$$

$$+ \sum_{k=1}^{r} [A_k(\eta(t))x(t) + G_k(\eta(t))\Delta (\varphi(t), \eta(t)) + B_k(\eta(t))u(t)]\, dw_k(t),$$

$$\varphi(t) = C(\eta(t))x(t),$$

where $x(t) \in \mathbf{R}^n$ is the state, $u(t) \in \mathbf{R}^m$ is the control variable, and $A_k(i) \in \mathbf{R}^{n \times n}$, $B_k(i) \in \mathbf{R}^{n \times m}$, $G_k(i) \in \mathbf{R}^{n \times m_1}$, $0 \le k \le r$, $C(i) \in \mathbf{R}^{p_1 \times n}$, $i \in \mathcal{D}$ are assumed to be known. The maps $y \to \Delta(y, i)$ are unknown functions including the uncertainties determined either by parameter variations or by truncation of nonlinear terms in the dynamic model. Denote by $\mathbf{\Delta}$ the class of admissible uncertainty

$$\mathbf{\Delta} = (\Delta(y, 1), \dots, \Delta(y, d)),$$

where $y \to \Delta(y, i) : \mathbf{R}^{p_1} \to \mathbf{R}^{m_1}$ are Lipschitz continuous functions with $\Delta(0, i) = 0$, $i \in \mathcal{D}$. In the following it is assumed that in (7.21) the whole state is available for measurement. The robust stabilization problem considered can be stated as follows: For a given $\rho > 0$ find a control law $u(t) = F(\eta(t))x(t)$ stabilizing the system (7.21) for all $\Delta \in \mathbf{\Delta}$ with $\|\Delta\| < \rho$. Recall that

$$\|\Delta\| = \sup_{y \ne 0, y \in R^{p_1}, i \in \mathcal{D}} \left\{ \frac{|\Delta(y, i)|}{|y|} \right\}.$$

Let $u(t) = F(\eta(t))x(t)$ be such that the zero solution of the system

$$dx(t) = [A_0(\eta(t)) + B_0(\eta(t))F(\eta(t))]\, x(t)dt$$

$$+ \sum_{k=1}^{r} [A_k(\eta(t)) + B_k(\eta(t))F(\eta(t))]\, x(t)dw_k(t)$$

is ESMS. When coupling this state feedback to (7.21) one obtains

$$dx(t) = \left\{ [A_0(\eta(t)) + B_0(\eta(t))F(\eta(t))]\, x(t) \right.$$
$$\left. + G_0(\eta(t))\Delta (\varphi(t), \eta(t)) \right\}dt \qquad (7.22)$$

$$+ \sum_{k=1}^{r} \left\{ [A_k(\eta(t)) + B_k(\eta(t))F(\eta(t))]\, x(t) \right.$$

$$\left. + G_k(\eta(t))\Delta (\varphi(t), \eta(t)) \right\}dw_k(t),$$

$$\varphi(t) = C(\eta(t))x(t).$$

Applying Theorem 24 of Chapter 6, we deduce that the zero solution of (7.22) is ESMS for arbitrary $\Delta \in \mathbf{\Delta}$ with $\|\Delta\| < \rho$ if the input–output operator $\mathcal{T}_F$ associated with the system

$$dx(t) = \{[A_0(\eta(t)) + B_0(\eta(t))F(\eta(t))]x(t) + G_0(\eta(t))v(t)\}\,dt$$

$$+ \sum_{k=1}^{r}\{[A_k(\eta(t))x(t) + B_k(\eta(t))F(\eta(t))]x(t) + G_k(\eta(t))v(t)\}\,dw_k(t),$$

$$z(t) = C(\eta(t))x(t)$$

satisfies the condition $\|\mathcal{T}_F\| < 1/\rho$. Therefore, in order to obtain a robust state feedback control with a given robustness radius $\rho > 0$ it is sufficient to solve the DAP with the level of attenuation $\gamma = 1/\rho$ for the following auxiliary system:

$$dx(t) = [A_0(\eta(t))x(t) + B_0(\eta(t))u(t) + G_0(\eta(t))v(t)]x(t) \qquad (7.23)$$

$$+ \sum_{k=1}^{r}[A_k(\eta(t))x(t) + B_k(\eta(t))u(t) + G_k(\eta(t))v(t)]\,dw_k(t),$$

$$z(t) = C(\eta(t))x(t).$$

From Theorem 1 applied for the system (7.23) one obtains the following theorem.

Theorem 6. *Assume that there exist* $Y = (Y(1), \ldots, Y(d)) \in \mathcal{S}_n^d$, $Y(i) > 0$, $\Gamma = (\Gamma(1), \ldots, \Gamma(d)) \in \mathcal{M}_{m,n}^d$ *satisfying the following LMIs:*

$$\begin{bmatrix} \mathcal{W}_{0,0}(Y,i) & \mathcal{W}_{0,1}(Y,i) & \cdots & \mathcal{W}_{0,r}(Y,i) & \mathcal{W}_{0,r+1}(Y,i) & \mathcal{W}_{0,r+2}(Y,i) \\ \mathcal{W}_{0,1}^*(Y,i) & \mathcal{W}_{1,1}(Y,i) & \cdots & \mathcal{W}_{1,r}(Y,i) & \mathcal{W}_{1,r+1}(Y,i) & \mathcal{W}_{1,r+2}(Y,i) \\ \vdots & \vdots & \ddots & \vdots & \vdots & \vdots \\ \mathcal{W}_{0,r}^*(Y,i) & \mathcal{W}_{1,r}^*(Y,i) & \cdots & \mathcal{W}_{r,r}(Y,i) & \mathcal{W}_{r,r+1}(Y,i) & \mathcal{W}_{r,r+2}(Y,i) \\ \mathcal{W}_{0,r+1}^*(Y,i) & \mathcal{W}_{1,r+1}^*(Y,i) & \cdots & \mathcal{W}_{r,r+1}^*(Y,i) & \mathcal{W}_{r+1,r+1}(Y,i) & \mathcal{W}_{r+1,r+2}(Y,i) \\ \mathcal{W}_{0,r+2}^*(Y,i) & \mathcal{W}_{1,r+2}^*(Y,i) & \cdots & \mathcal{W}_{r,r+2}^*(Y,i) & \mathcal{W}_{r+1,r+2}^*(Y,i) & \mathcal{W}_{r+2,r+2}(Y,i) \end{bmatrix} < 0,$$

$$(7.24)$$

$i \in \mathcal{D}$, *where*

$$\mathcal{W}_{0,0}(Y,i) = A_0(i)Y(i) + Y(i)A_0^*(i) + q_{ii}Y(i) + B_0(i)\Gamma(i)$$
$$+ \Gamma^*(i)B_0^*(i) + G_0(i)G_0^*(i),$$
$$\mathcal{W}_{0,k}(Y,i) = Y(i)A_k^*(i) + \Gamma^*(i)B_k^*(i) + G_0(i)G_k^*(i),\ 1 \le k \le r,$$
$$\mathcal{W}_{0,r+1}(i) = Y(i)C^*(i),$$
$$\mathcal{W}_{0,r+2} = \left[\sqrt{q_{i1}}Y(i) \ldots \sqrt{q_{i,i-1}}Y(i)\sqrt{q_{i,i+1}}Y(i) \ldots \sqrt{q_{id}}Y(i)\right],$$
$$\mathcal{W}_{l,k} = G_l(i)G_k^*(i),\ 1 \le l,k \le r,\ l \ne k,$$
$$\mathcal{W}_{l,l} = G_l(i)G_l^*(i) - Y(i),\ 1 \le l \le r,$$
$$\mathcal{W}_{l,r+1}(i) = 0,\ 1 \le l \le r,$$
$$\mathcal{W}_{l,r+2}(i) = 0,\ 1 \le l \le r+1,$$
$$\mathcal{W}_{r+1,r+1}(i) = -\gamma^2 I_{p_1},$$
$$\mathcal{W}_{r+2,r+2}(i) = diag\,(-Y(1) \ldots - Y(i-1) - Y(i+1) \ldots - Y(d)).$$

Then the control $u(t) = F(\eta(t))x(t)$ *with* $F(i) = \Gamma(i)Y^{-1}(i)$, $i \in D$, *provides a robust stability feedback gain.*   □

Remark 3. In order to maximize the robustness radius one can use the idea presented in Remark 2 but with the constraint (7.24) instead of (7.7).

7.2.3 A case study

In order to illustrate the theoretical developments concerning the DAP in the case when the state is measurable, we present in the following a case study for which some comparative aspects with the results provided by deterministic design approaches will be discussed.

Air-launched unmanned air vehicles (UAVs) are typically released with their wings folded in order to achieve a safe separation with respect to the launching aircraft. The vehicle's wings are deployed after several seconds when a glide slope maneuver is required. The wing deployment determines a "jump" of the aerodynamic coefficients leading to a transient of the angle of attack which must be minimized in order to prevent the loss of stability. The longitudinal short-period motion of the UAV has the following state-space equations:

$$\dot{x} = Ax + B\delta_{e_c} + Gv, \qquad (7.25)$$
$$z = Cx + D\delta_{e_c},$$

where the state vector is

$$x = \begin{bmatrix} w \\ q \\ \delta_e \\ \xi \end{bmatrix},$$

with w denoting the vertical component of the true airspeed, q is the pitch rate, δ_e is the internal state of the actuator, and ξ denotes the state of the integral action $\dot{\xi} = a_z - a_{z_c}$ introduced in order to obtain zero steady-state tracking error of the normal acceleration a_z with respect to its commanded piecewise constant value a_{z_c}. The control variable is the elevon command δ_{e_c} and the input vector v includes the external reference a_{z_c} and disturbances, namely:

$$v = \begin{bmatrix} a_{z_c} \\ d_{\dot{w}} \\ d_{\dot{q}} \end{bmatrix},$$

$d_{\dot{w}}$ and $d_{\dot{q}}$ denoting the disturbances in $\dot{w}$ and $\dot{q}$, respectively. The quality output z has two components

$$z = \begin{bmatrix} \beta\xi \\ \rho\delta_{e_c} \end{bmatrix},$$

where β and ρ are positive given weights. The matrix coefficients in (7.25) depend on the two flight conditions mentioned above, namely the situation when the UAV has the wings folded and the case when the wings are deployed, respectively. Therefore, in this case the Markov chain has two states, that is, $D = \{1, 2\}$. The numerical values corresponding to these two states are [108]:

$$A(1) = \begin{bmatrix} -0.1077 & 718.5340 & -31.3672 & 0 \\ -0.0219 & -0.7209 & -19.5316 & 0 \\ 0 & 0 & -30 & 0 \\ 0 & 2.8870 & 64.7283 & 0 \end{bmatrix},$$

$$A(2) = \begin{bmatrix} -0.4628 & 717.1890 & -16.7139 & 0 \\ -0.0333 & -0.7522 & -11.3638 & 0 \\ 0 & 0 & -30 & 0 \\ -0.2990 & 2.8210 & 39.1960 & 0 \end{bmatrix},$$

$$B(1) = B(2) = \begin{bmatrix} 0 \\ 0 \\ 30 \\ 0 \end{bmatrix},$$

$$G(1) = G(2) = \begin{bmatrix} 0 & 1 & 0 \\ 0 & 0 & 1 \\ 0 & 0 & 0 \\ -1 & 0 & 0 \end{bmatrix},$$

$$C(1) = C(2) = \begin{bmatrix} 0 & 0 & 0 & 20 \\ 0 & 0 & 0 & 0 \end{bmatrix},$$

$$D(1) = D(2) = \begin{bmatrix} 0 \\ 100 \end{bmatrix},$$

$\beta = 20$, $\rho = 100$. The transition rate matrix is

$$Q = \begin{bmatrix} -1 & 1 \\ 0.01 & -0.01 \end{bmatrix}.$$

The problem consists in determining a state feedback control $\delta_{e_c}(t) = F(\eta(t))x(t)$ such that the closed-loop system obtained when coupling it to (7.25), namely

$$\dot{x}(t) = [A(\eta(t)) + B(\eta(t))F(\eta(t))]x(t) + G(\eta(t))v(t),$$
$$z(t) = [C(\eta(t)) + D(\eta(t))F(\eta(t))]x(t),$$

is ESMS, and its associated input–output operator has norm less than a given $\gamma > 0$. Applying Corollary 2 we obtained for $\gamma = 20$,

$$\begin{aligned} F(1) &= [0.0290 & -2.7269 & -1.1120 & -1.5065], \\ F(2) &= [0.0110 & -0.7722 & -0.4793 & -0.2112]. \end{aligned} \qquad (7.26)$$

In order to compare these results with those provided by other standard design methods, we solved the same problem using two deterministic alternative approaches. The first one is the robust control (RC) design consisting in determining a unique "quadratically stabilizing" controller which stabilizes both systems corresponding to folded and unfolded wings situations. In this design we obtained, using again an LMI-based approach [9],

$$F_{RC} = [10.57 \quad -425.6 \quad -180.7 \quad -305.7]$$

for the minimum closed-loop disturbance attenuation level $\gamma = 33.43$.

The second deterministic method consists in designing separate H^∞ state feed-back zero-order controllers corresponding to each flight condition. This design will be abbreviated SDH, and it gives for $\gamma = 18.1$ and for $\gamma = 12.9$, respectively, the following gains corresponding to the two flight conditions considered:

$$F_{SDH}(1) = [0.0040 \quad -0.0825 \quad -0.7510 \quad -0.4253],$$
$$F_{SDH}(2) = [0.1212 \quad 1.2540 \quad -1.7674 \quad -1.6579].$$

Two comparison approaches have been used: the first is completely deterministic and the second is entirely stochastic. In the first method, the H^∞ norm of the closed-loop system for $i = 1$ and $i = 2$ has been determined for all three solutions obtained above. The results are presented in Table 7.1.

$\| T \|_\infty$	MJC	RC	SHD
$i = 1$	18.3	32.9	18.1
$i = 2$	15.7	22.5	12.9

Table 7.1. Deterministic comparison approach

One can see that for MJC and SHD design, the achieved H^∞ norms of the closed-loop system are very close to and much lower than those of the RC-feedback gain.

In the second method we computed the levels of attenuation corresponding to the three solutions using the stochastic framework. To this end, we determined the closed-loop system with the corresponding feedback gains. Regarding these systems as stochastic systems with Markov jumps, we applied Theorem 10 of Chapter 6 to computing the corresponding level of attenuation. The obtained results are presented in Table 7.2.

Method	MJC	RC	SHD
$\| T \|$	20	32.4	76.7

Table 7.2. Stochastic comparison approach

The fact that in the stochastic design case (MJC) the level of attenuation is signif-icantly lower is expected since the deterministic design (RC and SDH) does not take into consideration the parameter jumps.

The elements $P_{11}(t)$ and $P_{12}(t)$ of the transition probability matrix $P(t) = e^{Qt}$ as functions of time are illustrated in Figure 7.1a. In Figures 7.1b and 7.1c the time-responses of the angle of attack and of the elevon command to unit step acceleration are plotted. Inspecting these figures one can see that the angle of attack is similar for all three methods, but the MJC uses considerably less control effort than either RC or SDH design.

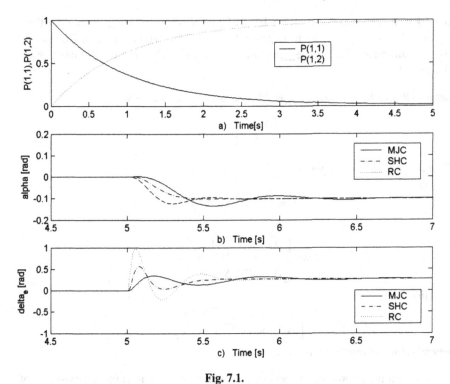

Fig. 7.1.

7.3 Solution of the DAP in the case of output measurement

In this section we consider the DAP with an imposed level of attenuation $\gamma > 0$ in the case when the output is available for measurement. Our approach is based on an LMI technique and it extends to this framework the well-known results in the deterministic context. As in the deterministic case the necessary and sufficient conditions guaranteeing the existence of a γ-attenuating controller are obtained using the following result (see [9]).

Lemma 7. *(Projection Lemma) Let* $\mathcal{Z} \in \mathbf{R}^{v \times v}$, $\mathcal{Z} = \mathcal{Z}^*$, $\mathcal{U} \in \mathbf{R}^{v_1 \times v}$, *and* $\mathcal{V} \in \mathbf{R}^{v_2 \times v}$, *with* v, v_1, v_2 *positive integers. Consider the following basic LMI:*

$$\mathcal{Z} + \mathcal{U}^* \Theta \mathcal{V} + \mathcal{V}^* \Theta^* \mathcal{U} < 0, \tag{7.27}$$

with the unknown variable $\Theta \in \mathbf{R}^{v_1 \times v_2}$. *Then the following are equivalent:*

  (i) *there exists* $\Theta \in \mathbf{R}^{v_1 \times v_2}$ *solving (7.27);*

  (ii)

$$\mathcal{W}_{\mathcal{U}}^* \mathcal{Z} \mathcal{W}_{\mathcal{U}} < 0 \tag{7.28}$$

and

$$\mathcal{W}_{\mathcal{V}}^* \mathcal{Z} \mathcal{W}_{\mathcal{V}} < 0, \tag{7.29}$$

where $\mathcal{W}_{\mathcal{U}}$ *and* $\mathcal{W}_{\mathcal{V}}$ *denote any bases of the null spaces* $Ker\mathcal{U}$ *and* $Ker\mathcal{V}$, *respectively.* □

Remark 4. It is known that if $\widetilde{W}$ is a basis of $Ker M$ where M is a given matrix, then any other basis of $Ker M$ can be expressed as $W = \widetilde{W}\Gamma$ with det $\Gamma \neq 0$. This shows that it is sufficient to check the conditions (7.28) and (7.29) for some suitable bases $W_{\mathcal{U}}$ and $W_{\mathcal{V}}$.

Lemma 8. *Let* $X, Y \in S_n$, $N \in \mathbf{R}^{n \times n_c}$ *and* $S \in S_{n_c}$ *with* $X > 0$ *and*

$$\begin{bmatrix} Y & N \\ N^* & S \end{bmatrix} > 0.$$

Then the following are equivalent:

(i)

$$X = \left(Y - NS^{-1}N^*\right)^{-1};$$

(ii)

$$\text{rank} \begin{bmatrix} X & I_n & 0 \\ I_n & Y & N \\ 0 & N^* & S \end{bmatrix} = n + n_c;$$

(iii)

$$\begin{bmatrix} Y & N \\ N^* & S \end{bmatrix}^{-1} = \begin{bmatrix} X & \star \\ \star & \star \end{bmatrix},$$

where $\star$ *denotes irrelevant entries.*

The next result provides necessary and sufficient conditions for the existence of a controller of type (7.2), solving the DAP for the system (7.1).

Theorem 9. *For a* $\gamma > 0$ *the following are equivalent:*

(i) *There exists a controller of order* $n_c > 0$ *which solves the DAP with the level of attenuation* $\gamma > 0$ *for the system (7.1).*

(ii) *There exist* $X = (X(1), \ldots, X(d)) \in S_n^d$, $X(i) > 0$, $i \in \mathcal{D}$, $Y = (Y(1), \ldots, Y(d)) \in S_n^d$, $Y(i) > 0$, $S = (S(1), \ldots, S(d)) \in S_n^d$, $S(i) > 0$, $N \in (N(1), \ldots, N(d))$, $N \in \mathcal{M}_{n,n_c}^d$ *such that*

$$\begin{bmatrix} V_0^*(i) & V_1^*(i) \end{bmatrix} \mathcal{N}_i (X) \begin{bmatrix} V_0(i) \\ V_1(i) \end{bmatrix} < 0, \tag{7.30}$$

$$\begin{bmatrix} \Pi_{0,0}(i) & \Pi_{0,1}(i) & -U_1^*(i)N(i) & \cdots & -U_r^*(i)N(i) & \Pi_{0,r+1}(i) \\ \Pi_{0,1}^*(i) & -\gamma^2 I_{m_1} & 0 & \cdots & 0 & 0 \\ -N^*(i)U_1(i) & 0 & -S(i) & \cdots & 0 & 0 \\ \vdots & \vdots & \vdots & \ddots & \vdots & \vdots \\ -N^*(i)U_r(i) & 0 & 0 & \cdots & -S(i) & 0 \\ \Pi_{0,r+1}^*(i) & 0 & 0 & \cdots & 0 & \Pi_{r+1,r+1}(i) \end{bmatrix} < 0, \tag{7.31}$$

$$\text{rank} \begin{bmatrix} X(i) & I_n & 0 \\ I_n & Y(i) & N(i) \\ 0 & N^*(i) & S(i) \end{bmatrix} = n + n_c, \tag{7.32}$$

where

$$\begin{bmatrix} V_0(i) \\ V_1(i) \end{bmatrix}$$

is a basis of $Ker\,[C_0(i)\quad D_0(i)]$,

$$\begin{bmatrix} U_0(i) \\ \vdots \\ U_{r+1}(i) \end{bmatrix}$$

is a basis of $Ker\,\big[B_0^*(i)\;\cdots\;B_r^*(i)\;D_{zu}^*(i)\big]$,

$$\mathcal{N}_i\,(X) = \begin{bmatrix} \mathcal{N}_{11}\,(X,i) & \mathcal{N}_{12}\,(X,i) \\ \mathcal{N}_{12}^*\,(X,i) & \mathcal{N}_{22}\,(X,i) \end{bmatrix},$$

$$\mathcal{N}_{11}\,(X,i) = A_0^*(i)X(i) + X(i)A_0(i) + \sum_{k=1}^{r} A_k^*(i)X(i)A_k(i)$$

$$+ \sum_{j=1}^{d} q_{ij} X(j) + C_z^*(i)C_z(i),$$

$$\mathcal{N}_{12}\,(X,i) = X(i)G_0(i) + \sum_{k=1}^{r} A_k^*(i)X(i)G_k(i) + C_z^*(i)D_{zv}(i),$$

$$\mathcal{N}_{22}\,(X,i) = -\gamma^2 I_{m_1} + D_{zv}^*(i)D_{zv}(i) + \sum_{k=1}^{r} G_k^*(i)X(i)G_k(i),$$

$$\Pi_{0,0}(i) = U_0^*(i)\big[A_0(i)Y(i) + Y(i)A_0^*(i) + q_{ii}Y(i)\big]U_0(i)$$

$$+ \sum_{k=1}^{r} U_0^*(i)Y(i)A_k^*(i)U_k(i) + U_0^*(i)Y(i)C_z^*(i)U_{r+1}(i)$$

$$+ U_{r+1}^*(i)C_z(i)Y(i)U_0(i) + \sum_{k=1}^{r} U_k^*(i)A_k(i)Y(i)U_0(i)$$

$$- \sum_{k=1}^{r} U_k^*(i)Y(i)U_k(i) - U_{r+1}^*(i)U_{r+1}(i),$$

$$\Pi_{0,1}(i) = \sum_{k=0}^{r} U_k^*(i)G_k(i) + U_{r+1}^*(i)D_{zv}(i),$$

$$\Pi_{0,r+1}(i) = U_0^*(i)[I_n \quad 0]\big[\sqrt{q_{i1}}\widetilde{Y}(i)\cdots\sqrt{q_{i,i-1}}\widetilde{Y}(i)\sqrt{q_{i,i+1}}\widetilde{Y}(i)\cdots\sqrt{q_{id}}\widetilde{Y}(i)\big],$$

$$\Pi_{r+1,r+1}(i) = -diag\,\big(\widetilde{Y}(1)\cdots\widetilde{Y}\,(i-1),\,\widetilde{Y}\,(i+1)\cdots\widetilde{Y}(d)\big),$$

$$\widetilde{Y}(i) = \begin{bmatrix} Y(i) & N(i) \\ N^*(i) & S(i) \end{bmatrix},\; i \in \mathcal{D}.$$

Proof. The outline of the proof is similar to the one in the deterministic framework. The stochastic feature of the considered system does not appear explicitly in the following developments of the proof. This feature appears only in the specific formulae of the Bounded Real Lemma. Therefore the proof is also accessible for readers who are not very familiar with stochastic systems.

(i) $\Rightarrow$ (ii) Assume that there exists a controller of form (7.2) stabilizing the system (7.1) such that $\|\mathcal{T}_{cl}\| < \gamma$. Using the implication (i) $\Rightarrow$ (ii) of Theorem 10 (Bounded Real Lemma) of Chapter 6 for the closed-loop system, we deduce that there exist

$$X_{cl} = (X_{cl}(1), \dots, X_{cl}(d)) \in \mathcal{S}_{n+n_c}^d, \quad X_{cl}(i) > 0$$

such that

$$\mathcal{N}_i(X_{cl}, \gamma) < 0, \tag{7.33}$$

where

$$\mathcal{N}_i(X_{cl}, \gamma) = \begin{bmatrix} (\mathcal{L}_{cl}^* X_{cl})(i) + C_{cl}^*(i) C_{cl}(i) & \mathcal{P}_i^*(X_{cl}) \\ \mathcal{P}_i(X_{cl}) & \mathcal{R}_i(X_{cl}) \end{bmatrix},$$

$$(\mathcal{L}_{cl}^* X_{cl})(i) = A_{0cl}^*(i) X_{cl}(i) + X_{cl}(i) A_{cl}(i)$$
$$+ \sum_{k=1}^{r} A_{kcl}^*(i) X_{cl}(i) A_{kcl}(i) + \sum_{j=1}^{d} q_{ij} X_{cl}(j),$$

$$\mathcal{P}_i(X_{cl}) = G_{0cl}^*(i) X_{cl}(i) + \sum_{k=1}^{r} G_{kcl}^*(i) X_{cl} A_{kcl}(i)$$
$$+ D_{cl}^*(i) C_{cl}(i),$$

$$\mathcal{R}_i(X_{cl}) = -\gamma^2 I_{m_1} + \sum_{k=1}^{r} G_{kcl}^*(i) X_{cl}(i) G_{kcl}(i).$$

Based on Schur complements arguments it is easy to see that (7.33) is equivalent to

$$\begin{bmatrix} (\mathcal{L}_0^* X_{cl})(i) & X_{cl}(i) G_{0cl}(i) & A_{1cl}^*(i) X_{cl}(i) & \cdots & A_{rcl}^*(i) X_{cl}(i) & C_{cl}^*(i) \\ G_{0cl}^*(i) X_{cl}(i) & -\gamma^2 I_{m_1} & G_{1cl}^*(i) X_{cl}(i) & \cdots & G_{rcl}^*(i) X_{cl}(i) & D_{cl}^*(i) \\ X_{cl}(i) A_{1cl}(i) & X_{cl}(i) G_{1cl}(i) & -X_{cl}(i) & \cdots & 0 & 0 \\ \vdots & \vdots & \vdots & \ddots & \vdots & \vdots \\ X_{cl}(i) A_{rcl}(i) & X_{cl}(i) G_{rcl}(i) & 0 & \cdots & -X_{cl}(i) & 0 \\ C_{cl}(i) & D_{cl}(i) & 0 & \cdots & 0 & -I_{p_1} \end{bmatrix} < 0, \tag{7.34}$$

where

$$(\mathcal{L}_0^* X_{cl})(i) = A_{0cl}^*(i) X_{cl}(i) + X_{cl}(i) A_{0cl}(i) + \sum_{j=1}^{d} q_{ij} X_{cl}(j).$$

Let us introduce the following notations:

$$\tilde{A}_k(i) = \begin{bmatrix} A_k(i) & 0 \\ 0 & 0 \end{bmatrix}, \ \tilde{G}_k(i) = \begin{bmatrix} G_k(i) \\ 0 \end{bmatrix}, \ 0 \le k \le r,$$

$$\tilde{B}_0(i) = \begin{bmatrix} 0 & B_0(i) \\ I_{n_c} & 0 \end{bmatrix}, \ \tilde{B}_k(i) = \begin{bmatrix} 0 & B_k(i) \\ 0 & 0 \end{bmatrix}, \ 1 \le k \le r,$$

$$\tilde{C}_0(i) = \begin{bmatrix} 0 & I_{n_c} \\ C_0(i) & 0 \end{bmatrix}, \ \tilde{C}_z(i) = [C_z(i) \quad 0],$$

$$\tilde{D}_{zu}(i) = [0 \quad D_{zu}(i)], \ \tilde{D}_0(i) = \begin{bmatrix} 0 \\ D_0(i) \end{bmatrix}, \ i \in \mathcal{D},$$

$$\Theta_c(i) = \begin{bmatrix} A_c(i) & B_c(i) \\ C_c(i) & D_c(i) \end{bmatrix}.$$

Using (7.4) one obtains

$$A_{kcl}(i) = \tilde{A}_k(i) + \tilde{B}_k(i)\Theta_c(i)\tilde{C}_0(i),$$
$$G_{kcl}(i) = \tilde{G}_k(i) + \tilde{B}_k(i)\Theta_c(i)\tilde{D}_0(i), \ 0 \le k \le r,$$
$$C_{cl}(i) = \tilde{C}_z(i) + \tilde{D}_{zu}(i)\Theta_c(i)\tilde{C}_0(i),$$
$$D_{cl}(i) = D_{zv}(i) + \tilde{D}_{zu}(i)\Theta_c(i)\tilde{D}_0(i), \ i \in \mathcal{D}.$$

With the above equations one can easily see that (7.34) can be written in the basic LMI form:

$$\mathcal{Z}(i) + \mathcal{U}^*(i)\Theta_c(i)\mathcal{V}(i) + \mathcal{V}^*(i)\Theta_c^*(i)\mathcal{U}(i) < 0, \ i \in \mathcal{D}, \qquad (7.35)$$

where

$$\mathcal{Z}(i) = \begin{bmatrix} (\tilde{\mathcal{L}}_0^* X_{cl})(i) & X_{cl}(i)\tilde{G}_0(i) & \tilde{A}_1^*(i)X_{cl}(i) & \cdots & \tilde{A}_r^*(i)X_{cl}(i) & \tilde{C}_z^*(i) \\ \tilde{G}_0^*(i)X_{cl}(i) & -\gamma^2 I_{m_1} & \tilde{G}_1^*(i)X_{cl}(i) & \cdots & \tilde{G}_r^*(i)X_{cl}(i) & D_{zv}^*(i) \\ X_{cl}(i)\tilde{A}_1(i) & X_{cl}(i)G_1(i) & -X_{cl}(i) & \cdots & 0 & 0 \\ \vdots & \vdots & \vdots & \ddots & \vdots & \vdots \\ X_{cl}(i)\tilde{A}_r(i) & X_{cl}(i)\tilde{G}_r(i) & 0 & \cdots & -X_{cl}(i) & 0 \\ \tilde{C}_z(i) & D_{zv}(i) & 0 & \cdots & 0 & -I_{p_1} \end{bmatrix},$$

$$\mathcal{U}(i) = [\tilde{B}_0^*(i)X_{cl}(i) \quad 0_{(m_2+n_c)\times m_1} \quad \tilde{B}_1^*(i)X_{cl}(i) \quad \cdots \quad \tilde{B}_r^*(i)X_{cl}(i) \quad \tilde{D}_{zu}^*(i)],$$
$$\mathcal{V}(i) = [\tilde{C}_0(i) \quad \tilde{D}_0(i) \quad 0_{(p_2+n_c)\times[p_1+r(n+n_c)]}], \ i \in \mathcal{D}, \qquad (7.36)$$

with

$$(\tilde{\mathcal{L}}_0^* X_{cl})(i) = \tilde{A}_{0cl}^*(i)X_{cl}(i) + X_{cl}(i)\tilde{A}_{0cl}(i) + \sum_{j=1}^{d} q_{ij}X_{cl}(j).$$

Therefore the existence of a stabilizing γ-attenuation controller for (7.1) is equivalent to the solvability of (7.35). Based on Lemma 7, (7.35) is solvable if and only if there exist

$$W^*_{\mathcal{U}(i)} \mathcal{Z}(i) W_{\mathcal{U}(i)} < 0, \tag{7.37}$$

$$W^*_{\mathcal{V}(i)} \mathcal{Z}(i) W_{\mathcal{V}(i)} < 0, \ i \in \mathcal{D}, \tag{7.38}$$

where $W_{\mathcal{U}(i)}$, $W_{\mathcal{V}(i)}$ denote bases of the null spaces of $\mathcal{U}(i)$ and $\mathcal{V}(i)$, respectively. It is easy to see that a basis of the null space of $\mathcal{U}(i)$ is

$$W_{\mathcal{U}(i)} = \mathcal{X}^{-1}(i) W_{\tilde{\mathcal{U}}(i)}, \tag{7.39}$$

where

$$\mathcal{X}(i) = diag \left(X_{cl} \ I_{m_1} \ X_{cl}(i) \cdots X_{cl}(i) \ I_{p_1} \right)$$

and $W_{\tilde{\mathcal{U}}(i)}$ is a basis of the null subspace of the matrix

$$\tilde{\mathcal{U}}(i) = \left[\tilde{B}^*_0(i) \ 0_{(m_2+n_c) \times m_1} \ \tilde{B}^*_1(i) \ \cdots \ \tilde{B}^*_r(i) \ \tilde{D}^*_{zu}(i) \right].$$

A basis of the null subspace of $\tilde{\mathcal{U}}(i)$ is

$$W_{\tilde{\mathcal{U}}(i)} = \begin{bmatrix} T_0(i) & 0 & 0 & \cdots & 0 \\ 0 & I_{m_1} & 0 & \cdots & 0 \\ T_1(i) & 0 & L & \cdots & 0 \\ \vdots & \vdots & \vdots & \ddots & \vdots \\ T_r(i) & 0 & 0 & \cdots & L \\ U_{r+1}(i) & 0 & 0 & \cdots & 0 \end{bmatrix}, \tag{7.40}$$

where

$$T_k(i) = \begin{bmatrix} U_k(i) \\ 0 \end{bmatrix}, \ 0 \le k \le r, \ L = \begin{bmatrix} 0 \\ I_{n_c} \end{bmatrix},$$

and

$$\begin{bmatrix} U_0(i) \\ \vdots \\ U_{r+1}(i) \end{bmatrix}$$

is a basis of the null subspace of the matrix

$$\left[B^*_0(i) \ B^*_1(i) \ \cdots \ B^*_r(i) \ D^*_{zu}(i) \right].$$

A suitable choice for $W_{\mathcal{V}(i)}$ is the following:

$$W_{\mathcal{V}(i)} = \begin{bmatrix} V_0(i) & 0 \\ 0 & 0 \\ V_1(i) & 0 \\ 0 & I_{p_1+r(n+n_c)} \end{bmatrix}, \tag{7.41}$$

where

$$\begin{bmatrix} V_0(i) \\ V_1(i) \end{bmatrix}$$

is a basis of the null subspace of the matrix $[C_0(i) \ D_0(i)]$.

Consider the partition of $X_{cl}(i)$:

$$X_{cl}(i) = \begin{bmatrix} X(i) & M(i) \\ M^*(i) & \tilde{X}(i) \end{bmatrix},$$

with $X(i) \in \mathbf{R}^{n \times n}$. Then by direct computations one obtains

$$\mathcal{W}_{\mathcal{V}(i)}^* \mathcal{Z}(i) \mathcal{W}_{\mathcal{V}(i)} = \begin{bmatrix} \Psi_{0,0}(i) & \Psi_{0,1}(i) & \cdots & \Psi_{0,r}(i) & \Psi_{0,r+1}(i) \\ \Psi_{0,1}^*(i) & -X_{cl}(i) & \cdots & 0 & 0 \\ \vdots & \vdots & \ddots & \vdots & \vdots \\ \Psi_{0,r}^*(i) & 0 & \cdots & -X_{cl}(i) & 0 \\ \Psi_{0,r+1}^*(i) & 0 & \cdots & 0 & -I_{p_1} \end{bmatrix}, \qquad (7.42)$$

where we denoted

$$\Psi_{0,0}(i) = V_0^*(i) \left[A_0^*(i) X(i) + X(i) A_0(i) + \sum_{j=1}^{d} q_{ij} X(j) \right] V_0(i)$$

$$+ V_0^*(i) X(i) G_0(i) V_1(i) + V_1^*(i) G_0^*(i) X(i) V_0(i) - \gamma^2 V_1^*(i) V_1(i),$$
$$\Psi_{0,k}(i) = \left(\begin{bmatrix} V_0^*(i) & 0 \end{bmatrix} \tilde{A}_k^*(i) + V_1^*(i) \tilde{G}_k^*(i) \right) X_{cl}, \quad 1 \le k \le r,$$
$$\Psi_{0,r+1}(i) = V_0^*(i) C_z^*(i) + V_1^*(i) D_{zv}^*(i).$$

Again using Schur complement arguments, it follows that condition (7.38) together with (7.42) is equivalent to

$$\Psi_{0,0}(i) + \sum_{k=1}^{r} \Psi_{0,k}(i) X_{cl}^{-1}(i) \Psi_{0,k}^*(i) + \Psi_{0,r+1}(i) \Psi_{0,r+1}^*(i) < 0.$$

Detailing the coefficients in the above inequality, (7.30) directly follows.

In order to detail the condition (7.37), one first computes

$$\mathcal{X}^{-1}(i) \mathcal{Z}(i) \mathcal{X}^{-1}(i) \qquad\qquad (7.43)$$

$$= \begin{bmatrix} (\mathcal{L}_0^* \tilde{Y})(i) & \tilde{G}_0(i) & \tilde{Y}(i) \tilde{A}_1^*(i) & \cdots & \tilde{Y}(i) \tilde{A}_r^*(i) & \tilde{Y}(i) \tilde{C}_z^*(i) \\ \tilde{G}_0^*(i) & -\gamma^2 I_{m_1} & \tilde{G}_1^*(i) & \cdots & \tilde{G}_r^*(i) & D_{zv}^*(i) \\ \tilde{A}_1(i) \tilde{Y}(i) & \tilde{G}_1^*(i) & -\tilde{Y}(i) & \cdots & 0 & 0 \\ \vdots & \vdots & \vdots & \ddots & \vdots & \vdots \\ \tilde{A}_r(i) \tilde{Y}(i) & \tilde{G}_r(i) & 0 & \cdots & -\tilde{Y}(i) & 0 \\ \tilde{C}_z(i) \tilde{Y}(i) & D_{zv}(i) & 0 & \cdots & 0 & -I_{p_1} \end{bmatrix}$$

where

$$(\mathcal{L}_0^* \tilde{Y})(i) = \tilde{A}_0(i) \tilde{Y}(i) + \tilde{Y}(i) \tilde{A}_0^*(i) + \sum_{j=1}^{d} q_{ij} \tilde{Y}(i) \tilde{Y}^{-1}(j) \tilde{Y}(i), \qquad (7.44)$$

$$\tilde{Y}(i) = X_{cl}^{-1}(i). \qquad\qquad (7.45)$$

We also introduce the notation

$$\tilde{Y}(i) = \begin{bmatrix} Y(i) & N(i) \\ N^*(i) & S(i) \end{bmatrix}, \quad Y(i) \in \mathbf{R}^{n \times n}.$$

Using (7.40), (7.43), (7.44), and (7.39), one obtains that (7.37) becomes

$$\begin{bmatrix} \tilde{\Pi}_{0,0}(i) & \Pi_{0,1}(i) & -U_1^*(i)N(i) & \cdots & -U_r^*(i)N(i) \\ \Pi_{0,1}^*(i) & -\gamma^2 I_{m_1} & 0 & \cdots & 0 \\ -N^*(i)U_1(i) & 0 & -S(i) & \cdots & 0 \\ \vdots & \vdots & \vdots & \ddots & \vdots \\ -N^*(i)U_r(i) & 0 & 0 & \cdots & -S(i) \end{bmatrix} < 0, \quad (7.46)$$

where

$$\tilde{\Pi}_{0,0}(i) = U_0^*(i) \left\{ A_0(i)Y(i) + Y(i)A_0^*(i) \right.$$

$$+ \sum_{j=1}^{d} q_{ij} [Y(i) \quad N(i)] \tilde{Y}^{-1}(j) \begin{bmatrix} Y(i) \\ N^*(i) \end{bmatrix} \right\} U_0(i)$$

$$+ \sum_{k=1}^{r} U_0^*(i)Y(i)A_k^*(i)U_k(i) + U_0^*(i)Y(i)C_z^*(i)U_{r+1}(i)$$

$$+ U_{r+1}^*(i)C_z(i)Y(i)U_0(i) + \sum_{k=1}^{r} U_k^*(i)A_k(i)Y(i)U_0(i)$$

$$- \sum_{k=1}^{r} U_k^*(i)Y(i)U_k(i) - U_{r+1}^*(i)U_{r+1}(i),$$

$$\Pi_{0,1}(i) = \sum_{k=0}^{r} U_k^*(i)G_k(i) + U_{r+1}^*(i)D_{zv}(i).$$

By Schur complement arguments one can see that (7.46) is equivalent to an extended LMI which coincides with (7.31). Taking into account that

$$rank \begin{bmatrix} X(i) & I & 0 \\ I & Y(i) & N(i) \\ 0 & N^*(i) & S(i) \end{bmatrix}$$

$$= rank \begin{bmatrix} X(i) - (Y(i) - N(i)S^{-1}(i)N^*(i))^{-1} & 0 & 0 \\ 0 & Y(i) - N(i)S^{-1}(i)N^*(i) & 0 \\ 0 & 0 & S(i) \end{bmatrix},$$

and $S(i) > 0$, $Y(i) - N(i)S^{-1}(i)N^*(i) > 0$, it follows that (7.45) gives

$$X(i) = (Y(i) - N(i)S^{-1}(i)N^*(i))^{-1}$$

from which (7.32) follows directly.

(ii) $\Rightarrow$ (i) Assume that there exist $X(i)$, $Y(i)$, $N(i)$, and $S(i)$ verifying (7.30)–(7.32). From (7.31) it follows that $\Pi_{r+1,r+1}(i) < 0$ and therefore

$$\widetilde{Y}(i) = \begin{bmatrix} Y(i) & N(i) \\ N^*(i) & S(i) \end{bmatrix} > 0.$$

Hence $\widetilde{Y}(i)$ is invertible. From Lemma 8 it results that $\widetilde{Y}^{-1}(i)$ has the structure

$$\begin{bmatrix} X(i) & \star \\ \star & \star \end{bmatrix},$$

where by $\star$ we denoted the irrelevant entries. From the developments performed to prove the implication (i) $\Rightarrow$ (ii), it follows that (7.37) and (7.38) are verified by

$$X_{cl}(i) = \widetilde{Y}^{-1}(i),$$

and hence (7.35) has a solution that guarantees the existence of a stabilizing and γ-attenuating controller. Thus the proof is complete. $\square$

Remark 5. In the case of the static output feedback ($n_c = 0$), in the above theorem we have to remove all variables n_c, $N(i)$, and $S(i)$, $i \in \mathcal{D}$.

Remark 6. According to the proof of the above result, the algorithm to determine a solution of the DAP is the following:
   Step 1 Solve the system of LMI (7.30) and (7.31) with the constraint (7.32).
   Step 2 Compute $\mathcal{Z}(i)$, $\mathcal{U}(i)$, and $\mathcal{V}(i)$, $i \in \mathcal{D}$, according to (7.36).
   Step 3 Solve the basic LMI (7.35) with respect to Θ_c. Then the solution of the DAP is given by the partition

$$\Theta_c(i) = \begin{bmatrix} A_c(i) & B_c(i) \\ C_c(i) & D_c(i) \end{bmatrix}.$$

Obviously, if $n_c = 0$, then $\Theta_c(i) = D_c(i)$.

In the following we shall emphasize the important particular cases when the system (7.1) is subjected only to Markovian jumping or to multiplicative white noise. In the situation when $A_k(i) = 0$, $B_k(i) = 0$, $G_k(i) = 0$, $1 \leq k \leq r$, $i \in \mathcal{D}$, the system (5.1) becomes

$$\dot{x}(t) = A_0(\eta(t))x(t) + G_0(\eta(t))v(t) + B_0(\eta(t))u(t),$$
$$z(t) = C_z(\eta(t))x(t) + D_{zv}(\eta(t))v(t) + D_{zu}(\eta(t))u(t), \qquad (7.47)$$
$$y(t) = C_0(\eta(t))x(t) + D_0(\eta(t))v(t).$$

The closed-loop system obtained by coupling a controller of form (7.2) to the system (7.47) has the following state-space realization:

$$\dot{x}_{cl}(t) = A_{0cl}(\eta(t))x_{cl}(t) + G_{0cl}(\eta(t))v(t), \qquad (7.48)$$
$$z(t) = C_{cl}(\eta(t))x_{cl}(t) + D_{cl}(\eta(t))v(t),$$

where the matrix coefficients are defined as in (7.4). The results in the preceding theorem lead for the particular system (7.47) to the following theorem.

Theorem 10. *For a $\gamma > 0$ the following are equivalent:*

(i) *There exists a controller of order $n_c \geq 0$ of type (7.2) which stabilizes the system (7.47) such that the input–output operator associated with the system (7.48) verifies $\|T_{cl}\| < \gamma$.*

(ii) *There exist $X = (X(1), \ldots, X(d)) \in S_n^d, Y = (Y(1), \ldots, Y(d)) \in S_n^d, S = (S(1), \ldots, S(d)) \in S_n^d, N \in (N(1), \ldots, N(d)), N \in M_{n,n_c}^d$ such that*

$$
\begin{bmatrix} V_0(i) \\ V_1(i) \end{bmatrix}^* \begin{bmatrix} A_0^*(i)X(i) + X(i)A_0(i) \\ +\sum_{j=1}^d q_{ij}X(j) + C_z^*(i)C_z(i) & X(i)G_0(i) + C_z^*(i)D_{zv}(i) \\ G_0^*(i)X(i) + D_{zv}^*(i)C_z(i) & -\gamma^2 I_{m_1} + D_{zv}^*(i)D_{zv}(i) \end{bmatrix}
$$
$$
\times [V_0(i) \quad V_1(i)] < 0, \tag{7.49}
$$

$$
\begin{bmatrix} \Pi_{0,0}(\gamma, i) & \Pi_{0,r+1}(\gamma, i) \\ \Pi_{0,r+1}^*(\gamma, i) & \Pi_{r+1,r+1}(\gamma, i) \end{bmatrix} < 0, \tag{7.50}
$$

$$
\text{rank} \begin{bmatrix} X(i) & I & 0 \\ I & Y(i) & N(i) \\ 0 & N^*(i) & S(i) \end{bmatrix} = n + n_c, \quad i \in \mathcal{D}, \tag{7.51}
$$

where

$$
\Pi_{0,0}(\gamma, i) = \begin{bmatrix} U_0(i) \\ U_{r+1}(i) \end{bmatrix}^*
$$
$$
\times \begin{bmatrix} A_0(i)Y(i) + Y(i)A_0^*(i) \\ +q_{ii}X(i) + \gamma^{-2}G_0(i)G_0^*(i) & Y(i)C_z^*(i) + \gamma^{-2}G_0(i)D_{zv}^*(i) \\ C_z(i)Y(i) + \gamma^{-2}D_{zv}(i)G_0^*(i) & -I_{p_1} + \gamma^{-2}D_{zv}(i)D_{zv}^*(i) \end{bmatrix}
$$
$$
\times [U_0(i) \quad U_{r+1}(i)],
$$

$\Pi_{0,r+1}(\gamma, i)$ *and* $\Pi_{r+1,r+1}(\gamma, i)$ *are as in Theorem 9,*

$$
\begin{bmatrix} U_0(i) \\ U_{r+1}(i) \end{bmatrix}
$$

is a basis of the null subspace of $[B_0^*(i) \quad D_{zu}^*(i)]$, *and*

$$
\begin{bmatrix} V_0(i) \\ V_1(i) \end{bmatrix}
$$

is a basis of the null subspace of $[C_0(i) \quad D_0(i)]$.   □

Remark 7. From the above theorem one can see that the necessary and sufficient conditions guaranteeing the solvability of the DAP involve the same unknown variables, namely $X(i), Y(i), S(i), N(i), i \in \mathcal{D}$, as in the general case of the system (7.1). It seems that this is the price paid to obtain a controller of order $n_c < n$. In the particular case when a full-order controller ($n_c = n$) is required, the rank condition (7.32) in the statement of Theorem 5 is removed (see Theorem 14 in Section 5.4).

Now consider the case when $\mathcal{D} = \{1\}$. Then the system (7.1) becomes

$$dx(t) = [A_0x(t) + G_0v(t) + B_0u(t)]\,dt$$

$$+ \sum_{k=1}^{r}[A_kx(t) + G_kv(t) + B_ku(t)]\,dw_k(t),$$

$$z(t) = C_zx(t) + D_{zv}v(t) + D_{zu}u(t), \tag{7.52}$$

$$y(t) = C_0x(t) + D_0v(t),$$

where the matrices A_k, B_k, G_k, $0 \leq k \leq r$, C_z, D_{zu}, D_{zv}, C_0, D_0 are given matrices of appropriate dimensions. The class of admissible controllers consists in deterministic controllers of the form

$$\dot{x}_c(t) = A_cx_c(t) + B_cy(t), \tag{7.53}$$

$$u(t) = C_cx_c(t) + D_cy(t).$$

The closed-loop system obtained when coupling (7.53) to (7.52) is

$$dx_c(t) = [A_{0cl}x_{cl}(t) + G_{0cl}v(t)]\,dt + \sum_{k=1}^{r}[A_{kcl}x_{cl}(t) + G_{kcl}v(t)]\,dw_k(t), \tag{7.54}$$

$$z(t) = C_{cl}x_{cl}(t) + D_{cl}v(t),$$

where the matrix coefficients are as in (7.4) with $d = 1$.

The next result provides a version of Theorem 9 for the particular case of the system (7.52).

Theorem 11. *For a given $\gamma > 0$ the following are equivalent:*

(i) There exist an $(n_c \geq 0)$-order controller stabilizing (7.53) such that the input–output operator associated with the system (7.54) verifies $\|T_{cl}\| < \gamma$.

(ii) There exist $X, Y \in \mathcal{S}_n$, $S \in \mathcal{S}_{n_c}$, $N \in \mathbf{R}^{n \times n_c}$ satisfying $X > 0$, $Y > 0$, $S > 0$, such that

$$\begin{bmatrix} V_0 \\ V_1 \end{bmatrix}^* \begin{bmatrix} A_0^*X + XA_0 \\ +\sum_{k=1}^{r} A_k^*XA_k + C_z^*C_z & XG_0 + \sum_{k=1}^{r} A_k^*XG_k \\ G_0^*X + \sum_{k=1}^{r} G_k^*XA_k & +C_z^*D_{zv} \\ +D_{zv}^*C_z & -\gamma^2 I_{m_1} + D_{zv}^*D_{zv} \\ & +\sum_{k=1}^{r} G_k^*XG_k \end{bmatrix}$$

$$\times [V_0 \quad V_1] < 0, \tag{7.55}$$

$$\begin{bmatrix} \Pi_{0,0}(Y) & \Pi_{0,1} & -U_1^*N & \cdots & -U_r^*N \\ \Pi_{0,1}^* & -\gamma^2 I_{m_1} & 0 & \cdots & 0 \\ -N^*U_1 & 0 & -S & \cdots & 0 \\ \vdots & \vdots & \vdots & \ddots & \vdots \\ -N^*U_r & 0 & 0 & \cdots & -S \end{bmatrix} < 0, \tag{7.56}$$

$$\text{rank} \begin{bmatrix} X & I & 0 \\ I & Y & N \\ 0 & N^* & S \end{bmatrix} = n + n_c, \tag{7.57}$$

where

$$\begin{bmatrix} V_0 \\ V_1 \end{bmatrix} \quad and \quad \begin{bmatrix} U_0 \\ \vdots \\ U_{r+1} \end{bmatrix}$$

are bases of the null subspaces of $[C_0 \quad D_0]$ and $\begin{bmatrix} B_0^* & B_1^* & \cdots & B_r^* & D_{zu}^* \end{bmatrix}$, respectively, and

$$\Pi_{0,0}(Y) = U_0^* \left[A_0 Y + Y A_0^* \right] U_0 + \sum_{k=1}^{r} U_k^* A_k Y U_0$$

$$+ \sum_{k=1}^{r} U_k^* A_k Y U_0 + U_0^* Y C_z^* U_{r+1} + U_{r+1}^* C_z Y U_0$$

$$- \sum_{k=1}^{r} U_k^* Y U_k - U_{r+1}^* U_{r+1},$$

$$\Pi_{0,1} = \sum_{k=0}^{r} U_k^* G_k + U_{r+1}^* D_{zv}.$$

The next result is well known in the deterministic case; however, for the sake of completeness, we shall briefly present it in the following lemma.

Lemma 12. *Let $X_{cl} \in \mathbf{R}^{n \times n_c}$ be partitioned as*

$$X_{cl} = \begin{bmatrix} X & M \\ M^* & \tilde{X} \end{bmatrix}, \ X \in \mathcal{S}_n, \ \tilde{X} \in \mathcal{S}_{n_c},$$

where $n_c \geq 1$. Assume that $X_{cl} > 0$ and consider the following partition of X_{cl}^{-1}:

$$X_{cl}^{-1} = \begin{bmatrix} Y & N \\ N^* & S \end{bmatrix}, \ Y \in \mathcal{S}_n, \ S \in \mathcal{S}_{n_c}.$$

Then we have

$$X \geq Y^{-1} > 0, \tag{7.58}$$

$$\operatorname{rank}\left(X - Y^{-1}\right) \leq n_c. \tag{7.59}$$

Conversely, if there exist $X \in \mathcal{S}_n$, $Y \in \mathcal{S}_n$ verifying conditions (7.58) and (7.59), then there exist $M \in \mathbf{R}^{n \times n_c}$, $\tilde{X} \in \mathcal{S}_{n_c}$, $N \in \mathbf{R}^{n \times n_c}$, $S \in \mathcal{S}_{n_c}$ such that

$$\begin{bmatrix} X & M \\ M^* & \tilde{X} \end{bmatrix} > 0$$

and

$$\begin{bmatrix} X & M \\ M^* & \tilde{X} \end{bmatrix} \begin{bmatrix} Y & N \\ N^* & S \end{bmatrix} = \begin{bmatrix} I & 0 \\ 0 & I \end{bmatrix}. \tag{7.60}$$

Proof. From $X_{cl} > 0$ it follows that $X > 0$, $\widetilde{X} > 0$, $S > 0$. From the condition $X_{cl} X_{cl}^{-1} = I$ one obtains that

$$X - Y^{-1} = Y^{-1} N \widetilde{X} N^* Y^{-1},$$

and therefore (7.58) immediately follows. The above conditions also leads to

$$rank\left(X - Y^{-1}\right) = rank(N) \leq n_c$$

and hence (7.59) results.

Conversely, let $X, Y \in S_n$ satisfying (7.58) and (7.59). Define $M \in \mathbf{R}^{n \times n_c}$ as the Cholesky factor:

$$X - Y^{-1} = MM^*$$

and

$$N = -YM,$$
$$\widetilde{X} = I_{n_c},$$
$$S = I_{n_c} + M^* Y M.$$

Then it follows that

$$X - M\widetilde{X}^{-1}M^* = Y^{-1} > 0,$$
$$S - N^* Y^{-1} N = I_{n_c} > 0.$$

Then (7.60) follows by direct computations and thus the proof is complete. □

The next result shows that it is possible to remove the unknown variables N and S, but in this case the condition (7.56) in Theorem 11 becomes nonlinear.

Theorem 13. *For a given $\gamma > 0$ the following are equivalent:*

(i) *There exists a stabilizing controller with $n_c > 0$ of form (7.53) solving the DAP for the system (7.52).*

(ii) *There exist $X, Y \in S_n$, $X > 0$, $Y > 0$ satisfying the following conditions:*

$$\begin{bmatrix} X & I \\ I & Y \end{bmatrix} \geq 0, \tag{7.61}$$

$$rank \begin{bmatrix} X & I \\ I & Y \end{bmatrix} \leq n + n_c, \tag{7.62}$$

$$\begin{bmatrix} V_0 \\ V_1 \end{bmatrix}^* \begin{bmatrix} A_0^* X + X A_0 & X G_0 + \sum_{k=1}^r A_k^* X G_k \\ + \sum_{k=1}^r A_k^* X A_k + C_z^* C_z & + C_z^* D_{zv} \\ G_0^* X + \sum_{k=1}^r G_k^* X A_k & -\gamma^2 I_{m_1} + D_{zv}^* D_{zv} \\ + D_{zv}^* C_z & + \sum_{k=1}^r G_k^* X G_k \end{bmatrix}$$
$$\times [V_0 \quad V_1] < 0, \tag{7.63}$$

$$U^* \Lambda(Y, \gamma) U < 0, \tag{7.64}$$

where $\begin{bmatrix} V_0 \\ V_1 \end{bmatrix}$ *is a basis of the null subspace of* $[C_0 \quad D_0]$,

$$U = \begin{bmatrix} U_0 \\ U_1 \\ \vdots \\ U_{r+1} \end{bmatrix}$$

is a basis of the null subspace of $[B_0^* \quad B_1^* \quad \cdots \quad B_r^* \quad D_{zu}^*]$, *and*

$$\Lambda = \begin{bmatrix} \Lambda_{0,0} & \cdots & \Lambda_{0,r} & \Lambda_{0,r+1} \\ \vdots & \ddots & \vdots & \vdots \\ \Lambda_{0,r}^* & \cdots & \Lambda_{r,r} & \Lambda_{r+1,r} \\ \Lambda_{0,r+1}^* & \cdots & \Lambda_{r+1,r}^* & \Lambda_{r+1,r+1} \end{bmatrix},$$

$$\Lambda_{0,0} = A_0 Y + Y A_0^* + \gamma^{-2} G_0 G_0^*,$$
$$\Lambda_{0,k} = Y A_k^* + \gamma^{-2} G_0 G_k^*, \ 1 \le k \le r,$$
$$\Lambda_{0,r+1} = Y C_z^* + \gamma^{-2} G_0 D_{zv}^*,$$
$$\Lambda_{l,k} = \gamma^{-2} G_l G_k^*, \ 1 \le l \ne k \le r,$$
$$\Lambda_{l,l} = \gamma^{-2} G_l G_l^* - X^{-1},$$
$$\Lambda_{l,r+1} = \gamma^{-2} G_l D_{zv}^*, \ 1 \le l \le r,$$
$$\Lambda_{r+1,r+1} = -I_{p_1} + \gamma^{-2} D_{zv} D_{zv}^*.$$

Proof. (i) $\Rightarrow$ (ii) If (i) in the statement is fulfilled, then using the implication (i) $\Rightarrow$ (ii) of Theorem 11 we deduce that there exist $X, Y \in \mathcal{S}_n$, $S \in \mathcal{S}_{n_c}$, $N \in \mathbf{R}^{n \times n_c}$ such that (7.55)–(7.57) are satisfied. One can see that (7.55) is just (7.63). On the other hand, (7.57) leads to

$$X = (Y - NS^{-1}N^*)^{-1}. \tag{7.65}$$

This means that X is the (1,1) block of the matrix

$$\begin{bmatrix} Y & N \\ N^* & S \end{bmatrix}^{-1}.$$

Applying Lemma 12 for

$$X_{cl} = \begin{bmatrix} Y & N \\ N^* & S \end{bmatrix}^{-1},$$

it follows that

$$X - Y^{-1} \ge 0 \tag{7.66}$$

and

$$rank\left(X - Y^{-1}\right) \le n_c.$$ (7.67)

It is obvious that (7.66) is equivalent to (7.61) and (7.67) is equivalent to (7.62). But (7.56) leads to

$$U_0^*\left(A_0Y + YA_0^*\right)U_0 + \sum_{k=0}^{r}\sum_{l=0}^{r}\gamma^{-2}U_k^*G_kG_l^*U_k$$

$$+ \sum_{k=1}^{r}\left(U_0^*YA_k^*U_k + U_k^*A_kYU_0\right) + U_0^*YC_z^*U_{r+1} + U_{r+1}C_zYU_0$$

$$+ \sum_{k=0}^{r}\gamma^{-2}\left(U_k^*G_kD_{zv}^*U_{r+1} + U_{r+1}^*D_{zv}G_k^*U_k\right)$$

$$- U_{r+1}^*\left(I_{p_1} - \gamma^{-2}D_{zv}D_{zv}^*\right)U_{r+1} - \sum_{k=1}^{r}U_k^*\left(Y - NS^{-1}N^*\right)U_k < 0. \quad (7.68)$$

Using (7.65), (7.68) becomes (7.64). Therefore there exist $X, Y \in S_n, X > 0, Y > 0$ verifying (7.61)–(7.64). Suppose now that (ii) holds. From (7.61) one deduces that $X \ge Y^{-1} > 0$ and $rank\left(X - Y^{-1}\right) \le n_c$. Then, according to Lemma 12, there exist $N \in \mathbf{R}^{n \times n_c}, M \in \mathbf{R}^{n \times n_c}, \tilde{X} \in \mathbf{R}^{n_c \times n_c}, S \in \mathbf{R}^{n_c \times n_c}$ such that

$$\begin{bmatrix} X & M \\ M^* & \tilde{X} \end{bmatrix} = \begin{bmatrix} Y & N \\ N^* & S \end{bmatrix}^{-1} > 0,$$ (7.69)

and therefore

$$X^{-1} = Y - NS^{-1}N^*.$$ (7.70)

Thus (7.64) becomes (7.68) and therefore (7.56) holds. Moreover, (7.69) and (7.70) imply (7.57). Taking into account that (7.63) is just(7.55), we conclude that if (ii) in the statement holds, then the condition (ii) in Theorem 11 is also verified. Then the implication (ii) $\Rightarrow$ (i) in Theorem 11 shows that (i) in the statement is fulfilled, and hence the proof is complete. □

Remark 8. In order to solve the system (7.61)–(7.64), one can suggest the following algorithm:

Step 1 Solve (7.63) with respect to X.

Step 2 Introduce X determined at Step 1 in (7.61), (7.62), and (7.64), and solve the obtained LMI system with respect to Y.

Now consider the particular case when in (7.52), $B_k = 0, k = 1, \ldots, r$. In this situation the base U becomes

$$U = \begin{bmatrix} \tilde{U}_0 & 0 \\ 0 & I_{nr} \\ \tilde{U}_{r+1} & 0 \end{bmatrix},$$

where

$$\begin{bmatrix} \tilde{U}_0 \\ \tilde{U}_{r+1} \end{bmatrix}$$

is a basis of the null subspace of the matrix $\begin{bmatrix} B_0^* & D_{zu}^* \end{bmatrix}$. Then condition (7.64) becomes

$$\begin{bmatrix} \tilde{\Pi}_{0,0} & \tilde{\Pi}_{0,1} & \cdots & \tilde{\Pi}_{0,r} \\ \tilde{\Pi}_{0,1}^* & \tilde{\Pi}_{1,1} & \cdots & \tilde{\Pi}_{1,r} \\ \vdots & \vdots & \ddots & \vdots \\ \tilde{\Pi}_{0,r}^* & \tilde{\Pi}_{1,1}^* & \cdots & \tilde{\Pi}_{r,r} \end{bmatrix} < 0, \tag{7.71}$$

where

$$\tilde{\Pi}_{0,0} = \begin{bmatrix} \tilde{U}_0 \\ \tilde{U}_{r+1} \end{bmatrix}^* \left(\begin{bmatrix} A_0 Y + Y A_0^* & Y C_z^* \\ C_z Y & -I_{p_1} \end{bmatrix} + \gamma^{-2} \begin{bmatrix} G_0 \\ D_{zv} \end{bmatrix} \begin{bmatrix} G_0^* & D_{zv}^* \end{bmatrix} \right) \begin{bmatrix} \tilde{U}_0 \\ \tilde{U}_{r+1} \end{bmatrix},$$

$$\tilde{\Pi}_{0,l} = \tilde{U}_0^* Y A_l^* + \gamma^{-2} \begin{bmatrix} \tilde{U}_0 \\ \tilde{U}_{r+1} \end{bmatrix}^* \begin{bmatrix} G_0 \\ D_{zv} \end{bmatrix} G_l^*, \quad 1 \le l \le r,$$

$$\tilde{\Pi}_{0,k} = \gamma^{-2} G_l G_k^*, \quad 1 \le l \ne k \le r,$$

$$\tilde{\Pi}_{l,l} = \gamma^{-2} G_l G_l^* - X^{-1}, \quad 1 \le l \le r.$$

By Schur complement arguments, (7.71) is equivalent to the extended inequality

$$\begin{bmatrix} \Lambda_{0,0}(Y, \Gamma) & \tilde{U}_0^* Y A_1^* & \cdots & \tilde{U}_0^* Y A_r^* & \tilde{U}_0^* G_0 + \tilde{U}_{r+1}^* D_{zv} \\ A_1 Y \tilde{U}_0 & -X^{-1} & \cdots & 0 & G_1 \\ \vdots & \vdots & \ddots & \vdots & \vdots \\ A_r Y \tilde{U}_0 & 0 & \cdots & X^{-1} & G_r \\ G_0^* \tilde{U}_0 + D_{zv}^* \tilde{U}_{r+1} & G_1^* & \cdots & G_r^* & -\gamma^2 I_{m_1} \end{bmatrix} < 0,$$

where

$$\Lambda_{0,0}(Y, \Gamma) = \begin{bmatrix} \tilde{U}_0 \\ \tilde{U}_{r+1} \end{bmatrix}^* \begin{bmatrix} A_0 Y + Y A_0^* & Y C_z^* \\ C_z Y & -I_{p_1} \end{bmatrix} \begin{bmatrix} \tilde{U}_0 \\ \tilde{U}_{r+1} \end{bmatrix}.$$

Taking the Schur complement of $diag(-X^{-1}, \ldots, -X^{-1})$ in the above inequality, one obtains

$$\begin{bmatrix} \Lambda_{0,0}(Y, \Gamma) & \tilde{U}_0^* G_0 + \tilde{U}_{r+1}^* D_{zv} \\ + \sum_{k=1}^r \tilde{U}_0 Y A_k^* X A_k Y \tilde{U}_0 & + \sum_{k=1}^r \tilde{U}_0^* Y A_k^* X G_k \\ G_0^* \tilde{U}_0 + D_{zv}^* \tilde{U}_{r+1} & -\gamma^2 I_{m_1} + \sum_{k=1}^r G_k^* X G_k \\ + \sum_{k=1}^r G_k^* X A_k Y \tilde{U}_0 & \end{bmatrix} < 0.$$

The above inequality together with (7.61), (7.62), and (7.63) are the necessary and sufficient conditions derived in [65].

In the final part of this section we shall discuss two problems of robust stabilization with respect to parametric uncertainty.

Consider the system described by

$$dx(t) = \left\{ \left[A_0(\eta(t)) + \widehat{G}_0(\eta(t)) \Delta_1(\eta(t)) \widehat{C}(\eta(t)) \right] x(t) \right. \tag{7.72}$$
$$+ \left[B_0(\eta(t)) + \widehat{B}_0(\eta(t)) \Delta_2(\eta(t)) \widehat{D}(\eta(t)) \right] u(t) \right\} dt$$

$$+ \sum_{k=1}^{r} \left\{ \left[A_k(\eta(t)) + \widehat{G}_k(\eta(t)) \Delta_1(\eta(t)) \widehat{C}(\eta(t)) \right] x(t) \right.$$

$$+ \left[B_k(\eta(t)) + \widehat{B}_k(\eta(t)) \Delta_2(\eta(t)) \widehat{D}(\eta(t)) \right] u(t) \right\} dw_k(t),$$

$$y(t) = C_0(\eta(t)) x(t),$$

where $x(t) \in \mathbf{R}^n$ denotes the state, $u(t) \in \mathbf{R}^m$ is the control variable, and $y \in \mathbf{R}^{p_2}$ is the measured output. The matrices $A_k(i) \in \mathbf{R}^{n \times n}$, $B_k(i) \in \mathbf{R}^{n \times m}$, $\widehat{G}_k(i) \in \mathbf{R}^{n \times \widehat{m}_1}$, $\widehat{B}_k(i) \in \mathbf{R}^{n \times \widetilde{m}_1}$, $0 \le k \le r$, $\widehat{C}(i) \in \mathbf{R}^{\widehat{p}_1 \times n}$, $\widehat{D}(i) \in \mathbf{R}^{\widetilde{p}_1 \times m}$, $C_0(i) \in \mathbf{R}^{p_2 \times n}$ are known matrices and $\Delta_1 \in \mathbf{R}^{\widehat{m}_1 \times \widehat{p}_1}$, $\Delta_2 \in \mathbf{R}^{\widetilde{m}_1 \times \widetilde{p}_1}$ are unknown matrices describing the parametric uncertainty. The robust stabilization problem we address has the following statement: Find a stabilizing controller of form (7.2) for the system (7.72) for arbitrary Δ_1, Δ_2 with $\max(|\Delta_1|, |\Delta_2|) < \rho$ for a prescribed $\rho > 0$, where $|\Delta_l| = \max_{i \in \mathcal{D}} |\Delta_l(i)|$, $l = 1, 2$. The closed-loop system obtained when coupling the controller (7.2) to (7.72) has the following state-space representation:

$$dx(t) = \left\{ [A_0(\eta(t)) + B_0(\eta(t)) D_c(\eta(t)) C_0(\eta(t))] x(t) \right.$$
$$+ B_0(\eta(t)) C_c(\eta(t)) x_c(t) + \left[\widehat{G}_0(\eta(t)) \Delta_1(\eta(t)) \widehat{C}(\eta(t)) \right.$$
$$+ \widehat{B}_0(\eta(t)) \Delta_2(\eta(t)) \widehat{D}(\eta(t)) D_c(\eta(t)) C_0(\eta(t)) \right] x(t)$$
$$+ \widehat{B}_0(\eta(t)) \Delta_2(\eta(t)) \widehat{D}(\eta(t)) C_c(\eta(t)) x_c(t) \right\} dt \tag{7.73}$$

$$\times \sum_{k=1}^{r} \left\{ [A_k(\eta(t)) + B_k(\eta(t)) D_c(\eta(t)) C_0(\eta(t))] x(t) \right.$$

$$+ B_k(\eta(t)) C_c(\eta(t)) x_c(t) + \left[\widehat{G}_k(\eta(t)) \Delta_1(\eta(t)) \widehat{C}(\eta(t)) \right.$$
$$+ \widehat{B}_k(\eta(t)) \Delta_2(\eta(t)) \widehat{D}(\eta(t)) D_c(\eta(t)) C_0(\eta(t)) \right] x(t)$$
$$+ \widehat{B}_k(\eta(t)) \Delta_2(\eta(t)) \widehat{D}(\eta(t)) C_c(\eta(t)) x_c(t) \right\} dw_k(t),$$
$$dx_c(t) = [B_c(\eta(t)) C_0(\eta(t)) x(t) + A_c(\eta(t)) x_c(t)] dt.$$

Denoting

$$G_k(i) = \left[\widehat{G}_k(i) \quad \widehat{B}_k(i) \right] \in \mathbf{R}^{n \times (\widehat{m}_1 + \widetilde{m}_1)}, 0 \le k \le r,$$

$$C_z(i) = \begin{bmatrix} \widehat{C}(i) \\ 0 \end{bmatrix} \in \mathbf{R}^{(\widehat{p}_1 + \widetilde{p}_1) \times n}, \tag{7.74}$$

$$D_{zu}(i) = \begin{bmatrix} 0 \\ \widehat{D}(i) \end{bmatrix} \in \mathbf{R}^{(\widehat{p}_1 + \widetilde{p}_1) \times m},$$

$$\Delta = \begin{bmatrix} \Delta_1(i) & 0 \\ 0 & \Delta_2(i) \end{bmatrix},$$

the system (7.73) can be rewritten in compact form as follows:

$$d\xi(t) = [A_{0cl}(\eta(t)) + G_{0cl}(\eta(t))\Delta(\eta(t))C_{cl}(\eta(t))]\,\xi(t)dt \tag{7.75}$$

$$+ \sum_{k=1}^{r} [A_{kcl}(\eta(t)) + G_{kcl}(\eta(t))\Delta(\eta(t))C_{cl}(\eta(t))]\,\xi(t)dw_k(t),$$

where $A_{kcl}(i)$ are defined as in (7.4) and

$$G_{kcl}(i) = \begin{bmatrix} G_k(i) \\ 0 \end{bmatrix},$$

$$C_{cl}(i) = [C_z(i) + D_{zu}(i)D_c(i)C_0(i) \quad D_{zu}(i)C_c(i)]$$

$$= \begin{bmatrix} \widehat{C}(i) & 0 \\ \widehat{D}(i)D_c(i)C_0(i) & \widehat{D}(i)C_c(i) \end{bmatrix}, i \in \mathcal{D}.$$

Therefore, the closed-loop system can be viewed as a perturbation of the system

$$d\xi(t) = A_{0cl}(\eta(t))\xi(t)dt + \sum_{k=1}^{r} A_{kcl}(\eta(t))\xi(t)dw_k(t)$$

obtained by coupling the controller (7.2) to the nominal system (7.72) obtained with $\Delta_1 = 0$, $\Delta_2 = 0$. Applying Corollary 21 of Chapter 6 to the system (7.75), it follows that a controller of type (7.2) stabilizes (7.72) for any Δ_1, Δ_2 with $\max(|\Delta_1|, |\Delta_2|) < \rho$ if the input–output operator $\mathcal{T}_{cl}$ associated with the fictitious system

$$d\xi_{cl}(t) = [A_{0cl}(\eta(t))\xi(t) + G_{0cl}(\eta(t))v(t)]\,dt$$

$$+ \sum_{k=1}^{r} [A_{kcl}(\eta(t))\xi(t) + G_{kcl}(\eta(t))v(t)]\,dw_k(t),$$

$$z(t) = C_{cl}(\eta(t))\xi(t)$$

verifies the condition $\|\mathcal{T}_{cl}\| < 1/\rho$. Then a stabilizing controller (7.2) providing the robustness radius ρ can be obtained as a solution of the DAP with $\gamma = \rho^{-1}$ corresponding to the two-input, two-output generalized system:

$$dx(t) = [A_0(\eta(t))x(t) + G_0(\eta(t))v(t) + B_0(\eta(t))u(t)]\,dt$$

$$+ \sum_{k=1}^{r} [A_k(\eta(t))x(t) + G_k(\eta(t))v(t) + B_k(\eta(t))u(t)]\,dw_k(t),$$

$$z(t) = C_z(\eta(t))x(t) + D_{zu}(\eta(t))u(t), \tag{7.76}$$

$$y(t) = C_0(\eta(t))x(t),$$

where $G_k(i)$, $0 \le k \le r$, $C_z(i)$, $D_{zu}(i)$, $i \in \mathcal{D}$ are defined as in (7.74). Then a robust stabilizing controller with the robustness radius ρ may be obtained, applying Theorem 9 to the system (7.76) for $\gamma = \rho^{-1}$.

The second robust stabilization problem with respect to parametric uncertainty considered in the final part of this section is the following: Find a stabilizing controller of type (7.2) for the system:

$$dx(t) = [A_0(\eta(t))x(t) + G_0(\eta(t))\Delta(\varphi(t), \eta(t)) + B_0(\eta(t))u(t)] \, dt$$

$$+ \sum_{k=1}^{r} [A_k(\eta(t))x(t) + G_k(\eta(t))\Delta(\varphi(t), \eta(t))$$

$$+ B_k(\eta(t))u(t)] \, dw_k(t), \tag{7.77}$$

$$y(t) = C_0(\eta(t))x(t) + D_0(\eta(t))v(t),$$

where $\varphi(t) = C(\eta(t))x(t)$ and Δ are unknown Lipschitz functions with $\Delta(0, i) = 0$ and

$$\sup_{i \in \mathcal{D}, z \in \mathbf{R}^{p_1}, z \neq 0} \frac{|\Delta(z, i)|}{|z|} < \rho. \tag{7.78}$$

When coupling a controller of type (7.2) to the system (7.77), the closed-loop system has the following state-space equation:

$$dx_{cl}(t) = [A_{0cl}(\eta(t))x_{cl}(t) + G_{0cl}(\eta(t))\Delta(\varphi(t), \eta(t))] \, dt \tag{7.79}$$

$$+ \sum_{k=1}^{r} [A_{kcl}(\eta(t))x_{cl}(t) + G_{kcl}(\eta(t))\Delta(\varphi(t), \eta(t))] \, dw_k(t),$$

where $A_{kcl}(i)$, $G_{kcl}(i)$ are defined as in (7.4), $0 \leq k \leq r$. Invoking Theorem 24 of Chapter 6 for the system (7.79), it follows that a controller (7.2) stabilizes (7.77) for any nonlinear perturbation Δ satisfying (7.78) if $\rho < 1/\|\mathcal{T}_{cl}\|$, where $\mathcal{T}_{cl}$ is the input–output operator of the system

$$d\xi(t) = [A_{0cl}(\eta(t))\xi(t) + G_{0cl}(\eta(t))v(t)] \, dt \tag{7.80}$$

$$+ \sum_{k=1}^{r} [A_{kcl}(\eta(t))\xi(t) + G_{kcl}(\eta(t))v(t)] \, dw_k(t),$$

$$z(t) = [C(\eta(t)) \quad 0] \xi(t).$$

Hence a robust stabilizing controller for (7.77) can be obtained by solving the DAP for $\gamma = 1/\rho$ for the system

$$dx(t) = [A_0(\eta(t))x(t) + G_0(\eta(t))v(t) + B_0(\eta(t))u(t)] \, dt$$

$$+ \sum_{k=1}^{r} [A_k(\eta(t))x(t) + G_k(\eta(t))v(t) + B_k(\eta(t))u(t)] \, dw_k(t), \tag{7.81}$$

$$z(t) = C(\eta(t))x(t),$$

$$y(t) = C_0(\eta(t))x(t) + D_0(\eta(t))v(t).$$

Solvability conditions for this DAP are provided by Theorem 9.

7.4 DAP for linear stochastic systems with Markovian jumping

In this section we shall investigate the γ-attenuation problem for linear stochastic systems of form (7.47) looking for strictly proper n-order controllers with $D_c(i) = 0$, $i \in \mathcal{D}$. More precisely, the class of considered controllers is given by

$$
\begin{aligned}
\dot{x}_c(t) &= A_c(\eta(t))x_c(t) + B_c(\eta(t))y(t), \\
u(t) &= C_c(\eta(t))x_c(t),
\end{aligned} \tag{7.82}
$$

where $A_c(i) \in \mathbf{R}^{n \times n}$, $B_c(i) \in \mathbf{R}^{n \times p_2}$, $C_c(i) \in \mathbf{R}^{m_2 \times n}$, $i \in \mathcal{D}$. When coupling the controller (7.82) to the system (7.47), one obtains

$$
\begin{aligned}
\dot{x}_{cl}(t) &= A_{cl}(\eta(t))x_{cl}(t) + G_{cl}(\eta(t))v(t), \\
z(t) &= C_{cl}(\eta(t))x_{cl}(t) + D_{cl}(\eta(t))v(t),
\end{aligned}
$$

where

$$
\begin{aligned}
A_{cl}(i) &= \begin{bmatrix} A_0(i) & B_0(i)C_c(i) \\ B_c(i)C_0(i) & A_c(i) \end{bmatrix}, \\
G_{cl}(i) &= \begin{bmatrix} G_0(i) \\ B_c(i)D_0(i) \end{bmatrix}, \\
C_{cl}(i) &= [C_z(i) \quad D_{zu}(i)C_c(i)], \\
D_{cl}(i) &= D_{zv}(i).
\end{aligned} \tag{7.83}
$$

The following result provides necessary and sufficient conditions that guarantee the existence of a solution of form (7.82) of the DAP.

Theorem 14. *For $\gamma > 0$ the following are equivalent:*
  (i) *There exists a controller of form (7.82) stabilizing (7.47) and solving the DAP with the level of attenuation γ.*
  (ii) *There exist $X = (X(1), \ldots, X(d)) \in \mathcal{S}_n^d$, $Y = (Y(1), \ldots, Y(d)) \in \mathcal{S}_n^d$, $F = (F(1), \ldots, F(d)) \in \mathcal{M}_{m_2 n}^d$, $K = (K(1), \ldots, K(d)) \in \mathcal{M}_{n p_2}^d$, which verify*

$$
X(i) > 0,
$$

$$
V(i) = \begin{bmatrix} V_{11}(i) & V_{12}(i) \\ V_{12}^*(i) & V_{22}(i) \end{bmatrix} < 0, \tag{7.84}
$$

$$
W(i) = \begin{bmatrix} W_{11}(i) & W_{12}(i) & W_{13}(i) \\ W_{12}^*(i) & W_{22}(i) & 0 \\ W_{13}^*(i) & 0 & W_{33}(i) \end{bmatrix} < 0, \tag{7.85}
$$

$$
\begin{bmatrix} Y(i) & I_n \\ I_n & X(i) \end{bmatrix} > 0, \tag{7.86}
$$

where

$$V_{11}(i) = A_0^*(i)X(i) + X(i)A_0(i) + K(i)C_0(i) + C_0^*(i)K^*(i)$$

$$+ \sum_{j=1}^{d} q_{ij} X(j) + C_z^*(i)C_z(i),$$

$$V_{12}(i) = X(i)G_0(i) + K(i)D_0(i) + C_z^*(i)D_{zv}(i),$$

$$V_{22}(i) = -\gamma^2 I_{m_1} + D_{zv}^*(i)D_{zv}(i),$$

and

$$W_{11}(i) = A_0(i)Y(i) + Y(i)A_0^*(i) + B_0(i)F(i) + F^*(i)B_0^*(i)$$

$$+ q_{ii} Y(i) + \gamma^{-2} G_0(i)G_0^*(i),$$

$$W_{12}(i) = Y(i)C_z^*(i) + F^*(i)D_{zu}^*(i) + \gamma^{-2} G_0(i)D_{zv}^*(i),$$

$$W_{13}(i) = \left[\sqrt{q_{i,1}}Y(i) \ldots \sqrt{q_{i,i-1}}Y(i)\sqrt{q_{i,i+1}}Y(i) \ldots \sqrt{q_{i,d}}Y(i) \right],$$

$$W_{22}(i) = -I_{p_1} + \gamma^2 D_{zv}(i)D_{zv}^*(i),$$

$$W_{33}(i) = -diag\,(Y(1) \ldots Y(i-1)\ Y(i+1) \ldots Y(d)).$$

Moreover, if (7.84)–(7.86) are feasible, then a controller of form (7.82) is given by

$$A_c(i) = \left[X(i) - Y^{-1}(i) \right]^{-1} \left\{ A_0^*(i) + X(i)A_0(i)Y(i) + X(i)B_0(i)F(i) \right.$$

$$+ K(i)C_0(i)Y(i) + C_z^*(i)\left[C_z(i)Y(i) + D_{zu}(i)F(i) \right]$$

$$+ \left[X(i)G_0(i) + K(i)D_0(i) + C_z^*(i)D_{zv}(i) \right]\left[\gamma^2 I_{m_1} - D_{zv}^*(i)D_{zv}(i) \right]^{-1}$$

$$\times \left[G_0^*(i) + D_{zv}^*(i)C_z(i)Y(i) + D_{zv}^*(i)D_{zu}(i)F(i) \right]$$

$$\left. + \sum_{j=1}^{d} q_{ij} Y(i)Y^{-1}(j) \right\} Y^{-1}(i), \tag{7.87}$$

$$B_c(i) = \left[Y^{-1}(i) - X(i) \right]^{-1} K(i),$$

$$C_c(i) = F(i)Y^{-1}(i).$$

Proof. (i) $\Rightarrow$ (ii) Assume that there exists a controller of form (7.82) such that the zero solution of the system (7.83) for $v(t) = 0$ is ESMS and $\|T_{cl}\| < \gamma$, where T_{cl} denotes the input–output operator associated with (7.83). Applying Corollary 15 in Chapter 6 for the system (7.83), we deduce that there exists $X_{cl} = (X_{cl}(1), \ldots, X_{cl}(d)) \in S_{2n}^d$, $X_{cl}(i) > 0$, $i \in \mathcal{D}$ such that

$$\Pi_i\,(X_{cl}) = \begin{bmatrix} \Pi_{i,11}\,(X_{cl}) & \Pi_{i,12}\,(X_{cl}) \\ \Pi_{i,12}^*(X_{cl}) & \Pi_{i,22}\,(X_{cl}) \end{bmatrix} < 0, \tag{7.88}$$

where we denoted

$$\Pi_{i,11}(X_{cl}) = A_{cl}^*(i)X_{cl}(i) + X_{cl}(i)A_{cl}(i) + \sum_{j=1}^{d} q_{ij}X_{cl}(j)$$

$$+ C_{cl}^*(i)C_{cl}(i),$$

$$\Pi_{i,12}(X_{cl}) = X_{cl}(i)G_{cl}(i) + C_{cl}^*(i)D_{cl}(i),$$

$$\Pi_{i,22}(X_{cl}) = -\gamma^2 I_{m_1} + D_{cl}^*(i)D_{cl}(i).$$

By a Schur complement reasoning, (7.88) leads to the following two conditions:

$$A_{cl}^*(i)X_{cl}(i) + X_{cl}(i)A_{cl}(i) + \sum_{j=1}^{d} q_{ij}X_{cl}(j) + C_{cl}^*(i)C_{cl}(i)$$

$$+ \left[X_{cl}(i)G_{cl}(i) + C_{cl}^*(i)D_{cl}(i)\right]\left[\gamma^2 I_{m_1} - D_{cl}^*(i)D_{cl}(i)\right]^{-1}$$

$$\times \left[G_{cl}^*(i)X_{cl}(i) + D_{cl}^*(i)C_{cl}(i)\right] < 0, \tag{7.89}$$

$$\gamma^2 I_{m_1} - D_{cl}^*(i)D_{cl}(i) > 0. \tag{7.90}$$

Consider the following partition of $X_{cl}(i)$:

$$X_{cl}(i) = \begin{bmatrix} X(i) & M(i) \\ M^*(i) & \tilde{X}(i) \end{bmatrix}$$

and

$$X_{cl}^{-1}(i) = \begin{bmatrix} Y(i) & N(i) \\ N^*(i) & S(i) \end{bmatrix},$$

where $X(i), Y(i) \in S_n^d$ and $M(i), N(i) \in \mathbf{R}^{n \times n}$. Without losing generality, one can assume that $M(i)$ is invertible for every $i \in \mathcal{D}$. Indeed, if $M(i)$ is not invertible for some $i \in \mathcal{D}$, then one can replace X_{cl} by

$$X_\varepsilon = X_{cl} + \begin{bmatrix} 0 & \varepsilon I_n \\ \varepsilon I_n & 0 \end{bmatrix} \text{ with some } \varepsilon > 0$$

such that $X_\varepsilon > 0$, $\Pi_i(X_\varepsilon) < 0$ for all $i \in \mathcal{D}$, and in addition $M_\varepsilon(i) = M(i) + \varepsilon I_n$ is invertible for every $i \in \mathcal{D}$. Since $X(i)N(i) + M(i)S(i) = 0$ it follows that $N(i) = -X^{-1}(i)M(i)S(i)$, and then $N(i)$ is invertible, too. Let us define

$$T(i) = \begin{bmatrix} Y(i) & I_n \\ N^*(i) & 0 \end{bmatrix}.$$

It is obvious that $T(i)$ is invertible and

$$T^{-1}(i) = \begin{bmatrix} 0 & (N^{-1}(i))^* \\ I_n & -Y(i)(N^{-1}(i))^* \end{bmatrix}.$$

Then we have

$$T^*(i)X_{cl}(i) = \begin{bmatrix} I_n & 0 \\ X(i) & M(i) \end{bmatrix} \tag{7.91}$$

and

$$T^*(i)X_{cl}(i)T(i) = \begin{bmatrix} Y(i) & I_n \\ I_n & X(i) \end{bmatrix}. \tag{7.92}$$

From (7.91) together with $X_{cl}(i) > 0$ one gets (7.86). By pre- and postmultiplication of (7.89) by $T^*(i)$ and $T(i)$, respectively, one obtains

$$T^*(i)\widehat{\Pi}_i(X_{cl})T(i) < 0, \tag{7.93}$$

where $\widehat{\Pi}_i(X_{cl})$ is the left-hand side of the inequality (7.89). Let

$$\Lambda(i) = T^*(i)\widehat{\Pi}_i(X_{cl})T(i) = \begin{bmatrix} \Lambda_{11}(i) & \Lambda_{21}^*(i) \\ \Lambda_{21}(i) & \Lambda_{22}(i) \end{bmatrix},$$

where by direct computations, based on (7.89)–(7.92), we have

$$
\begin{aligned}
\Lambda_{11}(i) &= A_0(i)Y(i) + B_0(i)C_c(i)N^*(i) + Y(i)A_0^*(i) + N(i)C_c^*(i)B_0^*(i) \\
&\quad + \left[G_0(i) + \left(Y(i)C_z^*(i) + N(i)C_c^*(i)D_{zu}^*(i)\right)D_{zv}(i)\right] \\
&\quad \times \left[\gamma^2 I_{m_1} - D_{zv}^*(i)D_{zv}(i)\right]^{-1} \\
&\quad \times \left[G_0^*(i) + D_{zv}^*(i)\left(C_z(i)Y(i) + D_{zu}(i)C_c(i)N^*(i)\right)\right] \\
&\quad + \left[Y(i)C_z^*(i) + N(i)C_c^*(i)D_{zu}^*(i)\right]\left[C_z(i)Y(i) + D_{zu}(i)C_c(i)N^*(i)\right] \\
&\quad + \sum_{j=1}^{d} q_{ij}\left[Y(i)X(j)Y(i) + N(i)M^*(j)Y(i) + Y(i)M(j)N^*(i)\right. \\
&\quad \left. + N(i)\widetilde{X}(j)N^*(i)\right], \\
\Lambda_{21}(i) &= A_0^*(i) + X(i)A_0(i)Y(i) + X(i)B_0(i)C_c(i)N^*(i) + M(i)B_c(i)C_0(i)Y(i) \\
&\quad + M(i)A_c(i)N^*(i) + \left[X(i)G_0(i) + M(i)B_c(i)D_0^*(i) + C_z^*(i)D_{zv}(i)\right] \\
&\quad \times \left[\gamma^2 I_{m_1} - D_{zv}^*(i)D_{zv}(i)\right]^{-1} \\
&\quad \times \left[G_0^*(i) + D_{zv}^*(i)C_z(i)Y(i) + D_{zv}^*(i)D_{zu}(i)C_c(i)N^*(i)\right] \\
&\quad + C_z^*(i)\left[C_z(i)Y(i) + D_{zu}(i)C_c(i)N^*(i)\right] \\
&\quad + \sum_{j=1}^{d} q_{ij}\left[Y(i)X(j) + N(i)M^*(j)\right], \\
\Lambda_{22}(i) &= A_0^*(i)X(i) + X(i)A_0(i) + M(i)B_c(i)C_0(i) + C_0^*(i)B_c^*(i)M(i) \\
&\quad + \left[X(i)G_0(i) + M(i)B_c(i)D_0^*(i) + C_z^*(i)D_{zv}(i)\right] \\
&\quad \times \left[\gamma^2 I_{m_1} - D_{zv}^*(i)D_{zv}(i)\right]^{-1} \\
&\quad \times \left[G_0^*(i)X(i) + D_0(i)B_c^*(i)M^*(i) + D_{zv}^*(i)C_z(i)\right] \\
&\quad + C_z^*(i)C_z(i) + \sum_{j=1}^{d} q_{ij}X(j).
\end{aligned}
$$

Let us introduce the following notation:

$$K(i) = M(i)B_c(i),$$
$$F(i) = C_c(i)N^*(i).$$

Thus one obtains

$$
\begin{aligned}
\Lambda_{11}(i) =\ & A_0(i)Y(i) + Y(i)A_0^*(i) + B_0(i)F(i) + F^*(i)B_0^*(i) && (7.94) \\
& + \left[G_0(i) + \left(Y(i)C_z^*(i) + F^*(i)D_{zu}^*(i)\right)D_{zv}(i)\right] \\
& \times \left[\gamma^2 I_{m_1} - D_{zv}^*(i)D_{zv}(i)\right]^{-1} \\
& \times \left[G_0^*(i) + D_{zv}^*(i)\left(C_z(i)Y(i) + D_{zu}(i)F(i)\right)\right] \\
& + \left[Y(i)C_z^*(i) + F^*(i)D_{zu}^*(i)\right]\left[C_z(i)Y(i) + D_{zu}(i)F(i)\right] \\
& + \sum_{j=1}^{d} q_{ij}\left[Y(i)X(j)Y(i) + N(i)M^*(j)Y(i) + Y(i)M(j)N^*(i)\right. \\
& \left. + N(i)\widetilde{X}(j)N^*(i)\right],
\end{aligned}
$$

$$
\begin{aligned}
\Lambda_{21}(i) =\ & A_0^*(i) + X(i)A_0(i)Y(i) + X(i)B_0(i)F(i) + K(i)C_0(i)Y(i) \\
& + M(i)A_c(i)N^*(i) + \left[X(i)G_0(i) + K(i)D_0^*(i) + C_z^*(i)D_{zv}(i)\right] \\
& \times \left[\gamma^2 I_{m_1} - D_{zv}^*(i)D_{zv}(i)\right]^{-1} && (7.95) \\
& \times \left[G_0^*(i) + D_{zv}^*(i)C_z(i)Y(i) + D_{zv}^*(i)D_{zu}(i)F(i)\right] \\
& + C_z^*(i)\left[C_z(i)Y(i) + D_{zu}(i)F(i)\right] \\
& + \sum_{j=1}^{d} q_{ij}\left[Y(i)X(j) + N(i)M^*(j)\right],
\end{aligned}
$$

$$
\begin{aligned}
\Lambda_{22}(i) =\ & A_0^*(i)X(i) + X(i)A_0(i) + K(i)C_0(i) + C_0^*(i)K^*(i) && (7.96) \\
& + \left[X(i)G_0(i) + K(i)D_0^*(i) + C_z^*(i)D_{zv}(i)\right] \\
& \times \left[\gamma^2 I_{m_1} - D_{zv}^*(i)D_{zv}(i)\right]^{-1} \\
& \times \left[G_0^*(i)X(i) + D_0(i)K^*(i) + D_{zv}^*(i)C_z(i)\right] \\
& + \sum_{j=1}^{d} q_{ij}X(j) + C_z^*(i)C_z(i).
\end{aligned}
$$

The condition (7.93) leads to

$$\Lambda_{11}(i) < 0, \tag{7.97}$$
$$\Lambda_{22}(i) < 0. \tag{7.98}$$

Using (7.96) and (7.98), by a Schur complement argument (7.84) directly follows. On the other hand, we may write

$$
\begin{aligned}
& Y(i)X(j)Y(i) + N(i)M^*(j)Y(i) + Y(i)M(j)N^*(i) + N(i)\widetilde{X}(j)N^*(i) \\
& = Y(i)\left[X(j) - M(j)\widetilde{X}^{-1}(j)M^*(j)\right]Y(i) + Y(i)M(j)\widetilde{X}^{-1}(j)M^*(j)Y(i) \\
& \quad + N(i)M^*(j)Y(i) + Y(i)M(j)N^*(i) + N(i)\widetilde{X}(j)N^*(i) \\
& = Y(i)Y^{-1}(j)Y(i) + \left[Y(i)M(j) + N(i)\widetilde{X}(j)\right]\widetilde{X}^{-1}(j) \\
& \quad \times \left[M^*(j)Y(i) + \widetilde{X}(j)N^*(i)\right].
\end{aligned}
$$

Then (7.97) and (7.94) lead to

$$A_0(i)Y(i) + Y(i)A_0^*(i) + B_0(i)F(i) + F_0^*(i)B_0^*(i)$$
$$+ \left[G_0(i) + \left(Y(i)C_z^*(i) + F^*(i)D_{zu}^*(i)\right)D_{zv}(i)\right]\left[\gamma^2 I_{m_1} - D_{zv}^*(i)D_{zv}(i)\right]^{-1}$$
$$\times \left[G_0^*(i) + D_{zv}^*(i)\left(C_z(i)Y(i) + D_{zu}(i)F(i)\right)\right]$$
$$+ \left[Y(i)C_z^*(i) + F^*(i)D_{zu}^*(i)\right]\left[C_z(i)Y(i) + D_{zu}(i)F(i)\right]$$
$$+ \sum_{j=1}^{d} q_{ij} Y(i)Y^{-1}(j)Y(i) < 0. \tag{7.99}$$

Again using Schur complement arguments, one can easily see that the above inequality together with (7.90) implies (7.85) in the statement. Thus the implication (i) $\Rightarrow$ (ii) is proved.

(ii) $\Rightarrow$ (i) Assume that there exist $X(i) > 0$, $Y(i) > 0$, $F(i)$, $K(i)$, $i \in \mathcal{D}$ verifying (7.84)–(7.86). From (7.86) we obtain that $X(i) - Y^{-1}(i) > 0$. Consider

$$X_{cl}(i) = \begin{bmatrix} X(i) & Y^{-1}(i) - X(i) \\ Y^{-1}(i) - X(i) & X(i) - Y^{-1}(i) \end{bmatrix}.$$

Then we have

$$X(i) - \left(Y^{-1}(i) - X(i)\right)\left(X(i) - Y^{-1}(i)\right)^{-1}\left(Y^{-1}(i) - X(i)\right)$$
$$= X(i) + Y^{-1}(i) - X(i) = Y^{-1}(i) > 0.$$

Therefore $X_{cl}(i) > 0$. Using (7.87), one obtains the closed-loop system

$$\dot{x}_{cl}(t) = \tilde{A}_{cl}(\eta(t))x_{cl}(t) + \tilde{G}_{cl}(\eta(t))v(t),$$
$$z(t) = \tilde{C}_{cl}(\eta(t))x_{cl}(t) + \tilde{D}_{cl}(\eta(t))v(t)$$

with the coefficients defined as in (7.83). Let

$$\tilde{\Pi}_i(X_{cl}) = \begin{bmatrix} \tilde{\Pi}_{i,11}(X_{cl}) & \tilde{\Pi}_{i,12}(X_{cl}) \\ \tilde{\Pi}_{i,12}^*(X_{cl}) & \tilde{\Pi}_{i,22}(X_{cl}) \end{bmatrix},$$

where

$$\tilde{\Pi}_{i,11}(X_{cl}) = \tilde{A}_{cl}^*(i)X_{cl}(i) + X_{cl}(i)\tilde{A}_{cl}(i) + \sum_{j=1}^{d} q_{ij}X_{cl}(j) + \tilde{C}_{cl}^*(i)\tilde{C}_{cl}(i),$$
$$\tilde{\Pi}_{i,12}(X_{cl}) = X_{cl}(i)\tilde{G}_{cl}(i) + \tilde{C}_{cl}^*(i)\tilde{D}_{cl}(i),$$
$$\tilde{\Pi}_{i,22}(X_{cl}) = -\gamma^2 I_{m_1} + \tilde{D}_{cl}^*(i)\tilde{D}_{cl}(i).$$

Then for

$$\tilde{T}(i) = \begin{bmatrix} Y(i) & I_n \\ Y(i) & 0 \end{bmatrix},$$

direct computations give

$$\widetilde{T}^*(i)\overline{\Pi}_i(X_{cl})\,\widetilde{T}(i) = \begin{bmatrix} \widetilde{\Lambda}_{11}(i) & 0 \\ 0 & \widetilde{\Lambda}_{22}(i) \end{bmatrix},$$

where

$$
\begin{aligned}
\overline{\Pi}_i(X_{cl}) &= \widetilde{\Pi}_{1,11}(X_{cl}) + \widetilde{\Pi}_{1,12}(X_{cl})\left(\gamma^2 I_{m_1} - \widetilde{D}^*_{cl}(i)\widetilde{D}_{cl}(i)\right)^{-1}\widetilde{\Pi}_{1,11}(X_{cl}), \\
\widetilde{\Lambda}_{11}(i) &= A_0(i)Y(i) + Y(i)A^*_0(i) + B_0(i)F(i) + F^*_0(i)B^*_0(i) \\
&\quad + \left[G_0(i) + \left(Y(i)C^*_z(i) + F^*(i)D^*_{zu}(i)\right)D_{zv}(i)\right] \\
&\quad \times \left[\gamma^2 I_{m_1} - D^*_{zv}(i)D_{zv}(i)\right]^{-1} \\
&\quad \times \left[G^*_0(i) + D^*_{zv}(i)\left(C_z(i)Y(i) + D_{zu}(i)F(i)\right)\right] \\
&\quad + \left[Y(i)C^*_z(i) + F^*(i)D^*_{zu}(i)\right]\left[C_z(i)Y(i) + D_{zu}(i)F(i)\right] \\
&\quad + \sum_{j=1}^{d} q_{ij}Y(i)Y^{-1}(j)Y(i),
\end{aligned}
$$

and $\widetilde{\Lambda}_{22}(i) = \Lambda_{22}(i)$ as defined in (7.96). From (7.84) and (7.85), by Schur complement arguments, it follows that

$$
\begin{aligned}
\widetilde{\Lambda}_{11}(i) &< 0, \\
\widetilde{\Lambda}_{22}(i) &< 0,
\end{aligned}
$$

respectively, and therefore $\overline{\Pi}_i(X_{cl}) < 0$. Moreover, from (7.84) $\gamma^2 I_{m_1} - D^*_{zv}(i)D_{zv}(i) > 0$, which coincides with the condition $\gamma^2 I_{m_1} - D^*_{cl}(i)D_{cl}(i) > 0$. This last condition together with $\overline{\Pi}_i(X_{cl}) < 0$ leads to an inequality of form (7.88) for $\widetilde{\Pi}_i(X_{cl})$, which shows that the controller (7.87) is a solution of the DAP and thus the proof is complete. $\square$

7.5 An H^∞-type filtering problem for signals corrupted with multiplicative white noise

In this section we consider a particular filtering problem in which the measured output is subjected to multiplicative white noise. Its solution is derived via an H^∞-type method based on the Bounded Real Lemma version proved in Theorem 10 of Chapter 6.

Consider the following linear stable system:

$$
\begin{aligned}
dx(t) &= [Ax(t) + Bu(t)]\,dt, & (7.100) \\
dy_1(t) &= C_1x(t)\,(dt + \sigma dw(t)), \\
y_2(t) &= C_2x(t),
\end{aligned}
$$

where $x(t) \in \mathbf{R}^n$ denotes the state, $u(t) \in \mathbf{R}^m$ is an input variable, $y_1 \in \mathbf{R}^{p_1 \times n}$ denotes the measured output, $y_2 \in \mathbf{R}^{p_2 \times n}$ is a quality output, $\sigma \in \mathbf{R}$, and $w(t)$ is a scalar standard Wiener process. Given $\gamma > 0$, the problem consists in determining an n_f-order deterministic filter where $n_f > 0$ is given, with the input y_1 and the output $y_f \in \mathbf{R}^{p_2}$, having the state-space equations

$$\dot{x}_f(t) = A_f x_f(t) + B_f y_1(t), \tag{7.101}$$
$$y_f(t) = C_f x_f(t),$$

such that the resulting system obtained by coupling it to (7.100) is ESMS, and the input–output operator

$$T : L_w^2\left([0, \infty), \mathbf{R}^m\right) \to L_w^2\left([0, \infty), \mathbf{R}^{p_2}\right)$$

from $u \mapsto z$, where $z(t) = y_2(t) - y_f(t)$, has the norm less than γ.

The solution of this problem is provided by the following result.

Theorem 15. *The filtering problem has a solution if and only if there exist the matrices* $P, X \in \mathcal{S}_n$, $\tilde{X} \in \mathcal{S}_{n_f}$, $P > 0$, $X > 0$, $\tilde{X} > 0$, and $\tilde{M} \in \mathbf{R}^{n \times n_f}$, *such that*

$$\begin{bmatrix} A^*P + PA + \sigma^2 C_1^* U^* \tilde{X} U C_1 & PB \\ B^T P & -\gamma^2 I \end{bmatrix} < 0, \tag{7.102}$$

$$\begin{bmatrix} A^*X + XA + MUC_1 + C_1^* U^* M^* & \\ +\sigma^2 C_1^* U^* \tilde{X} U C_1 + C_2^* C_2 & XB \\ B^*X & -\gamma^2 I \end{bmatrix} < 0, \tag{7.103}$$

$$\begin{bmatrix} X & M \\ M^* & \tilde{X} \end{bmatrix} > 0, \tag{7.104}$$

$$rank\left(\begin{bmatrix} P - X & M \\ M^* & -\tilde{X} \end{bmatrix}\right) = n_f, \tag{7.105}$$

where

$$U = \begin{cases} \begin{bmatrix} I_{p_1} \\ 0_{(n_f - p_1) \times p_1} \end{bmatrix} & \text{if } n_f \geq p_1 \text{ and} \\ \begin{bmatrix} I_{n_f} & 0_{n_f \times (p_1 - n_f)} \end{bmatrix} & \text{if } n_f < p_1. \end{cases} \tag{7.106}$$

Proof. When coupling the filter (7.101) to the system (7.100) one obtains the resulting system:

$$dx(t) = [Ax(t) + Bu(t)]\, dt,$$
$$dx_f(t) = \left[A_f x_f(t) + C_1 x(t)\right] dt + \sigma B_f C_1 x(t)\, dw(t),$$
$$z(t) = C_2 x(t) - C_f x_f(t),$$

or equivalently

$$
d\begin{bmatrix} x(t) \\ x_f(t) \end{bmatrix} = \left(\begin{bmatrix} A & 0 \\ B_f C_1 & A_f \end{bmatrix} \begin{bmatrix} x(t) \\ x_f(t) \end{bmatrix} + \begin{bmatrix} B \\ 0 \end{bmatrix} u(t) \right) dt
$$
$$
+ \begin{bmatrix} 0 & 0 \\ \sigma B_f C_1 & 0 \end{bmatrix} \begin{bmatrix} x(t) \\ x_f(t) \end{bmatrix} dw(t), \tag{7.107}
$$
$$
z(t) = \begin{bmatrix} C_2 & -C_f \end{bmatrix} \begin{bmatrix} x(t) \\ x_f(t) \end{bmatrix}.
$$

Let us introduce the following notation:

$$
\mathcal{A}_0 = \begin{bmatrix} A & 0 \\ B_f C_1 & A_f \end{bmatrix}, \; \mathcal{A}_1 = \begin{bmatrix} 0 & 0 \\ \sigma B_f C_1 & 0 \end{bmatrix}, \tag{7.108}
$$
$$
\mathcal{B}_0 = \begin{bmatrix} B_0 \\ 0 \end{bmatrix}, \; C = \begin{bmatrix} C_2 & -C_f \end{bmatrix}.
$$

Applying Theorem 10 of Chapter 6 for the resulting system (7.107), it follows that it is ESMS and its associated input–output operator has the norm less than γ if and only if there exists $\mathcal{X} > 0$ such that

$$
\begin{bmatrix} \mathcal{A}_0^* \mathcal{X} + \mathcal{X} \mathcal{A}_0 + \mathcal{A}_1^* \mathcal{X} \mathcal{A}_1 + C^* C & \mathcal{X} \mathcal{B}_0 \\ \mathcal{B}_0^* \mathcal{X} & -\gamma^2 I \end{bmatrix} < 0. \tag{7.109}
$$

Further, consider the partition of $\mathcal{X}$:

$$
\mathcal{X} = \begin{bmatrix} X & M \\ M^* & \widetilde{X} \end{bmatrix},
$$

where $X \in \mathbf{R}^{n \times n}$, $\widetilde{X} \in \mathbf{R}^{n_f \times n_f}$, and $M \in \mathbf{R}^{n \times n_f}$. Then using (7.108), the condition (7.109) becomes

$$
\mathcal{N}(X, M, \widetilde{X}, A_f, B_f, C_f) = \begin{bmatrix} \mathcal{N}_{11} & \mathcal{N}_{12} & \mathcal{N}_{13} & 0 \\ \mathcal{N}_{12}^* & \mathcal{N}_{22} & \mathcal{N}_{23} & \mathcal{N}_{24} \\ \mathcal{N}_{13}^* & \mathcal{N}_{23}^* & -\gamma^2 I_m & 0 \\ 0 & \mathcal{N}_{24}^* & 0 & -I_{p2} \end{bmatrix} < 0, \tag{7.110}
$$

where

$$
\mathcal{N}_{11} = A^* X + X A + M B_f C_1 + C_1^* B_f^* M^*
$$
$$
+ \sigma^2 C_1^* B_f^* \widetilde{X} B_f C_1 + C_2^* C_2,
$$
$$
\mathcal{N}_{12} = A^* M + C_1^* B_f^* \widetilde{X} + M A_f - C_2^* C_f, \tag{7.111}
$$
$$
\mathcal{N}_{13} = X B,
$$
$$
\mathcal{N}_{22} = A_f^* \widetilde{X} + \widetilde{X} A_f,
$$
$$
\mathcal{N}_{23} = M^* B,
$$
$$
\mathcal{N}_{24} = -C_f^*.
$$

Assume that B_f is full rank. This is not a restrictive assumption since in the case when the filtering problem stated above has a solution with B_f non–full rank, then one can always find a small enough perturbation of B_f such that the perturbed matrix $\widetilde{B}_f$ is full rank and verifies (7.110). Then, there exists a nonsingular transformation T such that

$$
T\widetilde{B}_f = \begin{cases} \begin{bmatrix} \Sigma \\ 0_{(n_f - p_1) \times p_1} \end{bmatrix} & \text{if } n_f \geq p_1 \text{ or} \\[2em] \begin{bmatrix} \Psi & 0_{n_f \times (p_1 - n_f)} \end{bmatrix} & \text{if } n_f < p_1, \end{cases}
$$

where Σ and Ψ are nonsingular. It follows that applying to $\widetilde{B}_f$ the nonsingular transformation

$$
\begin{bmatrix} \Sigma^{-1} & 0 \\ 0 & I \end{bmatrix} T \quad \text{if } n_f \geq p_1 \text{ or}
$$
$$
\Psi^{-1} T \quad \text{if } n_f < p_1,
$$

one obtains that $\widetilde{B}_f = U$ with U given by (7.106). Therefore, without losing generality, one can choose $B_f = U$.

The condition (7.110) can be expressed as

$$
\mathcal{Z} + \mathcal{P}^* \Omega \mathcal{Q} + \mathcal{Q}^* \Omega^* \mathcal{P} < 0, \tag{7.112}
$$

where we denoted

$$
\mathcal{Z} = \begin{bmatrix} \mathcal{N}_{11} & A^* M + C_1^* B_f^* \widetilde{X} & \mathcal{N}_{13} & 0 \\ M^* A + \widetilde{X} B_f C_1 & 0 & \mathcal{N}_{23} & 0 \\ \mathcal{N}_{11} & \mathcal{N}_{11} & -\gamma^2 I_m & 0 \\ 0 & 0 & 0 & -I_{p2} \end{bmatrix},
$$

$$
\mathcal{P} = \begin{bmatrix} M^* & \widetilde{X} & 0 & 0 \\ -C_2 & 0 & 0 & -I_{p2} \end{bmatrix}, \quad \mathcal{Q} = \begin{bmatrix} 0 & I_{n_f} & 0 & 0 \end{bmatrix}, \tag{7.113}
$$

$$
\Omega = \begin{bmatrix} A_f \\ C_f \end{bmatrix}.
$$

Using the Projection Lemma (Lemma 7), it follows that (7.112) has a solution Ω if and only if

$$
W_{\mathcal{P}}^* \mathcal{Z} W_{\mathcal{P}} < 0, \tag{7.114}
$$
$$
W_{\mathcal{Q}}^* \mathcal{Z} W_{\mathcal{Q}} < 0, \tag{7.115}
$$

where $W_{\mathcal{P}}$ and $W_{\mathcal{Q}}$ denote bases of the null subspaces of $\mathcal{P}$ and $\mathcal{Q}$, respectively. Further, perform the partition of $\mathcal{X}^{-1}$ according to the partition of $\mathcal{X}$:

$$
\mathcal{X}^{-1} = \begin{bmatrix} Y & N \\ N^* & \widetilde{Y} \end{bmatrix}.
$$

With these notations, one obtains that

$$
W_{\mathcal{P}}^* = \begin{bmatrix} Y & N & 0 & -YC_2^T \\ 0 & 0 & -I_{n_f} & 0 \end{bmatrix},
$$

$$
W_{\mathcal{Q}} = \begin{bmatrix} I_n & 0 & 0 \\ 0 & 0 & 0 \\ 0 & I_m & 0 \\ 0 & 0 & I_{p_2} \end{bmatrix}.
$$

Direct algebraic computations using $Y^{-1} - X = MN^*Y^{-1}$ show that (7.114) is equivalent to (7.102), where $P = Y^{-1}$, and (7.115) is equivalent to (7.103). The rank condition (7.105) follows directly from the relationship between $\mathcal{X}$ and $\mathcal{X}^{-1}$, and it shows that $Y^{-1} = X - M\widetilde{X}^{-1}M^*$. Thus the proof is complete.   □

If the necessary and sufficient conditions in Theorem 15 are fulfilled, then a solution of the filtering problem can easily be obtained by solving the basic LMI (7.112) with respect to Ω.

In the following, we present a numerical example illustrating the above result. The *Instrumental Landing System* (ILS) is radioelectronic equipment that provides aircrafts with on-board, on-line information concerning the aircraft's position relative to some glideslope references in the landing phase of the flight. The glideslope signal is expressed as

$$
i_{gs} = K i_0, \tag{7.116}
$$

where the multiplicative factor K depends on the glideslope sensitivity and i_o denotes the nominal signal. The offset in the glideslope sensitivity depends on the performance category of the ILS. If σ denotes the mean square deviation of K, then $P(|K(t) - K_0| < 3\sigma) \geq 0.997$, where K_0 denotes the nominal value of the multiplicative factor. This probability increases when $\sigma \to 0$. Then, taking $\sigma = 0.06$, for which $3\sigma = 0.18$, one can obtain a maximum deviation from the glideslope sensitivity of 18%, in conformance with international standards (Category II of ILS). Therefore, the multiplication factor K in (7.116) can be replaced by

$$
K = K_0 + \sigma\xi, \tag{7.117}
$$

where ξ is a white noise with unitary covariance. If the altitude dynamics is approximated by $\dot{x} = Ax + Bu$ with $i_0 = Cx$, then according to (7.116) and (7.117), the glideslope measured signal is $i_{gs} = (K_0 + \sigma\xi) Cx$. Thus one obtains a stochastic system of form (7.100) with the output subjected to multiplicative white noise, for which a deterministic filter is designed. For $A = -1/30$, $B = 50/30$, $C_1 = C_2 = 1$, and $K_0 = 1$, using the result stated in Theorem 15, we obtained for the level of attenuation $\gamma = 5$, the following solution of the system of inequalities (7.102–7.105): $X = 1.9457$; $M = -0.6692$; $\widetilde{X} = 0.3132$; $P = 0.5161$. Solving the LMI (7.112),

$$
\Omega = \begin{bmatrix} -0.4073 \\ 0.4450 \end{bmatrix},
$$

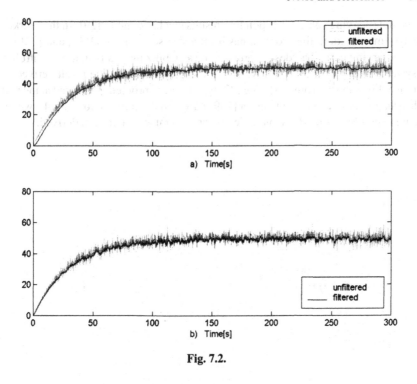

Fig. 7.2.

and therefore the solution of the filtering problem is given by

$$\dot{x}_f = -0.4073x_f + y_1,$$
$$y_f = 0.4045x_f.$$

In Figure 7.2a the unfiltered and the filtered signals are plotted. For comparison, we further determined a Kalman filter for the attitude dynamics by tuning the covariance matrices Q_0 and R_0 corresponding to the control and to the output additive white noise perturbations. For $Q_0 = 100$ and $R_0 = 0.1$, the resulting Kalman filter provides the results shown in the Figure 7.2b, where the filtered and unfiltered signals are represented.

Analyzing the numerical results illustrated in the above figure, one concludes, as is expected, that a filter designed using the specific multiplicative feature of the stochastic perturbation provides better results with respect to those given by Kalman filters that are suitable in the case of additive stochastic perturbations.

Notes and references

Most of the results derived in this chapter are presented for the first time. State feedback H^∞ control for linear systems with multiplicative white noise has been studied in several works. Among them we cite [99], [9], and the references therein.

For the time-varying case, corresponding results can be found in [29]. In the Markovian systems situation, the problem has been addressed in [41], [108], and [32] for the time-varying case. The design problem of a stabilizing γ-attenuating controller for systems with state-dependent white noise is given in [64]. The result derived in Section 5.4 is inspired from [25]. The H^{∞} type filtering problem presented at the end of this chapter has been considered in [109] based on the formulation in [54], where deterministic filters with the same order as the generator systems are derived.

Bibliography

1. H. Abou-Kaudil, G. Freiling, G. Jank, Solution and asymptotic behavior of coupled Riccati equations in jump linear systems, *IEEE Trans. Automat. Control*, 39 (1994), 1631–1636.
2. H. Abou-Kaudil, G. Freiling, V. Ionescu, G. Jank, *Matrix Riccati Equations in Control Systems Theory*, Birhäuser, Basel, 2003.
3. M. Ait-Rami, L. El-Ghaoui, *Robust stabilization of jump linear systems using linear matrix inequalities*, IFAC Symposium in Robust Control, 1994.
4. M. Ait-Rami, X.Y. Zhou, Linear matrix inequalities, Riccati equations and indefinite stochastic linear quadratic controls, *IEEE Trans. Automat. Control*, 45(6) (2000), 1131–1143.
5. L. Arnold, *Stochastic Differential Equations: Theory and Applications*. John Wiley, New York, 1974.
6. L. Arnold, *Random Dynamical Systems*, Springer-Verlag, Berlin, 1988.
7. R. Belmann, *Introduction to Matrix Analysis*, McGraw-Hill, New York, 1960.
8. J.M. Bismut, Linear qudratic optimal control with random coefficients, *SIAM J. Control Optim.*, 14 (1976), 419–444.
9. S. Boyd, L. El-Ghaoui, E. Feron, V. Balakrishnan, *Linear Matrix Inequalities in Systems and Control Theory*, SIAM, Philadelphia, PA, 1994.
10. R.W. Brockett, Parametrically stochastic linear differential equations. *Math. Program. Stud.* 5 (1979), 8–21.
11. H. Bunke, *Gewöhnliche Differential-gleichungen mit zufäligen Parametern*, Academie Verlag, Berlin, 1972.
12. H.F. Chen, On stochastic observability, *Scientia Sinica*, 20 (1977), 305–323.
13. H.F. Chen, On the stochastic observability and controllability, *Proceedings of the 7th IFAC World Congress*, 1978, 1115–1162.
14. S. Chen, X. Li, Y. Zhou, Stochastic linear quadratic regulators with indefinite control weight costs, *SIAM J. Control Optim.*, 36(5) (1998), 1685–1702.
15. O.L.V. Costa, J.B.R. do Val, J.C. Geromel, Continuous-time state-feedback H_2 control of Markovian jump linear systems via convex analysis, *Automatica*, 35 (1999), 259–268.
16. K.L. Chung, *Markov Chains with Stationary Transition Probabilities*, Springer-Verlag, Berlin, 1967.
17. R.F. Curtain, *Stability of Stochastic Dynamical Systems*, Lecture Notes in Math., vol. 294, Springer-Verlag, Berlin, 1972.
18. G. Da Prato, A. Ichikawa, Stability and quadratic control for linear stochastic equation with unbounded coefficients. *Boll. Un Mat.*, 6 (1985), 989–1001.

19. G. Da Prato, A. Ichikawa, Quadratic control for linear periodic systems. *Appl. Math. Optim.*, 18 (1988), 39–66.
20. G. Da Prato, A. Ichikawa, Quadratic control for linear time-varying stochastic systems. *SIAM J. Control Optim.*, 28 (1990), 359–381.
21. G. Da Prato, J. Zabczyk, *Stochastic Equations in Infinite Dimension*, Cambridge Univ. Press, Cambridge, UK, 1992.
22. T. Damm, *Rational Matrix Equations in Stochastic Control*, Lecture Notes in Control and Inform Sci. 297, Springer-Verlag, Berlin, 2004.
23. T. Damm, D. Hinrichsen, Newton's Method for a rational matrix equation occuring in stochastic control, *Linear Algebra Appl.*, 332/334 (2001), 81–109.
24. T. Damm, D. Hinrichsen, Newton's Method for Concave Operators with Resolvent Derivatives in Ordered Banach Spaces, *Linear Algebra Appl.*, 363 (2003), 43–64.
25. D.P. de Farias, J.C. Geromel, J.B.R. do Val, O.L.V. Costa, Output feedback control of Markov jump linear systems in continuous-time, *IEEE Trans. Automat. Control*, 45(2) (2000), 944–948.
26. J.L. Doob, *Stochastic Processes*, John Wiley, New York, 1967.
27. J.L. Doob, *Measure Theory*, Graduate Texts in Mathematics, Springer-Verlag, Berlin, 1994.
28. V. Dragan, G. Freiling, A. Hochhaus, T. Morozan, A class of nonlinear differential equations on the space of symmetri matrices, *Electronic J. Differential Equations*, 96 (2004), 1–48.
29. V. Dragan, T. Morozan, Global solutions to a game-theoretic Riccati equation of stochastic control, *J. Differential Equations*, 138(2) (1997), 328–350.
30. V. Dragan, T. Morozan, Systems of matrix rational differential equations arising in connection with linear stochastic systems with Markovian jumping, *J. Differential Equations*, 194 (2003), 1–38.
31. V. Dragan, T. Morozan, The linear quadratic optimization problem and tracking problem for a class of linear stochastic systems with multiplicative white noise and Markovian jumping, *IEEE Trans. Automat. Control*, 49 (2004), 665–676.
32. V. Dragan, T. Morozan, Game-theoretic coupled Riccati equations associated to controlled linear differential systems with jump Markov perturbations, *Stoch. Anal. Appl.*, 19(5) (2001), 715–751.
33. V. Dragan, T. Morozan, Stability and robust stabilization to linear stochastic systems described by differential equations with Markov jumping and multiplicative white noise, *Stoch. Analy. Appl.*, 20(1) (2002), 33–92.
34. V. Dragan, T. Morozan. Stochastic observability and applications, *IMA J. Math. Control Inform.*, 21 (2004), 323–344.
35. V. Dragan, T. Morozan, A. Stoica, Iterative procedure for stabilizing solutions of differential Riccati type equations arising in stochastic control, in *Analysis and Optimization of Differential Systems*, Eds. V. Barbu, I. Lasiecka, C. Varsan, Kluwer Academic Publishers, Dordrecht, the Netherlands, 2003, 133–144.
36. V. Dragan, T. Morozan, A. Stoica, Optimal H^2 state-feedback control for linear systems with Markovian jumps, *Revue Roumaine des Sciences Techniques*, T 48 (2003), 523–537.
37. V. Dragan, T. Morozan, A. Stoica, Stochastic version of Bounded Real Lemma and applications, *Math. Rep.*, 6(56) (2004), 389–430.
38. V. Dragan, T. Morozan, A. Stoica, H^2 Optimal control for linear stochastic systems, *Automatica*, 40 (2004), 1103–1113.
39. V. Dragan, A. Halanay, T. Morozan, Optimal stabilizing compensator for linear systems with state dependent noise, *Stoch. Anal. Appl.*, 10(5) (1992), 557–572.

40. V. Dragan, A. Halanay, A. Stoica, A small gain theorem for linear stochastic systems, *Systems Control Lett.*, 30 (1997), 243–251.

41. V. Dragan, A. Halanay, A. Stoica, A small gain and robustness for linear systems with jump Markov perturbations, *Proceedings of Mathematical Theory of Networks and Systems (MTNS)*, Padova, Italy, 6–10 July, ThE11, 779–782.

42. V. Dragan, A. Halanay, A. Stoica, The γ-attenuation problem for systems with state-dependent noise, *Stoch. Anal. Appl.*, 17(3) (1999), 395–404.

43. N. Dunford, J. Schwartz, *Linear operators: Part I*. Interscience Publishers, New York, 1958.

44. A. El Bouhtouri, A.J. Protchard, Stability radii of linear systems with respect to stochastic perturbations, *Systems Control Lett.*, 19 (1992), 29–33.

45. L. El-Ghaoui, State-feedback control of systems with multiplicative noise via linear matrix inequalities, *Systems Control Lett.*, 24 (1995), 223–228.

46. K. El Hadri, *Robustesse de la stabilité et stabilization d'une classe de systèmes a sauts Markoviens assujettis a des perturbations structurées incertaines*, Ph.D. Thesis, Université Chouaib Doukkali, El Jadida, Morocco, 2001.

47. M. Ehrhardt, W. Kliemann, *Controllability of Linear Stochastic Systems*, Report no. 50, Institut of Dynamische Systems, Bremen, August 1981.

48. X. Feng, K.A. Loparo, Y. Ji, H.J. Chizeck, Stochastic stability properties of jump linear systems. *IEEE Trans. Automat. Control*, 37(1) (1992), 38–53.

49. M.D. Fragoso, O.L.V. Costa, *A Unified Approach for Mean Square Stability of Continuous-Time Linear Systems with Markov Jumping Parameters and Additive Disturbances*. LNCC Internal Report no. 11/1999 (1999).

50. M.D. Fragoso, O.L.V. Costa, C.E. de Souza, A new approach to linearly perturbed Riccati equations arising in stochastic control, *Appl. Math. Optim.*, 37 (1998), 99–126.

51. G. Freiling, A. Hochhaus, On a class of rational matrix differential equations arising in stochastic control, *Linear Algebra Appl.*, 379 (2004), 43–68.

52. A. Friedman, *Stochastic Differential Equations and Applications*, vol. I, Academic Press, New York, 1975.

53. Z. Gajic, I. Borno, Liapunov iterations for optimal control of jump linear systems at steady state, *IEEE Trans. Automat. Control*, 40 (1995), 1971–1975.

54. E. Gershon, D.J.N. Limebeer, U. Shaked, I. Yaesh, H_∞ filtering of continuous-time linear systems with multiplicative noise, *Proceedings of the 3rd IFAC Symposium on Robust Control Design (ROCOND)*, Prague, Czech Republic, 2000.

55. I. Ghihman, A. Skorohod, *Introduction to Stochastic Processes* (in Russian), Nauka, Moskow, 1965.

56. I. Ghihman, A. Skorohod, *Stochastic Differential Equations*, Springer-Verlag, Berlin, 1972.

57. C.H. Guo, Iterative solution of a matrix Riccati equation arising in stochastic control, *Oper. Theory Adv. Appl.*, 130 (2001), 209–221.

58. A. Halanay, *Differential Equations, Stability, Oscillations, Time Lag*, Academic Press, New York, 1966.

59. P. Halmos, *Measure Theory*, Van Nostrand, New York, 1951.

60. A. Haurie, A. Leizarowitz, Overtracking optimal regulation and tracking of piecewise diffusion linear systems, *SIAM J. Control Optim.*, 30(4) (1992), 816–832.

61. U.G. Hausmann, Optimal stationary control with state and control dependent noise, *SIAM J. Control Optim.*, 9 (1971), 184–198.

62. U.G. Haussmann, Stability of linear systems with control dependent noise, *SIAM J. Control*, 11(2) (1973), 382–394.

63. E. Hille, R. Phillips, Functional analysis and semigroups, *Amer. Math. Soc. Coll. Pub.* 31 (1957).
64. D. Hinrichsen, A.J. Pritchard, Stochastic H^∞, *SIAM J. Control Optim.*, 36 (1998), 1504–1538.
65. D. Hinrichsen, A.J. Pritchard, Stability radii of systems with stochastic uncertainty and their optimization by output feedback, *SIAM J. Control Optim.*, 34 (1996), 1972–1998.
66. A. Ichikawa, Optimal control of linear stochastic evolution equation with state and control dependent noise, *Proc. IMA Conf. on Recent Theoret. Develop. in Control*, Leicester, Academic Press, 1978, 383–401.
67. A. Ichikawa, Dynamic programming approach to stochastic evolution equation, *SIAM J. Control Optim.* 17 (1979), 162–174.
68. A. Ichikawa, Filtering and control of stochastic differential equations with unbounded coefficients, *Stoch. Anal. Appl.*, 4 (1986), 187–212.
69. N. Ikeda, S. Watanabe, *Stochastic Differential Equations*, North-Holland, Amsterdam, 1981.
70. Y. Ji, H.J. Chizeck, Controllability, stabilizability and continuous–time Markovian jump linear quadratic control, *IEEE Trans. Automat. Control*, 35(7) (1990), 777–788.
71. Y. Ji, H.J. Chizeck, Jump linear quadratic Gaussian control: Steady–state solution and testable conditions, *Control Theory and Advanced Technology*, 7(2) (1991), 247–270.
72. R.E. Kalman, Contributions to the theory of optimal control, *Bull. Soc. Math. Mexicana*, 5 (1960), 102–119.
73. I. Kats, N.N. Krasovskii, On stability of systems with random parameters (in Russian), *P.M.M.*, 24 (1960), 809–823.
74. R.Z. Khasminskii, *Stochastic Stability of Differential Equations*. Sÿthoff and Noordhoff Alpen aan den Rijn, NL, 1980.
75. J. Klamka, L. Socha, Some remarks about stochastic controllability, *IEEE Trans. Automat. Control*, 27 (1977), 880–881.
76. J. Klamka, L. Socha, Stochastic controllability of dynamical systems (in Polish), *Podstawy Sterowonia*, 8 (1978), 191–200.
77. H. Kushner, *Stochastic Stability and Control*, Academic Press, New York, 1967.
78. G.S. Ladde, V. Lakshmikantham, Random Differential Inequalities, Academic Press, New York, 1980.
79. Levit, V.A. Yakubovich, Algebraic criteria for stochastic stability of linear systems with parametric excitation of white noise type, *Prikl. Mat.*, 1 (1972), 142–147.
80. M. Lewin, On the boundedness measure and stability of solutions of an Itô equation perturbed by a Markov chain, *Stoch. Anal. Appl.*, 4(4) (1986), 431–487.
81. A. Lipser, A.N. Shiryaev, *Statistics of Random Processes*, Nauka, Moscow, 1974 (Springer-Verlag, Berlin, 1977).
82. K.A. Loparo, Stochastic stability of coupled linear systems: a survey of method and results. *Stoch. Anal. Appl.*, 2 (1984), 193–228.
83. X. Mao, Stability of stochastic differential equations with Markov switching, *Stoch. Process. Appl.*, 79 (1999), 45–67.
84. M. Mariton, *Jump Linear Systems in Automatic Control*, Marcel Dekker, New York and Basel, 1990.
85. M. Mariton, Almost sure and moments stability of Jump linear systems, *Systems Control Lett.*, 11 (1988), 393–397.
86. T. Morozan, Optimal stationary control for dynamic systems with Markov perturbations, *Stoch. Anal. Appl.*, 1(3) (1983), 299–325.
87. T. Morozan, Stabilization of some control differential systems with Markov perturbations (in Romanian), *Stud. Cerc. Mat.*, 37(3) (1985), 282–284.

88. T. Morozan, Stochastic uniform observability and Riccati equations of stochastic control. *Rev Roumaine Math. Pures Appl.*, 38(9) (1993), 771–781.

89. T. Morozan, Stability and control for linear systems with jump Markov perturbations. *Stoch. Anal. Appl.*, 13(1) (1995), 91–110.

90. T. Morozan, Stability radii of some stochastic differential equations, *Stochastics Stochastic Reports*, 54 (1995), 281–291.

91. T. Morozan, Stability radii of some time-varying linear stochastic differential systems, *Stoch. Anal. Appl.*, 15(3) (1997), 387–397.

92. T. Morozan, Parametrized Riccati equations for controlled linear differential systems with jump Markov perturbations, *Stoch. Anal. Appl.*, 14(4) (1998), 661–682.

93. T. Morozan, Parametrized Riccati equation and input–output operators for time-varying stochastic differential equations with state dependent noise, *Stud. Cerc. Mat.*, 1 (1999), 99–115.

94. T. Morozan, Linear quadratic and tracking problems for time-varying stochastic differential systems perturbed by a Markov chain. *Revue Roumaine Math. Pures Appl.*, 25(6) (2001), 783–804.

95. I.P. Natanson, *Theory of Functions of a Real Variable* (in Russian). Gost, Moscow, 1950.

96. J. Neveu, *Bases mathématiques du calcul des probabilités*, Masson, Paris, 1970.

97. J. Neveu, Notes sur l'intégrale stochastique. Cours de 3ème cycle. Lab. de Probabilités, Paris VI, 1972.

98. B. Oksendal, *Stochastic Differential Equations*. Springer-Verlag, Berlin, 1998.

99. I.R. Petersen, V.A. Ugrinovskii, A.V. Savkin, *Robust Control Design Using H^∞ Methods*, Springer-Verlag, Berlin, 2000.

100. A.J. Pritchard, J. Zabczyk, Stability and stabilizability of infinite dimensional systems, *SIAM Rev.*, 23 (1981), 25–52.

101. T. Sasagawa, LP-stabilization problem for linear-stochastic control systems with multiplicative noise. *J. Optim. Theory Appl.*, 61(3) (1989), 451–471.

102. T. Sasagawa, Stabilization of linear stochastic systems with delay by using a certain class of controls, Trans. 8th Prague Conference on Inform. Theory, *Statistical Decision Functions* (1979), 295–300.

103. L. Schwartz, *Analyse mathématique*, vol. I, Hermann, Paris, 1967.

104. A.N. Shiryayev, *Probability*, Springer-Verlag, New York, 1984.

105. Y. Shunahara, S. Aihara, K. Kishino, On the stochastic observability and controllability of nonlinear systems. *IEEE Trans. Automat. Control*, 19 (1974), 49–54.

106. Y. Shunahara, S. Aihara, K. Kishino, On the stochastic observability and controllability of nonlinear systems. *Internat. J. Control*, 22 (1975), 65–82.

107. A. Skorohod, *Random Processes with Independent Increments* (in Russian), Nauka, Moscow, 1964.

108. A. Stoica, I. Yaesh, Robust H_∞ control of wings deployment loops for an uninhabited air vehicle—The Jump Markov model approach, *J. Guidance, Control Dynamics*, 25 (2002), 407–411.

109. A. Stoica H^∞ *filtering of sygnals subjected to multiplicative white noise*, Proceedings of CDC Barcelona, 21–26 July 2002.

110. J.C. Taylor, *An Introduction to Measure and Probability*, Springer-Verlag, Berlin, 1996.

111. G. Tessitore, Some remarks on the Riccati equation arising in an optimal control problem with state and control dependent noise, *SIAM J. Control Optim.*, 30(3) (1992), 714–744.

112. C. Tudor, *Procesos Estochasticas*, Sociedad Matematica Mexicana, 1994.

113. J.L. Willems, Mean square stability criteria for linear white noise stochastic systems. *Problems Control Inform. Theory*, 2(3–4) (1973), 199–217.

114. J.L. Willems, *Stochastic Problems in Dynamics*, Ed. B.L. Clarkson, Pitman, London, 1977.
115. J.L. Willems, J.C. Willems, Feedback stabilization for stochastic systems with state and control dependent noise. *Automatica*, 12 (1976), 277–283.
116. J.L. Willems, J.C. Willems, Robust stabilization of uncertain systems, *SIAM J. Control*, 21 (1983), 395–409.
117. W.M. Wonham, Random differential equations in control theory. *Probabilistic Methods in Appl. Math.*, vol. 2, Academic Press, New York, 1970, 131–132.
118. G.G. Yin, Q. Zhang, *Continuous-Time Markov Chains and Applications. A Singular Perturbation Approach*, Springer-Verlag, Berlin, 1997.
119. J. Zabczyk, *An Introduction to Probability Theory. Control Theory and Topics in Functional Analysis*, vol. I, IAEA, Vienna, 1976.
120. J. Zabczyk, *Stochastic Control of Discrete-Time Systems. Control Theory and Topics in Functional Analysis*, vol. III, IAEA, Vienna, 1976.
121. J. Zabczyk, Controllability of stochastic linear systems. *Systems Control Lett.*, 1(1) (1981), 25–31.

Index